Using Words and Things

This book offers a systematic framework for thinking about the relationship between language and technology, and an argument for interweaving thinking about technology with thinking about language. The main claim of philosophy of technology—that technologies are not mere tools and artefacts not mere things, but crucially and significantly shape what we perceive, do, and are—is rethought in a way that accounts for the role of language in human technological experiences and practices. Engaging with work by Wittgenstein, Heidegger, McLuhan, Searle, Ihde, Latour, Ricoeur, and many others, the author critically responds to, and constructs a synthesis of, three "extreme," idealtype, untenable positions: (1) Only humans speak and neither language nor technologies speak, (2) Only language speaks and neither humans nor technologies speak, and (3) Only technology speaks and neither humans nor language speak. The construction of this synthesis goes hand in hand with a narrative about subjects and objects that become entangled and constitute one another. *Using Words and Things* thus draws in central discussions from other subdisciplines in philosophy, such as philosophy of language, epistemology, and metaphysics, to offer an original theory of the relationship between language and (philosophy of) technology centered on use, performance, and narrative, and taking a transcendental turn.

Mark Coeckelbergh is Professor of Philosophy of Media and Technology at the Department of Philosophy, University of Vienna, and (part-time) Professor of Technology and Social Responsibility at De Montfort University, UK. His publications include *Growing Moral Relations* (2012), *Human Being @ Risk* (2013), *Environmental Skill* (2015), *Money Machines* (2015), *New Romantic Cyborgs* (2017), and numerous articles in the area of philosophy of technology.

Routledge Studies in Contemporary Philosophy

For a full list of titles in this series, please visit www.routledge.com

83 **Time and the Philosophy of Action**
   Edited by Roman Altshuler and Michael J. Sigrist

84 **McTaggart's Paradox**
   R. D. Ingthorsson

85 **Perspectives on Ignorance from Moral and Social Philosophy**
   Edited by Rik Peels

86 **Self-Reflection for the Opaque Mind**
   An Essay in Neo-Sellarsian Philosophy
   T. Parent

87 **Facts and Values**
   The Ethics and Metaphysics of Normativity
   Edited by Giancarlo Marchetti and Sarin Marchetti

88 **Aesthetic Disinterestedness**
   Art, Experience, and the Self
   Thomas Hilgers

89 **The Social Contexts of Intellectual Virtue**
   Knowledge as a Team Achievement
   Adam Green

90 **Reflective Equilibrium and the Principles of Logical Analysis**
   Understanding the Laws of Logic
   Jaroslav Peregrin and Vladimír Svoboda

91 **Philosophical and Scientific Perspectives on Downward Causation**
   Edited by Michele Paolini Paoletti and Francesco Orilia

92 **Using Words and Things**
   Language and Philosophy of Technology
   Mark Coeckelbergh

Children's Games by Pieter Bruegel the Elder, 1560. Kunsthistorisches Museum, Vienna.

Using Words and Things

Language and Philosophy of Technology

Mark Coeckelbergh

Routledge
Taylor & Francis Group

LONDON AND NEW YORK

First published 2017
by Routledge

2 Park Square, Milton Park, Abingdon, Oxfordshire OX14 4RN
52 Vanderbilt Avenue, New York, NY 10017

Routledge is an imprint of the Taylor & Francis Group, an informal business

First issued in paperback 2020

Library of Congress Cataloging-in-Publication Data
A catalog record for this book has been requested

ISBN: 978-1-138-69416-3 (hbk)
ISBN: 978-0-367-59502-9 (pbk)

Typeset in Sabon
by Apex CoVantage, LLC

Contents

Acknowledgments   ix

1   Introduction: Words and Things   1

PART I
Humans Speak (Subjects versus Objects)   21

2   Speaking with and about Technology   23

3   Giving Meaning to Technology: A Searlean Social
    Ontology of Technological Artefacts   50

PART II
Language Speaks (Subjects Change Objects)   81

4   Language and the Social Construction of Artefacts   83

5   All about Language: Postmodern Interpretations,
    or the Muting of Humans and Technology   106

PART III
Technology Speaks (Objects Change Subjects)   139

6   What Technology Tells Us (to Do) (Part 1): Media,
    Artefacts, Networks   141

7   What Technology Tells Us (to Do) (Part 2): Narrative
    Technologies, or Interpreting and Materializing Ricoeur   211

PART IV
Humans, Language, and Technology Speak
(Subjects and Objects Entangled) 251

8 Using and Performing with Words and Things 253

Index 289

Acknowledgments

Many thanks to Andrew Weckenmann, Allie Simmons, and their colleagues from Routledge for supporting and working on this book project. I warmly thank Michael Funk, Stefan Koller, Wessel Reijers, Peter Rantasa, Martin Kusch, and Hans Bernhard Schmid for our discussions about topics relevant to this book, including Wittgenstein, use, performance, Searle, collective intentionality, Ricoeur, narrative, and voice. I am also very grateful to Yoni Van Den Eede and Martin Kusch for reading and commenting on an earlier draft of this manuscript, to the anonymous reviewers for reading and appreciating my initial book proposal, and to Agnes Buchberger for her assistance with formatting and correcting the final version of the manuscript. Finally, with this essay on language and technology I wish to honor Carl Mitcham, Don Ihde, Langdon Winner, and other "first generation" contemporary philosophers of technology, without whom this field would not exist in its present form, and to whose pioneering work this book can read as a respectful footnote.

1 Introduction
Words and Things

1.1. Introduction to the Topic: The Ancient Divide Between Words and Things

It is common to distinguish between words and things, humans and nature, language and technology. This is not only a modern habit, as for instance Latour (1993, 2004) suggests. Already in ancient-Greek times, words and things were conceptually separated, which then of course raised the question as to how they are linked; for instance, when we name things: Is the link a matter of convention, or are some words naturally linked to things?

One way to link them, I will argue in this book, is to focus on the notion of use and performance. Influenced by the later Wittgenstein, I will compare the use of language to the use of (other) tools. I will argue that, just as the use of language becomes crystallized in languages and language games, the use of technologies becomes crystallized in what I will call "technology games," and ultimately in forms of life—which in turn shape our lives and our use of languages and tools. Thus, I will argue that if we consider their use and performance in a social and cultural context, there is a strong similarity between words and things, and that in this sense language can be understood as a technology. I will also propose different ways to bridge the gap between words and things, such as a revised material hermeneutics that departs from Ihde but takes into account the role of language, a Latourian call for giving voice to things, and the concept of "narrative technologies" influenced by Ricoeur. But let me start with ancient-Greek ways of thinking about the problem concerning the relation between words and things, and then introduce some paths to address the problem that will be developed in this book.

Two useful starting points are Plato's *Cratylus* and Aristotle's *De Interpretatione*. In *Cratylus*, the question is whether names are natural and conventional: Cratylus argues that they are natural (Plato 1997, 383), whereas Hermogenes argues they are conventional (384c-d): 'No name belongs to a particular thing by nature, but only because of the rules and usage of those who establish the usage and call it by that name' (384d). Interesting in these arguments for my purposes here is the Socratic view that a name is an *instrument*. Socrates says that 'it is by using names that people say things' (387c6). The use of names is compared to weaving, and the making of names to carpentry (387–388). Socrates defends the naturalist view, but at the same time recognizes the aspects of instrumentality and use when he sees naming as an activity. He even explicitly says that a name 'is also a sort of tool' (388a) and compares naming to various crafts (388). In particular, he argues that names are tools to divide things according to their natures, just as for the weaver 'a shuttle is a tool for dividing warp and woof' (388b). Hence, names can be seen as what one could call ontological instruments, and ontology is the practice and craft of naming.

This comparison to instruments and crafts is interesting with regard to dealing with the gap between words and things, since it makes a clear link between the use of words and technological practice. I will argue that this connection is more than metaphorical, or rather metaphorical in a rich, McLuhanian sense of metaphor. Words can also be seen as instruments and tools, and the use of words can be compared to technological practice, to craft, which involves the mastery of a technique, and hence know-how and tacit knowledge: The use of words also requires technique and know-how. In the 20th century, Wittgenstein will argue against Augustine that meaning is not in words—Augustine thought that words belong to objects, stand for objects, and that this is their meaning—but in the *use* of words. Focusing on this "use" line of thinking will guide my reading of Wittgenstein in terms of technological practice and will help me to develop a Wittgensteinian approach to the relation between language and technology, and the view that language is a technology.

Aristotle's discussion of nouns is also relevant in this context. In the second chapter of *De Interpretatione* (*On Interpretation*), Aristotle argues that the meaning of nouns is conventional ('a name is a spoken sound significant by convention'), and later, in Chapter 3 and 4, he suggests that syllables and verbs only gain their full meanings when they are combined with other linguistic components. If we, perhaps against Aristotle, radicalize and generalize this thought, this suggestion makes us think of what is known as semantic holism, which holds that parts of language can only be understood in relation to a larger (linguistic) whole. Consider Wittgenstein's view in the *Investigations* (1953) that 'to understand a sentence means to understand a language'—and he adds: 'To understand a language means to have mastered a technique' (§199, 87e). Moving towards later-20th-century work in philosophy of language, we may also be reminded of Davidson's holistic view of meaning. According to Davidson, words and sentences have only meaning in the context of a whole language (Davidson 1967, 308).

Davidson's view built on Frege's thesis that words have only meaning in the context of a proposition, the so-called "context principle," which asks 'never to ask for the meaning of a word in isolation, but only in the context of a proposition' (Frege 1884). This also influenced Wittgenstein, who wrote in the *Investigations*: 'Naming is not yet a move in a language-game—any more than putting a piece in its place on the board is a move in chess. One may say: with the mere naming of a thing, nothing has yet been done. Nor *has* it a name except in a game. This was what Frege meant too when he said that a word has a meaning only in the context of a sentence' (§49, 28e). Could Wittgenstein's (and Davidson's) semantic holism be applied to the meaning of *technologies* and their use?

Influenced by Wittgenstein, I will introduce the term "technology games" and a form of life later in this book in order to suggest a kind of "technological holism." And if meaning of words is conventional, is the meaning of artefacts also conventional, as for instance Searle's view and the social construction of artefacts (see Chapters 3 and 4) seem to imply? Moreover, Wittgenstein's comparison with the mastery of a technique reminds us of the comparisons with crafts in *Cratylus*: It suggests that the use of language and the use of tools are both a matter of mastery. Yet Aristotle also claims, in the first chapter, that spoken words are symbols of mental experience, or more precisely: 'Spoken sounds are symbols of affections in the soul, and written marks symbols of spoken sounds.' Wittgenstein will question this mentalist and symbolist view, and contribute to the view that language is more than an instrument; it has its own kind of autonomy, which in turn shapes our thinking (and, keeping in mind Aristotle, one could add: our affections). Again the analogy with technology is interesting here: Similar to words, (other) technologies are also always more than instruments; they make their own contributions to meaning.

Thus, on this reading of Plato and Aristotle, words and things are conceptually separated, but on a closer reading of the their texts, they are also at the same time linked in various ways, and there are already some elements here that parachute us into 20th-century thinking about language: Plato already compared the use of words and the use and making of things, arguments for conventionalism were already discussed in Plato and in Aristotle, in Aristotle we already find the seeds of a holism that Wittgenstein and Davidson will radicalize, and in response to Aristotle we can already raise the question as to how the instrumentality of language and technology must be understood: Do we only use language, or does language also use us? Although I will criticize Wittgenstein's neglect of the material object, I will sympathize with his view that language is all about use, and that when our use becomes crystallized in language, language assumes its own autonomy and then *uses us*, constituting what Wittgenstein called "language games" and a "form of life" (Wittgenstein 1953). Thus, everyday language use takes on a specific form, language use gets literally formalized in a language and a culture, and this in turn shapes everyday language use. Similarly, I will argue, uses of technologies become formalized in technology games and ultimately in a form of life, which in turn shape our use of these technologies. Like linguistic holism, then, there is also a kind of "technological holism." Moreover, we could consider how both "wholes" are or can be connected.

Perhaps there are more bridges between words and things, between language and technology, and between culture and materiality.

In Plato's time, however, essentialism, naturalism, and universalism prevailed, and the analogy with technological practice remained an analogy. Technology was generally seen as belonging to a separate, entirely nonlinguistic realm. *Logos* seemed to be a way to gain independence from the variable and context-dependent, a road to the perfect rather than the imperfect, a way of getting in line with the natural—understood by Plato in terms of eternal forms rather than transient materiality. This rendered technological practices and their materiality amoral and apolitical. As Hannah Arendt articulated in *The Human Condition* (1958), the ancient Greeks defined the sphere of freedom and politics as having to do with speaking, whereas the nonpolitical was the sphere of the household, defined as the sphere of necessity. In other words, politics was about having, using, and giving voice, about using language; things, crafts, and arts, by contrast, belong to the sphere of necessity, the sphere of slavery and unfreedom. Free men and philosophers, so it was assumed, should speak, deliberate, and discuss; if they are concerned with making things, then they should speak about the principles that govern this (*techne*), about the forms; they should not get their hands dirty.

Today, this view and hierarchy might have been eroded a little, but the very divide between words and things still influences our thinking and shapes our practices. Consider, for instance, the way we organize research and higher education. At universities, scholars and students occupying themselves with language and humans (humanities and social sciences) are usually separated from those who deal with things and with nature (science and engineering). This conceptual separation often has a physical and geographical counterpart: People are put in different buildings or even different parts of the city. There are those swimming in language and living in libraries where they engage with books and texts (humanities, social sciences), and there are those steeped in the wet and dirty world of living things and material artefacts (natural and engineering sciences). There are the human, "subjective," and political academic activities (in the seminar room), and there are the "objective" and apolitical academic activities that deal with nature (in the lab). Already at primary-school level, language learning is traditionally separated from learning about things.[1] Similarly, in occupations and professions, there are tends to be a line between those who deal with words (lawyers, teachers, politicians, writers, etc.) and those who make things (engineers, workers, craftspeople, etc.). There are people who are trained to be good at speaking (using their voice, writing), and people who are educated to be good at making things (using their hands, using tools), and usually these groups are separated.

Metaphysically speaking, one could say that objects and subjects are separated. There are human subjects on the one side, and nonhuman objects on the other side. Insofar as one is making things, one does not have subjectivity;

subjectivity can only come from language. In so far as one is speaking, one is not touched by natural or artificial things. There is the art of speaking and there is the art of making, and never shall they meet.

As Latour has argued, this is a typical modern way of speaking words and doing things. Although *in practice* we may be far less modern than this metaphysical picture suggests (Latour 1993), this is the way we continue to understand ourselves and make sense of our activities, and as such it shapes our thinking and doing. It is also largely the way we continue to build our world—our physical buildings and our social institutions. But this excursion to ancient thinking shows that the modern dichotomies and categories have roots in ancient Platonic and Aristotelian thinking, or rather: in ancient ontological-linguistic techniques, in a particular ontology understood as the use and mastery of names that divide up being into words and things.

Insofar as these ontological and cultural distinctions still hold sway, then, thinking about technology seems to have little to do with thinking about language. Thinking about technology appears to concern thinking (that is, use of language, logos, words) *about* technology (material artefacts, tools). The words are not changed by the things, and the things are not changed by the words. Subjects and objects are neatly separated. One can talk and write *about* technology, and of course one can program a computer to utter words. But according to this approach to language and technology, these are not the words of the computer. The linguistic agency and hence the subjectivity remain entirely on the side of the human.

But what if we cross the line? *What if the line has already been crossed?* What if we consider, for instance, a science such as archaeology, which connects tools to culture, making a seemingly magical jump from things to words? What if we consider computer sciences, which seem to do things with words *and* things, making an equally magical shift from code to material processes and back? What if in contemporary ICT, *techne* and *logos* come closer or are even united, for instance in the programming *and* making of very material robots? What if contemporary biotechnology merges code/text, living things, and engineered things? What if there never *was* a line between words and things? What if *all* sciences have always done a hermeneutics of words and things? What if the history of technology is also a history of words and if the history of culture is fundamentally always also a history of technology? What if we rewrite the history of cultures and civilizations in terms of the ways we use language, understood as the use of language-technologies and other media/technologies we have? Walter Ong, a student of McLuhan, famously defined writing as a technology like other technologies—a technology which enabled a major shift from orality to literacy, impacting human consciousness and culture: Writing and later print technology changed the way we think, personality, and social structures (Ong 1982). *What if we define language itself as a technology?* What if we consider even the human voice as a technology (Young 2015, 6), which then itself can be extended by, and mediated through, other technologies? And what if the claim of contemporary postphenomenological philosophers of technology, that subjects and object co-constitute one another, can be interpreted as implying that words co-constitute things and vice versa? What if a material hermeneutics à la Ihde (e.g., Ihde 1990) is enriched by the recognition that words and things are intensely entangled? What if Latour's critique of a purification that separates culture and nature (1993, 2004) can be applied to words and things, as I suggested in the previous pages? What if technologies themselves can have a "narrative" function, as I will argue in Chapter 7?

Such questions open up an exciting area of research for philosophers of technology at the intersection of thinking about technology and thinking about language. Of course in philosophy of technology and neighboring fields such as science and technology studies (STS), *some* work has been done that explicitly or implicitly touches upon the topic of language and technology. For instance, Winner (2014) already made references to Wittgenstein, and STS scholars have done discourse analysis. But this book develops a more systematic project. I will show how we can (re-)interpret and revise some classic and contemporary thinking about technology, and indeed thinking about *language*, as attempts to say something about technology *and* language. I will use work by Wittgenstein, Heidegger, McLuhan, Latour, Ihde, and Ricoeur in order to try to better understand the relationship(s) between language and technology. When writing this book, I became particularly fascinated by the work of the later Wittgenstein, especially the Wittgenstein of the *Investigations*. I will interpret and use that work throughout this book as opening up the way for a fruitful comparison between use of language and use of technology—indeed, for conceptually integrating them, by *understanding both language and technology in terms of use* of tools, as performances, and as technological practices. I will also use Ricoeur in a perhaps unexpected way to think about "narrative technologies," thus rendering at least two "philosophers of language" that are usually not considered by philosophers of technology, useful to thinking about technology.

Note again that, outside philosophy, there is, and there always has been, a lot of research that crosses the lines between words and things, humans and nature, and language and technology. This is also true for contemporary work in the domain of high tech, inside and outside academia. For instance, in the domains of social robotics and artificial intelligence, there is at least some exchange between the worlds of social science (and occasionally humanities) and computer science, as an "engineering" science. And computers seem to develop their own "voice," as they get more "human" and literally speak far more often (e.g., as robots). In this kind of research, linguistics often meets computer science. However, it is far from clear, for instance, how the relation between language and technology should be conceptualized in such interdisciplinary projects, if computers really can have a "voice." And it would be interesting, for instance, to explore what exactly the thinking of Wittgenstein, Latour, or Ricoeur implies for practices that connect language and technology. More generally, there are currently few interfaces between, on the one hand, philosophy of technology, which often stays on the technology side of the technology/language divide, and on the other hand, multidisciplinary scientific and technology research practices that cross this line. And as suggested before, perhaps our scientific and other practices have *always* crossed the line in various ways, even if we have difficulties in seeing that hybridity through our modern glasses. However, for philosophers of technology to meaningfully and successfully engage with these other fields, and indeed to interpret many everyday practices inside and outside science and academia, it is important to sharpen one's conceptual tools and to expand one's theoretical toolbox *as a philosopher of technology*. Although I will give some examples and, for instance, use the case of social media to develop my views, the emphasis in this book is on making a significant and substantial contribution to the construction of a conceptual framework that enables us to address issues concerning language and technology. So far, such a framework is lacking in the still relatively young discipline of philosophy of technology.

To conclude, at this point in the development of the field, philosophy of technology can do with *a systematic study of how technology relates to language* and vice versa, which addresses the ancient problem of the divide between words and things from a new angle (philosophy of technology), and rethinks contemporary philosophy of technology in ways that account for the role of language.

Ideally, such a project would also spark off and involve rethinking contemporary philosophy of language in a way that accounts for the role of technology. As this section already illustrates and as I will show in the next chapters, a focus on technology also throws new light on standard philosophy of language texts. However, my main focus in this book will be on rethinking *philosophy of technology*.

1.2. Aims and Rationale of This Book

This book aims to make a significant contribution to this project of rethinking philosophy of technology in order to account for the role of language, and enquires into, more specifically, attempts to conceptualize and map **the relationship(s) between technology and language**. Let me further explain the rationale and aim of the book.

At first sight, contemporary philosophy of technology seems to largely adhere to the old ontological lines described in the beginning of this introduction, lines which separate language and technology, words and things. While earlier philosophers of technology such as Heidegger were much occupied with issues concerning language *and* with issues concerning technology, language has received far less attention from contemporary philosophers of technology. After the turn away from what Mitcham called "humanities philosophy of technology" (Mitcham 1994), the main focus was and is on "technology itself": on artefacts, things. Apparently it was assumed that one had to *choose* between studying either words or things, and the latter was chosen. Analytic philosophy of technology focused on topics such as engineering science and design, and the metaphysical status of artifacts. Others were influenced by continental philosophy, for example phenomenology, but took an "empirical turn," (Achterhuis 2001) which has largely resulted in neglect of issues concerning language. Consider, for instance, the material hermeneutics of Ihde (Ihde 1990). The focus is on materiality, on things (Verbeek 2005). The same is true for so-called engineering ethics. Moreover, philosophers of technology influenced by critical theory (e.g., Feenberg 1999) and political economy quite naturally focus on the material structure of production in society and on power issues related to artefacts (e.g., Winner 1980—although Winner is also an exception, since he used Wittgenstein; I will return to this). And philosophers influenced by, for instance, Merleau-Ponty and/or by feminist theory who focus on embodiment (Ihde again) and the body often see the body as entirely distinct from, or even opposed to, language. In feminist discourse there has been a discussion about whether culture and a particular way of thinking can be "inscribed" in the body. Here, Judith Butler's response to Foucault is interesting since it questions the pre-cultural and pre-discursive body and, more generally, discusses the relation between body and culture (Butler 1989). But these kinds of discussions seem to be missing in philosophy of technology. Thus, although there are some exceptions, language and its relation to technology was and is not on the agenda of contemporary *philosophy of technology*, which tends to be focused on things, and sometimes also on bodies. The (perhaps largely unintended) result of these developments is that **contemporary philosophy of technology lacks systematic reflection on the relationship between technology and *language*.**

At the same time, work in 20th century philosophy of language—both in the analytic and continental tradition—has overemphasized language to the neglect of the material object (and embodiment). For instance, in the work of Wittgenstein, Austin, and Searle, the focus has been on the use of language, on words. The use of technology and material artefacts has remained out of sight. Similarly, in the French poststructuralist and postmodern tradition, influenced by a linguistically oriented semiotics, the focus has been on signs. Material objects could only appear as signs or symbols. More generally, until today, philosophers of language have shown little interest in the growing field of philosophy of technology and its insights into how to understand technology. Thus, there is **a gap between philosophy of technology and philosophy of language.**

This book aims to start closing this gap by offering a systematic and sustained reflection on the relationship between technology and language. Engaging with the work of thinkers such as Wittgenstein, Searle, Heidegger, McLuhan, Ihde, Winner, Latour, and Ricoeur, it **offers a systematic framework** for thinking about language and technology, and an argument for interweaving thinking about technology with thinking about language for their mutual benefit. The result is that the main claim of contemporary **philosophy of technology**—that technologies are not mere tools and artefacts not mere things, but crucially and significantly shape what we perceive, do, and are—**is rethought in a way that accounts for the role of language in human technological experiences and practices**, which I will mainly understand in terms of use and performance.

The book achieves this aim by integrating thinking about language use with thinking about technology use (starting from Wittgenstein and commenting on Heidegger), and by reinterpreting and revising Ihde, McLuhan, Latour, and Ricoeur in ways that also establish a more intimate relation between language and technology, linking use of technology to performance and narrative. In terms of argumentative structure, it achieves its aim by learning from, critically responding to, and constructing a synthesis of, three "extreme" untenable positions: (1) The claim that only **humans** speak and that neither language nor technologies speak, (2) The claim that only **language** speaks and that neither humans nor technologies speak, and (3) The claim that only **technologies** speak and that neither humans nor language speak. The concluding position, which constructs the view that there is an entanglement between subject and object, and which attempts to give "voice" to humans, language, and technology, can then be represented as the centroid of a triangle formed by connecting the three positions:

technology

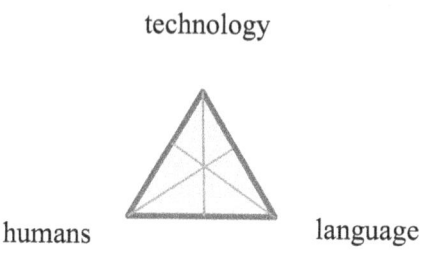

humans   language

Figure 1.2.1 Who or what speaks? Triangle of positions.

Note that the three positions are indeed "extreme" and are not meant to correspond entirely and exactly to existing theoretical positions. They are what (with Weber) we might call **"ideal types."** Ideal types are not meant to correspond exactly to positions in the literature, which are typically more nuanced than presented here, but help to bring order in what otherwise would be a far more chaotic field of positions. I use generalization and even exaggeration (hyperboles) to bring out particular dangers associated with some real positions in the field, which are interpreted as being on a slippery slope towards the ideal type position. One could also call the outline of these positions a map (Coeckelbergh 2015), which is always a generalization and simplification. The map is not the territory. If a map is supposed to be a perfect and true representation, then the framework presented in this book fails as a map. But as a conceptual *instrument*, it enables me to navigate, construct, define, interpret, and revise real, existing positions, which is pivotal in constructing my own synthesis. My discussions of the "extreme" positions can hence be regarded as necessary *detours* on the way to an integrated framework.

The structure of the book follows the structure of this argument, with the first three parts representing my constructions of, and responses to, the three extreme positions, and the final part representing the synthesis. In addition, the book connects this argument concerning positions with **a narrative about language and technology** getting closer and more intimate, and **a narrative about subjects and objects** becoming entangled: Whereas in the view discussed in the first part, subjects and objects are kept apart and are not altered by one another, this changes in the next parts when subjects and objects start touching one another, and when subjects change objects and objects change subjects. In the concluding chapter, an attempt is made to "balance the triangle": A view is developed according to which subjects and objects are entangled and constitute each other, and which conceptually integrates the use of language with the use of technology in a way that gives voice to humans, language, and technology.

The result is a book that rewrites philosophy of technology in a way that engages with influential work in philosophy of technology (e.g., Ihde and McLuhan) and related fields (e.g., Latour), but also draws in central discussions from **other subdisciplines of philosophy**, such as philosophy of language, epistemology, metaphysics, philosophical anthropology, and so on—offering original interpretations and revisions of Wittgenstein, Searle, and Ricoeur. This renders the book not only relevant to philosophers of technology but also to other philosophers, who might find my readings of "their" texts helpful, as these texts are reread and reinterpreted with a focus on technology and in light of some central concerns within philosophy of technology.

1.3. Overview of the Chapters

The structure of the book reflects the argument concerning who/what speaks and the narrative about subjects and objects.

Part I articulates and critically discusses the position that **humans speak,** whereas language and technology are instruments, and that there are subjects and objects, but that these are not intrinsically linked; they do not mutually shape one another.

The present chapter, Chapter 1, has introduced the problem of the divide between words and things, and in response to Plato and Aristotle, it has already announced a Wittgensteinian-Heideggerian approach to addressing that problem. **Chapter 2** continues in that direction. First it articulates instrumental views of language and technology: they are mere tools, instruments for human purposes. Indeed, one view of the relationship between language and technology is that there is no intrinsic connection between the two. Humans decide about ends; all the rest, indeed everything nonhuman, belongs to the category of means. Language is a tool for communication, and in the use of media, the message and humans remain unchanged. There are subjects and objects, and both remain unchanged in the encounter. As instruments, language and technology are neutral. Then I start criticizing this instrumental view of language and technology by using Wittgenstein and Heidegger. Language and technology are indeed instruments, but they are more than instruments. They co-constitute what Wittgenstein (1953) calls our form of life and, as Heidegger (1977) has argued, they also shape our thinking. I use Wittgenstein's comparison between use of language and use of technology: Turning the metaphor around, I apply his view of language to technology. I argue that technologies are part of what I call "technology games" and are a form of life. Thus, while Wittgenstein's upfront focus is on use of language, we can use Wittgenstein to think about technology. Moreover, I show that, paradoxically, through Wittgenstein's emphasis on *use* and instrumentality, we can go beyond an instrumental understanding of *both* language and technology. I also comment on Dewey to develop my arguments.

In the next section, I consider the view that there are narratives and discourse *about* technology. There are different ways of talking about and interpreting technology, and we can acknowledge that these ways are not neutral. Yet, although language and technology get a little "closer" in this view, there remains a gap between the two. The subject does not touch or change the object; at most it coats and colors it. Again, Heidegger and Wittgenstein are used to show that an alternative, noninstrumental view is possible, which connects technology and language in more intimate ways and goes beyond detached analysis of discourses and narratives. I ask if philosophy of technology is critical enough of its own thinking, understood as use of language. Influenced by Wittgenstein, I question the "theoretical" attitude and argue that we should critically examine speaking about technology, and also speaking about technology by philosophers of technology, since that speaking is also a language game and is linguistically and culturally embedded. The language we use shapes how we talk and think about technology. At the same time, however, technology also shapes our language and thinking. Technology is, to speak with a Heideggerian term, a "mode of existence." Guided by Heim's discussion of computers and language (1992), I use Heidegger's comments on the typewriter (1982) and on "language machines" to argue that current information technologies change our relation to the world.

Another, far less radical way to conceptualize a more active role of language and "closer" relation between language and technology, which also focuses on the *use* of language, is Searle's social ontology and its concepts of declarations and status functions. This is the view I articulate and critically discuss in Chapter 3. What specific technological artefacts "are" depends on what we declare and agree them to be. Like in the discourses *about* technology view, language again plays the role of coater and painter. There is a physical-material object—which itself remains unchanged—and this object is coated with subjective meaning by language users. But the role of language is no longer merely instrumental here: Language also constitutes the meaning of the object; the object is "touched" by the subject.

Yet this chapter also criticizes Searle's one-sided focus on language to the point of downplaying of the role of material objects, his physical-world/social-world dualism, and his exclusive focus on declaration. It attempts to give technology a more central role. Continuing the Wittgensteinian-Heideggerian argument started in the previous chapter and learning from Austin's work on performatives and speech acts, a different view is sketched, which integrates language use and technology use, highlights the role of material objects in Searle's *own* work, considers other relations and functions next to declaration, replaces Searle's contractarian view of the social with one according to which there is already a social-technological "given" within which concrete uses of words and tools are embedded, and suggests the concept of performance to further open up this direction of thought. Wittgenstein's towel example is used to bring together language use and technology use, and thinking itself is understood as an activity that includes artefacts and body parts as tools. Technology starts "speaking." This position is further developed in the last chapters.

Part II first considers the position that **language speaks**, whereas humans and technological artefacts do not. Here, subjects change objects, with "subjects" understood in linguistic terms, as standing for language.

Chapter 4 starts with considering theory about the social construction of artefacts. If artefacts are "socially constructed," (Pinch and Bijker 1984) then, in its radical version, this means that there are neither facts nor artefacts apart from the social; what artefacts are and how they are use is the outcome of a societal process in which meanings and uses come to be shared by a social group or society. Thus, again, meaning is given to artefacts, but, in contrast to the Searlean view, the subject—in the form of the social—is more in the foreground and the object retreats to the background. However, here, use and the role of *language* remain undertheorized.

This problem is overcome when social constructivism is reinterpreted and revised in a Wittgensteinian way and when we focus more on use: the use of technology (also in design) and the use of *language*. Meaning emerges through use, although that use is always connected to larger patterns and to social interests. If the design and interpretation of artefacts is a social and cultural matter, then we should also include language, which is interwoven with the social and cultural. In this sense, there is something like *the social-linguistic construction of artefacts*. The meaning, "rhetoric," and "grammar" of artefacts is connected to the meaning, rhetoric, and grammar of the practice and of the society and culture they belong to, and this implies that their meanings are connected to language. This also implies that in innovation and technology development, there are dynamics of *linguistic-social-technological* change. To invent a technology is also to invent a language and a world. Use of language turns out to be a crucial component in innovation processes.

I also point out other views and discussions in sociology of knowledge relevant to social constructivism, in particular Winch's interpretation of Wittgenstein. I argue that social constructivism, by taking distance from these kinds of more linguistically oriented discussions and approaches in sociology of knowledge, has come to neglect the role of language. This also holds for philosophers of technology who have taken a material turn and have used social constructivism in a way that neglects language. This differs from Latour's work with Akrich: While partly turning away from "old-fashioned" philosophy and sociology of science, their approach retains an interest in semiotics and even what we may interpret as a Wittgensteinian interest in performativity and a non-representational view of language and knowledge. I propose to use Latour to further think about the relation between words and things; perhaps artefacts and language are much more intertwined than usually supposed. Latour's work also hints at the aspect of narrativity, which I will discuss in Chapter 7.

But first, in Chapter 5, I articulate postmodern and poststructuralist interpretations of Heidegger, Wittgenstein, and McLuhan, which tend towards the extreme claim that only language speaks. Humans become mere vehicles of language and the wider social-cultural whole. For thinking about technology, this view implies that now the subject completely eclipses the object. Materiality disappears; only semiotics and grammar (understood in a purely linguistic way) remain. There are signs, but they no longer refer to objects. Only the subject remains—no longer in human form, but in the form of language. This position helps us to acknowledge the important role language plays in thinking about technology. It makes us understand why—in spite of the lessons of Heidegger and contemporary philosophers of technology such as Ihde—we continue to think about technology as an instrument: We speak about it as an instrument, since *language* speaks about it as an instrument. The very word "technology" misleads us. However, in its radical, postmodern version, this view mutes both the human and technology. There is no longer a human speaker; humans are media, used by language for its speech. And technology can only appear here as language. Its materiality has been rendered invisible. This is problematic and has led many philosophers of technology to turn away from these directions of thinking.

However, as the previous chapter also shows, another interpretation of Heidegger, Wittgenstein, and McLuhan is possible; in this chapter, I continue this route. I also argue that one should not throw away the postmodern child with the bathwater, as postphenomenology did with its empirical turn. I show that in work by Derrida, Lyotard, Baudrillard, and other poststructuralist and postmodern thinkers, there are interesting elements we can use for thinking about technology, including Derrida's use of the *bricolage* metaphor, Lyotard's Wittgensteinian concepts, and Mersch's suggestion that not everything can be said. Moreover, apart from the "message" or the "content" of these thinkers, they also offer some interesting approaches, some of which may be compatible with postphenomenology or, as Smith has argued, may even be interpreted as constituting *another version* of post-phenomenology from which those who today claim postphenomenology as a school may learn. In particular, I argue that contemporary philosophy of technology may want to incorporate *transcendental* arguments, do more to acknowledge difference, and study the phenomenology and epistemology of concrete, unique encounters with technology. Moreover, these are also helpful additions to, and revisions of, the Wittgensteinian approach developed so far.

In the next and final part of the book, I engage more directly and more extensively with postphenomenology (especially Ihde), next to Latour and (more) narrative approaches. **Part III** articulates the position that **technology speaks**, but at the expense of humans and language.

Chapter 6 considers various ways of constructing and (re-)interpreting the position that technology tells us things (McLuhan and media ecology, but also the material hermeneutics of Ihde's postphenomenology) and tells us what to do (Latour and Akrich). In response to Wittgenstein, Searle, and the postmoderns, this position brings back the material object. First, I discuss McLuhan's view, including Postman's media ecology's reading of McLuhan which puts more emphasis on language, but which remains too dualist. I point to interesting ways in which not only McLuhan's *Understanding Media* but also *Laws of Media* (1988) can help us to connect language and technology, seeing language as technology and bringing in insights from "the grammarians."

Then I focus on Ihde, whose turn to material artefacts was also a turn away from language. Understandably tired of the perversions of the linguistic turn and rejecting postmodern interpretations of Heidegger of the 1970s and 1980s, mistaken for Heidegger's own view, postphenomenology stressed the materiality of technology and (re)turned to the things. Technological artefacts, in their very materiality, mediate our relation to the world (e.g., Ihde 1990). Things "do" things (Verbeek 2005). But on the way, postphenomenology lost attention for the role of language. In this chapter, I read language back into Ihde and show that his phenomenology and hermeneutics, in particularly what has become his mediation theory, can be usefully expanded in order to account for the role of *language* in mediating our relation to the world—often together with technology. Moreover, using Wittgensteinian and other insights gained so far, and adding a transcendental argument, I show how postphenomenology could be revised with, and at least partly replaced by, a Wittgensteinian-Heideggerian approach that puts human-technology-language relations in a larger social-cultural context, recognizing that there are games and forms of life that are given and that make possible and structure our use of, and experience with, technology and language.

The next section explores another route to rethinking technology in a way that accounts for the role of language and puts technology in a social context: interpreting, using, and critically discussing Latour. First, I return to Akrich and Latour, and show how this work can bring in a semiotic approach that mixes language and technology. This is about what technology tells us to *do*, on the script of artefacts. Technological artefacts can have prescriptive messages—to use a term from Austin: They can perform elocutionary speech acts. As Latour and Akrich (1992) argue, prescriptions are inscribed into artefacts by humans and hence there is a sense that things make people do things by telling them what to do. It is material semiotics, which moves from signs to things and back (Akrich and Latour 1992). The metaphor is the theatre: Things have a script, and this script makes us, human actors, do things. This "performative" and process dimension of Latour is also stressed in Pickering's interpretation of Latour's posthumanism (Pickering 1995). Furthermore, in Latour's nonmodern view developed in *We Have Never Been Modern* (1993) and *Politics of Nature* (2004), things also "speak". But here humans do not use things to speak and script; things use humans and technologies to speak. There is a collective of humans and nonhumans, and artefacts are liberated from their muteness; culture/nature and human/nonhuman dualism is overcome. Yet, although the collective contains humans and even discourse, Latour stresses things and their speech; what humans and language say retreats to the background. In response, I try to read more language into Latour's text and attempt to make sense of various kinds of "speaking."

By offering these interpretations, I show that Latour's nonmodern, posthumanist view and Ihde's postphenomenology can charitably be interpreted as attempting to provide a synthesis of the three extreme positions identified in the beginning of the book. Yet, I argue that this synthesis and integration only partly succeeds if it stays too close to the authors; although they do not totally neglect language, both Ihde and Latour pay not enough attention to *language* in their central conceptualizations of our relations to technology and of the social; they do not sufficiently clarify its role vis-à-vis technology. They also miss insights from structuralist, "grammarian," and transcendental arguments—the latter which do not need to be remote or dismissive of the material at all. Moreover, while Ihde, Latour, and their postphenomenological readers and interpreters (e.g., Verbeek) can be interpreted as inventing a new language to talk about things and, hence, as having made original contributions to thinking about technology, as philosophers they remain insufficiently critical of their own use of language, which shapes their discourse about what things do. This is partly so because they reject a *transcendental* perspective (or what Heidegger would call the "ontological" dimension): a rejection which blinds them to the transcendental role of both language and technology in structuring our speech and our thinking, that is, our use of words, tools, and media. In addition, although in his earlier work Latour was influenced by a narrative semiotics and although Ihde knows Ricoeur's work, both do not sufficiently conceptualize the *narrative* dimension of our relations to technologies (and indeed of human praxis and existence). Furthermore, Latour's nonphenomenological approach does not do enough justice to *experience* (it is constructed from a third-person view), *concrete language use*, and technology use by embodied humans in specific practices. From a Wittgensteinian and Heideggerian perspective, one could say that his view is too theoretical.

Chapter 7 then further explores the position that technology tells us things and tells us what to do, but this time from a different angle that *does* start from (thinking about) language and pays a lot more attention to other structural elements of language and meaning that did not receive much attention so far: temporality and, in particular, narrativity. Indeed, another way to conceptualize the *logos* of technology is to use Ricoeur's theory of narrativity and apply it to technology. Ricoeur's theory can be used to say that technology can be "read" like a text (see also Ihde), against a broader cultural context, and that stories can be told *about* the ways technology figures into our lives (Kaplan 2006; see also Chapters 2, 4, and 5). But technology can also be seen as actively configuring actors and events into a plot: then they become what Reijers and I have called narrative technologies (Coeckelbergh and Reijers 2016). In this chapter, I use and further develop this view by combining it with the Wittgensteinian focus on use and performance, and by comparing it with Latour's view. In our use of, and performance with, technology, technology cowrites, co-narrates, and coauthors our stories; it "speaks" in the sense that it structures our narrative and narrative time. Furthermore, the social is not an unstructured collective, but has a diachronic aspect and narrative structure. Again technology takes up a role usually ascribed to humans, and again the metaphor of the theatre is used, but in contrast to Ihde's postphenomenology and Latour's work on the nonmodern, the temporal dimension and the narrative structure of technology use and of the social life are emphasized.

This approach does not necessarily mean that humans are muted. Humans can also in turn interpret and organize technological artefacts, or co-write the narrative and codirect the configuration of characters and events. Furthermore, while in the posthumanist accounts discussed in Part II and Part III human experience seems to be missing, including the experience of technology and the experience of language, as language users and as beings-addressed, this is not necessarily the case in the approach I propose. There is "individual" experience, but human experience and performance is always also at the same time social and linked to collective (crystallization of) experience and performance, since it is related to and structured by language and technology games, and presupposes a form of life.

That being said, it remains questionable how much the "narrative technologies" view developed here really involves language. Is language (and narrative) used as a metaphor, or is the narrative process really linguistic in nature? A more comprehensive and balanced view is needed. Perhaps a synthesis can be constructed by combining the insights gained so far from using, revising, and responding to work by Wittgenstein, Heidegger, McLuhan, Ricoeur, Latour, and Ihde. At the end of Chapter 7 I already ask what it means for humans to acquire and have a voice, through language and through technology. I also discuss contemporary social media in order to show what the proposed approach means for understanding technology and its relation to language and humans. Finally, I raise the question of power (here I also comment on Foucault) and argue that, so far, the aspect of embodiment is still missing and should be added to the framework. In this context, I use again a transcendental argument.

Part IV attempts to further develop an integrated approach, a position that gives **humans, language, and technology** a voice. It tries to balance the triangle: Can we define a position that gives humans and technology a voice, humanizes technology and language, and materializes humans and language? Assembling work done in the previous chapters and integrating insights gained along the way, Chapter 8 attempts a synthesis of the extreme positions: it tries to give humans, language, and technology their due, to let them all "speak." It lists a number of conceptual building blocks for a framework to think about the relations between humans, language, and technology, and then articulates an integrated, balanced view that gives voice to humans, language, and technology. It does so by focusing on what it means (a) to use and perform with words and things, and (b) to understand language and technology as entangled. The chapter ends with a reflection on the role of art in thinking about technology.

Let me offer a preview of the integrated view by summarizing its contributions and responses to (1) philosophy of *technology* in the tradition of the empirical turn, (2) philosophy of *language*, and (3) (post)*human*ism, reflecting the three positions on the triangle presented earlier (technology, language, humans).

On the one hand, the integrated view pays more attention to *language* than, say, Ihde or Verbeek. If the human-technology relation is conceptualized in terms of the entanglement and—in line with Verbeek—mutual constitution of subjects and objects, and if the use and performance of language and the use and performance of technology are conceptually linked and integrated, then the picture that emerges is one which halts the material turn (and the linguistic turn) where it is in danger of going too far, or rather one in which the material turn is redirected in a way that takes into account the role of language. The view that only humans speak is no longer tenable, and neither is the postmodern deletion of the human in favor of the text. But this book shows that the empirical turn's one-sided if not obsessive focus on the artefact, the thing, the object, tends to throw away the baby with the bathwater. Contemporary philosophy of technology can learn a lot from more human-centered and language-centered accounts if it wants to give the full range of human experience with technology its due, and if it wants to further elaborate and deepen its approaches and theories, for instance by transforming its phenomenological and hermeneutic approaches in a way that integrates thinking about use and performance, transcendental arguments, process thinking, and narrative theory. Technology may do much more than instrumentalists think it does, but this noninstrumentality or more than instrumentality should not only be conceptualized in terms of mediation (understood as an "in between") and in a way that is divorced from use and performance, including use of language and performance with language; it is also part of human experience and practice that we use technology and perform with technology. I will argue that, paradoxically, taking seriously this use dimension of technological experience and practice makes possible conceptualizing the noninstrumentality of technology. Moreover, it is also part of technological experience and practice that we speak about technology, that we give meaning to objects, that language structures and makes possible particular uses and performances with technology, and thus more generally that language does more than we commonly think it does. It is important to better understand and conceptualize the relations between language and technology in human experience and practice. Moreover, so far, and perhaps ironically, most philosophers of technology have spoken about technology without sufficiently criticizing their own instrument: language. A truly *philosophical* thinking about technology that awakens from its dogmatic slumber is aware of this problem, tries to remedy it, and critically discusses the relationship between language and (thinking about) technology.

On the other hand, in this book, philosophies of language such as Wittgenstein's and Ricoeur's are used in a new way that takes into account *technology*. As said, currently there is an unfruitful gap between philosophy of language and philosophy of technology. Much philosophy of language is still done without reflecting on technologies and technological artefacts, and most philosophers of technology—both in the "analytic" and "continental" traditions—tend to conveniently ignore lessons from philosophy of language. By applying Wittgenstein's (but also Austin's and Searle's) thinking to use of technology and performance with technology, by taking seriously McLuhan's suggestion that language is technology (and vice versa) and his "grammatical" approach, by "reading language back into Latour and into Ihde," and by interpreting Ricoeur in a way that highlights the temporal dimension and narrative functions of technology, the present book makes a contribution to bridging this unfruitful gap.

Finally, a posthumanist view of the *human* emerges which acknowledges that humans are not the only speakers, that there is a sense in which language and technology always already "co-speak" with us. When we focus on the concrete phenomenology and hermeneutics of use and performance with technologies, we see that language and technology not only mediate our embodied relations to these technologies but also function as conditions of possibility for our human speaking. It turns out that the human-speech subject and acts, which are so dear to humanists, are constituted by and emerge from a larger linguistic and technological structure, whole and *milieu*, which makes possible our speech in the first place and pre-shapes and pre-structures what we do with words and (other) things. As we are entangled and enmeshed in language and technology, our speech acts and our narratives are always cowritten and codirected by the linguistic and technological words, gestures, and grammars that surround us and by the games, social-material processes, and form(s) of life that carry us, from which our own individual and collective voices emerge.

Note

1 For instance, in the subjects taught at German and Austrian primary schools, a distinction is made between *Sprache* (language), Mathematics, and *Sachunterricht* (literally: teaching about *things*), which is concerned with geography, history, biology, economy, physics, technology, etc. The latter is seen as *sachbezogen*, which means: concerned with things. Although, in contemporary education, efforts are made to connect language and mathematics with learning about (other) things, these names still bear witness to an older way of thinking, which neatly separates learning about words, numbers, and things.

References

Achterhuis, Hans, ed. 2001. *American Philosophy of Technology: The Empirical Turn*. Bloomington/Indianapolis: Indiana University Press.

Akrich, Madeleine and Bruno Latour. 1992. "A Summary of a Convenient Vocabulary for the Semiotics of Human and Nonhuman Assemblies." In *Shaping Technology/Building Society: Studies in Sociotechnical Change*, edited by Wiebe E. Bijker and John Law, 259–64. Cambridge, MA: MIT Press.

Arendt, Hannah. 1958. *The Human Condition*. Chicago, IL: University of Chicago Press.

Aristotle. 1984. "De Interpretatione. (On Interpretation.)" In Vol. 1 of *The Complete Works of Aristotle*, edited by Jonathan Barnes, 25–38. Princeton, NJ: Princeton University Press.

Butler, Judith. 1989. "Foucault and the Paradox of Bodily Inscriptions." *The Journal of Philosophy* 86(11): 601–7.

Coeckelbergh, Mark. 2015. "Language and Technology: Maps, Bridges, and Pathways." *AI & Society* (online first). doi:10.1007/s00146-015-0604-9.

Coeckelbergh, Mark and Wessel Reijers. 2016. "Narrative Technologies: A Philosophical Investigation of Narrative Capacities of Technologies by Using Ricoeur's Narrative Theory." *Human Studies* 39(3): 325–46. doi:10.1007/s10746-016-9383-7.

Davidson, Donald. 1967. "Truth and Meaning." *Synthese* 17(3): 304–23.

Feenberg, Andrew. 1999. *Questioning Technology*. New York/London: Routledge.

Frege, Gottlob. 1980 (1884). *The Foundations of Arithmetic*. Translated by John Langshaw Austin. Second Revised Edition. Evanston, IL: Northwestern University Press.

Heidegger, Martin. 1977. "The Question Concerning Technology." In *The Question Concerning Technology and Other Essays*, edited by Martin Heidegger, translated by William Lovitt, 3–35. New York: Harper & Row.

Heidegger, Martin. 1982. *Parmenides*. Frankfurt: Vittorio Klostermann.

Ihde, Don. 1990. *Technology and the Lifeworld: From Garden to Earth*. Bloomington: Indiana University Press.

Kaplan, David M. 2006. "Paul Ricoeur and the Philosophy of Technology." *Journal of French and Francophone Philosophy* 16(1–2): 42–56.

Latour, Bruno. 1993. *We Have Never Been Modern*. Cambridge, MA: Harvard University Press.

Latour, Bruno. 2004. *Politics of Nature: How to Bring the Sciences into Democracy*. Cambridge, MA: Harvard University Press.

McLuhan, Marshall and Eric McLuhan. 1988. *Laws of Media: The New Science*. Toronto: University of Toronto Press.

Mitcham, Carl. 1994. *Thinking Through Technology: The Path Between Engineering and Philosophy*. Chicago, IL: University of Chicago Press.

Ong, Walter J. 2002 (1982). *Orality and Literacy: The Technologizing of the Word*. London/New York: Routledge.

Pickering, Andrew. 1995. *The Mangle of Practice: Time, Agency, and Science*. Chicago, IL: University of Chicago Press.

Plato. 1997. "Cratylus." Translated by C.D.C. Reeve. In *Plato, Complete Works*, edited by John M. Cooper and D.S. Hutchinson, 101–56. Indianapolis/Cambridge: Hackett Publishing Company.

Verbeek, Peter-Paul. 2005. *What Things Do*. University Park, PA: Pennsylvania State University Press.

Winner, Langdon. 1980. "Do Artifacts Have Politics?" *Daedalus* 109(1): 121–36.

Winner, Langdon. 2014. "Technologies as Forms of Life." In *Ethics and Emerging Technologies*, edited by Ronald Sandler, 48–60. Basingstoke/New York: Palgrave Macmillan.

Wittgenstein, Ludwig. 1969. *On Certainty*. Translated by Denis Paul and Elisabeth Anscombe. Oxford: Basil Blackwell.

Wittgenstein, Ludwig. 2009 (1953). *Philosophische Untersuchungen/Philosophical Investigations*, Revised 4th edition. Translated by Elisabeth Anscombe, P.M.S. Hacker, and Joachim Schulte. Malden, MA/Oxford/Chichester: Wiley-Blackwell.

Young, Miriam. 2015. *Singing the Body Electric: The Human Voice and Sound Technology*. Farnham: Ashgate.

Part I

Humans Speak (Subjects versus Objects)

2 Speaking with and about Technology

2.1. Using Technology and Language to Speak: Technology and Language as Mere Tools, and a Heideggerian-Wittgensteinian Criticism of This View

Instrumentalism and its Critics: Heidegger on Technology

It is common to assume an instrumental view of both language and technology. Technology users assume that technology is an instrument. Technology is seen as a means to an end, an instrument (*instrumentum*) used by humans to reach an aim. Heidegger writes about technology:

> The current conception of technology, according to which it is a means and a human activity, can therefore be called the instrumental and anthropological definition of technology. Who would even deny that it is correct? It is in obvious conformity with what we are envisioning when we talk about technology. The instrumental definition of technology is . . . uncannily correct . . . the instrumental conception of technology conditions every attempt to bring man into the right relation to technology.
>
> (Heidegger 1977, 5)

With the terms "human activity" and "anthropological," Heidegger suggests that we commonly assume that technology is a matter of human agency, and the instrumental view implies that the effects of technology can be reduced to its intended effects: to what humans want to achieve with them. This is why Heidegger (1977) links the conception of technology to mastery, in particular the will to master (5). This includes wanting to control technology. Furthermore, the instrumental conception of technology holds that technologies themselves are value neutral and have no meaning beyond what is intended by humans. What matters are the ethics and meaning given to it by humans through their use. Moreover, in this view, the social and cultural context of technology is not intrinsically connected to the technology. The social and cultural context matters to the values and meaning we give (with) technology. But it does not "touch" technology and vice versa.

Interestingly, language seems to be in the same boat—an observation which supports the argument that language can helpfully be understood as a technology. What Heidegger says about technology could be applied to language, of which we also tend to hold an instrumental and anthropological view: Language is usually seen as a means for human communication and it is a human activity. Indeed, it is also commonly assumed that we have full control of language. Knowing a language is to master it, and it seems that if we know a language, we can do with it what we want. To put it in the language of media: As language users we commonly assume that language is a mere tool, that medium and message are basically independent from one another. On the one hand, there is what we want to say. On the other hand, there is the tool or medium we use to say it. The instrumental conception of language assumes that the language does not influence what we say, that the medium does not influence our message. Like in the case of (other) technology, it is assumed that the message depends only on *human* thought, human intention, human goals, human values, and so on. Like other technologies, language is seen as a tool created by the human mind to get across its message. Human subjectivity and human thinking are seen as independent from their linguistic tools.

By itself, it is not wrong to say that technology and language are instruments. But they are also more than instruments. The instrumental conception of technology has been criticized by philosophers of technology and media. Heidegger (1977) explicitly argued against it and McLuhan (1964) argued that technologies actively shape human consciousness and culture, and that the medium shapes the message, even "is" the message. (In the next section and chapters I will say more about these alternative, more radical ways of conceptualizing technology.) Similarly, the instrumental conception of language has been criticized, for instance by Heidegger and Wittgenstein, among many others. In these views, language is much more than a mere tool and instead shapes human subjectivity, thinking, goals, etc. The medium shapes the message. Let me say more about language in Heidegger and Wittgenstein, and then build a bridge to technology.

For the later Heidegger, language always discloses the world in a particular way. We always think and perceive within language. It renders possible a particular understanding of the world. Heidegger even says that "language speaks:" 'language is not a work of human beings: Language speaks. Humans speak only insofar as they co-respond to language' (Heidegger 1967, 57). When we speak, we can only do so on the basis of a language which precedes our speaking and thinking. Even the invention of a "new" language starts from language as we already speak it: Heidegger writes that, 'humans may be able to invent artificial speech constructions and signs, but they are able to do so only in reference to and from out of an already spoken language' (57). We do not fully control language; it also has a life of its own and conditions us. It is what Heidegger calls "the house of being" in which human beings dwell (254). We can only know and experience beings based on a prior understanding of them, which is already given in language.

Language makes possible that things show up for us in the first place. Truth, meaning, and value are shaped by language.

Similarly, for Heidegger technology is not merely an instrument. As I mentioned, he admits that it is of course correct to say that it is an instrument. This instrumental definition conditions our thinking. But, he argues, it is also a way of revealing and a way of thinking: 'Technology is therefore no mere means. Technology is a way of revealing' (Heidegger 1977, 12). There are several kinds of revealing. For example, Heidegger argues that there is the bringing forth of *poiesis* in ancient crafts as a bringing into appearance (10), so to speak, helping it to burst open. Today, by contrast, we are under the spell of modern technology/thinking/language: The revealing in modern technology is a challenging (14) and we think of nature as a "standing reserve" (17) for industrial production; that is, for us. The world is revealed as a "standing reserve," a resource for our use. Whether or not Heidegger is right about ancient and modern technology, it is interesting that he sees technology as more than an instrument; it is also a way of thinking and conditions our speaking. It is also 'no merely doing' (19), since we do not fully control that and how it influences us. It 'sets upon man' (20), Heidegger writes, and he concludes: 'The merely instrumental, merely anthropological definition of technology is therefore in principle untenable' (21). If we 'represent technology as an instrument,' he argues, we miss its essence (32). Again, we do not need to accept Heidegger's particular views, for instance his use of the term the "essence" of technology, to recognize that technology's role is more than merely instrumental and that it shapes our thinking and speaking (and that this is not fully under human control). To further develop this view, we can apply Heidegger's view of *language* to technology: Like language, technology shapes what precedes our thinking and speaking. Like language, it is part of the transcendental conditions of speaking and thinking; it shapes our prior understanding of things. We cannot isolate technology from mind and culture, confine it to a separate sphere of the material or the natural; it spills over, and contaminates our thinking and our culture.

But why and how exactly can technology play this role? Paradoxically, to understand this, we have to take seriously its instrumentality, and in particular its *use* by human beings—but then with "use" understood in a way that downplays the agency of humans and acknowledges that technology can have its own kind of "agency," in the sense that it shapes our thinking and our practices (we may call this its "hermeneutic agency" and "practical agency"). For this philosophical operation, I propose that we turn to Wittgenstein, in particular to his view of language, which I will interpret in a way that renders it useful for thinking about technology.

Technology Games: Wittgenstein's Toolbox

For the later Wittgenstein of the *Philosophical Investigations* (1953), language is a tool, but also more than a tool. It is not simply a tool we use to represent the world or a structure we impose on the world; *as* a tool, it is—to use McLuhan's words—not only the medium but also the message. Used as a tool, it is *more* than a tool. Against what he takes to be Augustine's view in the *Confessions*—that the meaning of a word is the object it refers to—Wittgenstein turns to the *use* of words and the way this use shapes meaning. But this use is always related to a larger whole: to language and to the social-cultural context of the language user, which is always also a practical context—the context of the lifeworld. A crucial point made by Wittgenstein is that the use of words is firmly linked to everyday practice and experience, in particular to what he calls "language-games," and a "form of life." He writes:

> I shall also call the whole, consisting of language and the activities into which it is woven, a "language-game."
>
> (Wittgenstein 1953, §7, 8e)

When we use language, we follow certain rules. We play a game. When we learn to use language, we learn these rules. But they are often not explicit. We usually learn by watching and by participating in a game, as when we learn a board game (§31, 18e-19e). Learning language, then, is like learning a skill. Moreover, our language games and this learning always have a social and cultural context. Wittgenstein calls this a "form of life:" 'to imagine a language means to imagine a form of life' (§19, 11e). Thus, meaning depends on language use, and this language use is an activity and a game, which is part of a larger cultural whole—a form of life. To illustrate that 'the speaking of language is part of an activity, or of a form of life,' Wittgenstein mentions a number of activities, such as giving orders, drawing, reporting, etc. (§24, 15e). The point is that meaning is tied to these (social) activities; it is not attached to a particular sign (e.g., a word).

Indeed, language itself is woven into our activities (§7, 8e), such as giving orders, describing something, acting in a play, singing, etc. Words are also deeds (§546, 155e). Against the "logicians"—including the early Wittgenstein—the later Wittgenstein emphasizes the diversity of use: the diversity of 'the tools of language and of the ways they are used' (§24, 15e). With every use, the meaning changes. Meaning is not 'an aura the word brings along with it and retains in every kind of use' (§117, 53e). What matters to meaning, then, is not only the message—what you say—but also how you say it, in which context, and what you do with it: what matters to meaning is how language is used in context. For instance, in some contexts a particular language game is not appropriate. Wittgenstein gives the example of believing that you will burn yourself on the hotplate: in practice, when there is an actual risk of burning yourself, giving reasons for this belief seems out of place; we do not need reasons for this belief (§477, 143e) and we are careful not to burn ourselves. He also argues again that one can learn a game without learning the rules (§31, 18e-19e). Thus, in everyday life and practice, knowledge need not be explicit and we do not need to give reasons for it.

In *On Certainty* (1969), Wittgenstein stresses that we learn from experience and after a while have a "picture of the world." We trust without having grounds (§600, 79e). He suggests that a single belief cannot be justi-fied and that we do not need that kind of justification. For instance, we do not need a theory of induction. He writes that, 'the squirrel does not infer by induction that it is going to need stores next winter as well. And no more do we need a law of induction to justify our actions and our predictions' (§287, 37e). Of course one can be uncertain about many things. But, Wittgenstein argues, that will not make a difference to your life (§338, 43e): he suggests that we will just go on living as we do, and in this sense we can only doubt on the basis of non-doubting behavior (§354, 46e)—in other words, it is super-fluous. We do not need "knowledge" or "beliefs" for our language games and our activities. For instance, we do not need the knowledge that there are physical objects (§479, 63e); experience is first. He gives the example of taking hold of a towel: we do this 'without having doubts' (§510, 67e). Similarly, mastering a language is possible without "knowing;" instead, one 'must be *able to do* certain things:' to master a language game is to be able to do certain things (§534, 71e). In other words, if we need knowledge at all in language use and other use (indeed, in daily life), then it is a know-how we need. Moreover, even if we doubt, this is only possible on the basis of knowl-edge and beliefs that are already there in our culture; we trust these and rely on them. (Employing a transcendental argument, which I will introduce later in this book, one could say: There is already a "grammar" and a form of life which *makes possible* particular games, such as doubting.) Wittgenstein uses the metaphor of "hinges:" 'our doubts depend on the fact that some proposi-tions are exempt from doubt, are as it were like hinges on which those turn' (§341, 44e) and 'If I want the door to turn, the hinges must stay put' (§343, 44e). This implies that we live our lives while 'being content to accept many things' (§344, 44e), without having certainty. Descartes's radical doubt then, is only possible on the basis of this experience of use and of doing things in the lifeworld, in which we 'accept many things.' Wittgenstein stresses the deed and the performance: 'certain things are *in deed* not doubted' (§342, 44e; Wittgenstein's emphasis). It is only by alienating ourselves from con-crete lived performance and activity that we can have doubts.

Again the comparison with technologies seems appropriate. Technology, like language, is something we use, it is linked to activities, it is done in a particular context, its use and know-how is learned by doing, and it is part of a form of life. It also relies on trust in know-how that is already there in our culture. This includes trust in technology. Wittgenstein writes in *On Certainty*: 'If I make an experiment I do not doubt the existence of the apparatus before my eyes' (§337, 43e). *In* use and performance, *in* the technological practice, there is no doubt. Meaning and knowledge arises in use and this use is connected to knowledge and beliefs that are there in our culture, that are part of our form of life, in which we trust, on which we rely, and which (as I will argue) *make possible* that use.

The meaning of a technology is then not an aura attached to the artefact, which is then retained in every kind of use, but varies with the use of the tech-nology and hence with context. Ihde calls this "multistability:" technologies 'may be variantly embedded; the "same" technology in another cultural context becomes quite a "different" technology' (Ihde 1990, 144). According to him, 'a technological object, whatever else it is, becomes what it is "is" through its uses' (70). This thought can be supported and developed by using Wittgenstein. We learn the meaning of technology through its use in a particular context and as it is linked to particular activities. Technology use is always part of a game, practice, and larger whole that gives it meaning. Following Wittgenstein, we can understand that the use of artefacts is woven into activities and a larger whole. If it is true, as Wittgenstein writes in *On Certainty*, that 'our talk gets its meaning from the rest of our proceedings' (§229, 30e), then something similar can be said about our use of technology, which is also related to our (other) activities and our entire culture. When we look at the practice and the culture, we will see the meaning and "logic" of technology. We will see the "technology-game" that is being played (to play with the term "language-games" from the *Investigations* [§654, 175e]). The use of technologies is also diverse, even within the same culture. And we also trust it without having grounds; our technology use is like the activities of the squirrel, and we trust that it will work. As practical, game-playing, and cultural squirrels, we do not need a theory or rules when it comes to tool use. We obtain know-how from experience, and we trust and rely on a larger whole of related knowledge available in our culture. Using a tool does not necessarily involve thinking about it at all; a "theoretical" attitude even gets in the way of proper use and understanding of tools as tools. As Heidegger put it in *Being and Time*: Usually when we use tools, they are not 'present-at-hand' (*vorhanden*), but 'ready-to-hand' (*zuhanden*); the meaning is in the use and in the act, as we are involved in the world:

> The less we just stare at the thing called hammer, the more actively we use it, the more original our relation to it becomes and the more undisguisedly it is encountered as what it is, as a useful thing. The act of hammering itself discovers the specific "handiness" of the hammer. . . . When we just look at things "theoretically," we lack an understanding of handiness. . . . Handiness is not grasped theoretically at all. . . . What is peculiar to what is initially at hand is that it withdraws, so to speak, in its character of handiness in order to be really handy.
>
> (Heidegger 1927, 65)

Similarly, language also withdraws from view when we use it. Usually we do not look at it theoretically at all. As we actively use it and are involved in what we are doing, it mediates our relations with others and with the world. Moreover, sometimes we have a knowledge and an understanding without that we can make this knowledge and understanding explicit by using language. Wittgenstein, commenting on Augustine's famous question about time in the *Confessions*, writes about 'something that one knows when nobody asks one, but no longer knows when one is asked to explain it' (§89, 47e). When we use technology, we often have a know-how—we know how to use it—without being able to explain exactly how it works or even what we do (e.g., as when a child uses a tablet computer even before it knows what a computer is; perhaps this kind of use of technology is not the exception but the rule).

Finally, *language itself can be understood as a technology*. And indeed Wittgenstein himself, perhaps influenced by his engineering background, makes the comparison between language use and technology use, albeit to say something about language rather than about technology. He writes that language is a tool we use; it is an instrument: 'Language is an instrument. Its concepts are instruments' (§569, 159e). He even compares language to tools in a toolbox, which have diverse functions:

> Think of the tools in a toolbox: there is a hammer, pliers, a saw, a screwdriver, a rule, a glue-pot, glue, nails and screws.—The functions of words are as diverse as the functions of these objects. (And in both cases there are similarities.)
>
> (Wittgenstein 1953, §11, 9e)

Wittgenstein makes the analogy to stress that what we *do* with language is important: its use. Signs are dead, unless we use them—that use gives them life, breath (§432, 135e). And he compares the invention of a language to the invention of a device (§492, 145e).

But *like all comparisons, **this metaphor also works in the other direction; we can turn around the comparison***: we can use the use of language as a metaphor for the use of technology (technological artefacts). Using technology is like using language. And then we can **apply Wittgenstein's view of language to technology**: With Wittgenstein, we can focus on *use* for understanding technology. We can apply what Wittgenstein says about language to technology: The use of technology is part of "technology games", it is part of a context of use, of a practice, and ultimately of a form of life. *Use is the breath that gives life to technology*. Artefacts do not have a fixed aura of meaning, but get their meaning through use, and the meaning of a technology is as diverse as its uses. Moreover, the use of technology is also a matter of trust and know-how: It is about learning to use it, experiencing it, and then trusting it. Then we have the know-how of the squirrel; this is usually sufficient. Only apprentices (users in training) may need explicit instruction, and experts will discuss theory. But in daily use, we trust the technology if we have learned to use it as part of activities, games, and a form of life.

Thus, while Wittgenstein focuses on use of language and, perhaps in an overreaction against the Augustinian view, downplays the material object, we can *bring back the object and technology by turning around his metaphor of the toolbox*, and by applying his view of *language* to *technology*, thereby fully benefiting from the richness of his philosophy of language, but also from his epistemology, which highlights know-how, practice and the lifeworld.

Note that this passage in the *Investigations* is probably not the first time Wittgenstein makes a comparison between technology and language in his thinking. In *Wittgenstein flies a Kite* (2005), Susan Sterrett has argued that Wittgenstein's view of language in the *Tractatus* is influenced by his experience with technologies in his youth and in his work as an aeronautical engineer. For instance, she has suggested that kites and scale models of propellers inspired him to the *Tractatus* view that language represents facts and provides a model of the world. Nordmann (2002) has also commented on the engineering background of the *Tractatus*. We may presume that also the later work was influenced by his experience of technology, and not only experience of the toolbox but also of modern technology. Indeed, Wittgenstein refers to *machines* in the *Investigations*(§194, 84e and §359, 120e); he also refers to automata: Echoing Descartes, he asks if we can imagine people around us as automata (§420, 133e).

Other interesting comparisons with technology in the *Investigations* are Wittgenstein's references to measurement instruments in order to say something about language. In fact, the text is full of references to measurement. For instance, according to Wittgenstein, communication by means of language requires agreement in judgment (and, ultimately, in a form of life, as his previous paragraph says—see also below). He refers to measurement when he writes:

> It is not only agreement in definitions, but also (odd as it may sound) agreement in judgments that is required for communication by means of language. This seems to abolish logic, but does not do so.—It is one thing to describe methods of measurement, and another to obtain and state results of measurement. But what we call "measuring" is in part determined by a certain constancy in results of measurement.
>
> (Wittgenstein 1953, §242, 94e–95e)

Wittgenstein distinguishes here between describing the method of measurement, and measuring. This metaphor is then applied to *language*, which requires definitions (the meaning of words, say, the unit of measurement), but also judgment based on experience, in particular the experience of obtaining constancy in results when one says something. *Again we can turn this around*, and stress the need for performativity and constancy in *technology* as use, experience, and practice in order for meaning and knowledge (or rather: know-how) of the technology to arise. The meaning and knowledge of a tool, for instance a hammer, is not only and not so much determined by agreement on what the word means (this would require what Heidegger and Wittgenstein call a "theoretical" attitude), but also and especially by the act and practice of hammering and (the constancy of) its results, which gives us trust in it. I really know a particular app, for instance, not so much if I know what that app "is meant for" (i.e., we agree that the name of the app "stands" for this or that function; this is its "theoretical" meaning), but also and especially if I have *used* it and *experienced* what it does when I use it, if I have experienced a certain constancy in the results. Based on this experience, I trust it. More generally, we know the meaning of the technology not so much if we know the name or share a definition about the technology, but especially if we share *experience* with *using* it in specific contexts and situations, which each time again involves practical judgment and know-how. This shared and repeated practice and experience, and not so much agreement on "what the tool is for," gives us knowledge and meaning of the technology. In other words, it is the hermeneutics of its *use* and of *acting*, *performing* with it, that is important to the meaning of technology, rather than (merely) agreement about the meaning of a name, about a definition, about naming and defining an artefact. (This contrasts with the Searlean account I will present in the next chapter.)

This excursion into instruments also further supports the insight that technology is always more than a tool as we commonly understand it; it is bound up with communal practice and social wisdom. I will soon make an excursion to Dewey to stress this. However, let me first return to Wittgenstein's view of language in order to elaborate why, according to Wittgenstein, *language* is more than a tool, and then see how we can apply it to technology. We need to know more about the concepts of language game and form of life in the context of Wittgenstein's general view of language, and hence about what may be called the "cultural" interpretation of Wittgenstein.

From Instrumentality to More-than-Instrumentality; The "Cultural" and "Use" Interpretation of Wittgenstein

Let me revisit and rephrase Wittgenstein's view of language. For Wittgenstein, in contrast to Heidegger, there is not a sharp distinction between the instrumental and non-instrumental aspects of language. Paradoxically, its instrumentality takes us to its non-instrumentality.

To understand what this means, and to better understand the terms "language games" and "forms of life," let us return to Wittgenstein's text. As said, Wittgenstein's point was that words and language do not have meaning in isolation, but should be seen in terms of their *use* within a larger social-linguistic whole. But to understand this, we must start with the understanding of language as a tool, an instrument. *As it is used*, the tool—language—is bound up with social and communal activities and practice. It is because it *is* a tool and an instrument that it is part of that practice and—we may add—shapes that practice. Learning a language is learning a particular activity and game. This is why Wittgenstein writes that inventing a language is not only like inventing a device but also like inventing a game (§492, 145e). Language is a game; there are language games. Like technologies, language is an instrument that we can use for particular purposes. But, Wittgenstein argues in the *Investigations*, this "invention" (see again §492) is always a social and communal matter. If language is invented at all, it is like inventing a game. Meaning must be understood by looking at the use of words within a language game.

This means that we cannot restrict the locus of meaning to individual intention, thought, feeling, and subjectivity—they themselves depend on language. Words describing intentions, thoughts, feelings, etc. are only meaningful in the context of (specific) language games and forms of life; that is, the society and culture in which they are used. I do not first have meanings in my mind, which then are expressed in context. Instead, Wittgenstein argues, 'language itself is the vehicle of thought' (§329, 113e). For Wittgenstein, the language game is primary; feelings and thoughts are ways of interpreting the language game (§656, 175). Language games are embedded in forms of life and cultures. Language is about use, but that use is anchored and limited by the particular practices, games, and cultures that are given to us. Language is a tool, and instrument, but it is a deeply practical and social one. Meaning is public; it is about what languages users do and about the way *we* live our lives.

This view can be further unpacked by stressing the "social" or "cultural" interpretation of Wittgenstein. As Peter Winch (1958) argued, 'it is a central feature of what Wittgenstein writes about language that this can only be seen for what it is if looked at in the more general context of behavior in which it is embedded' (Winch 1958, xiii). He already pointed to Wittgenstein's concepts of following a rule and "forms of life" (40). Connecting Wittgenstein's view of language and epistemology with sociology, he argued that all meaningful behavior is rule governed (52). Winch was also influenced by Wittgenstein's criticism of Frazer. Frazer, a 19th century Scottish anthropologist, had interpreted indigenous magic practices as primitive science, whereas Wittgenstein called attention to the meaning of the practices themselves, instead of "explaining" them by referring to modern science. Instead of making these people look stupid, Wittgenstein suggested, one should try to understand the practices in their own terms and in their cultural context (Wittgenstein 1967). In other words (or at least this is how some readers of Wittgenstein interpreted it), the indigenous people had their own language games. There are different forms of life; these can be described, but justification (from the outside) is not the way to approach them. (Of course this then raises the questions of if and how people from these different "conceptual schemes"—to use a term from Davidson—can reach mutual understanding. For a helpful overview and interpretation of some of the "rationality" debates in the 1960s and 1970s see Springs 2008). Thus, according to this reading of Wittgenstein, language is essentially a social and cultural matter, since it is connected to language games and forms of life. Or as Susan Hekman (2002) summarizes the social-cultural interpretation of Wittgenstein's view of language: 'For the later Wittgenstein, language is exclusively a social activity: meanings are defined by use; justification is conventional and relative to language games; private languages are impossible' (166).[1]

This cultural-Wittgensteinian view of language can now be applied to technology. Similarly, one can now argue, (other) technologies must also be linked to our activities, practices, and their social and cultural context, and indeed to our *thinking*. This is so since we use technologies always in the context of social activities and culture. In this sense, the instrumentality of technology (its use) takes us to its noninstrumentality. Influenced by Wittgenstein, we can say that using a technology is like using a language: Technological artefacts are dead, unless use gives them life, breath. But this use is not merely "individual," since it is always linked to activities and larger social-cultural wholes. Technologies are part of what I call "technology games:" Using a technology and learning to use a technology are like learning and playing a game, and the use of technology is always a social and communal matter. Technologies are embedded in forms of life and cultures, and their use is shaped and constrained by those cultures. What technologies "are" and mean, then, depends on the hermeneutics of their use *in practice* and on a social and cultural context. (This also means, of course, that there are different cultural-technological forms of life, different technological cultures—which raises the question as to how commensurable they are and how "we" should interpret "them.")

An Excursion to Pragmatism: Dewey

This social-cultural interpretation of Wittgenstein's view of language can be helpfully compared to, and seems at least compatible with, Dewey's view of language, which also starts from instrumentality and use, but links this use to wider social meanings, and which can also be "turned around" and applied to technology.

In *Experience and Nature* (1929), Dewey argues that language must be understood as a social instrument, which makes possible social institutions and the social life. He writes that language is 'the instrument of social cooperation and mutual participation' and clearly understands it as a tool when he writes about 'language and other tools' that enable social participation—a concept used by Dewey to connect mind with nature, the ideal with the spiritual (vi). He argues against theories that 'oppose subject and object' (162) and says that language joins the gap between spiritual and empirical, 'existence and essence' (167). Like Wittgenstein, Dewey was skeptical about the language of "mental" or "inner" states; instead, he sees language as having a key role in shaping the message. He sees mind as emergent from interaction and as bound up with the emergence of language: ' "mind" is an added property assumed by a feeling creature, when it reaches that organized interaction with other living creatures which is language, communication' (258). Against solipsism, Dewey argues that 'the world of inner experience is dependent upon an extension of language which is a social product and operation,' 'a natural function of human association' (173). He compares words with money: In line with Simmel, he says that they are things, but at the same time they 'embody relationships' (173) and transform and create new interactions and transactions. Language changes the social; it is not a mere medium. And it does not come into existence 'by intent and mind but by over-flow, by-products, in gestures and sound. The story of language is the story of the *use* made of these occurrences' (175; Dewey's emphasis). For instance, in daily life we use water, thus "water" is normally connected to concrete uses in human (social) life; the scientific use of the sign "H_2O" abstracts from this use and experience, and 'would be totally meaningless, a mere sound, not an intelligible name' (194) if it was disconnected from that experience. Thus, Dewey here has a deeply social and embodied understanding of language use: Language makes possible the social and is itself emergent *from* the social and its embodied and experiential occurrences. He stresses sharing and cooperation:

> The heart of language is not "expression" of something antecedent, much less expression of antecedent thought. It is communication; the establishment of cooperation in an activity in which there are partners, and in which the activity of each is modified and regulated by partnership. . . . Meaning is not indeed a psychic existence.
>
> (Dewey 1929, 179)

Forms of language create a 'sense of sharing and merging in a whole' (184). Like Wittgenstein, he also argues against nominalism, which regarded words as expressions of mental states. Interesting in anticipation of my discussion of Searle in the next chapter is especially Dewey's insistence that a 'sound, gesture, or written mark' does 'not become a word by declaring a mental existence; it becomes a word by gaining meaning; and it gains meaning when its use establishes a genuine community of action. . . . Nominalism ignores organisation, and thus makes nonsense of meanings' (184–185). Language is thus all about interaction between two beings and 'presupposes an organized group to which these creatures belong, and from whom they have acquired their habits of speech' (185).

These views accord with Wittgenstein's social, "cultural," and anti-nominalist view of language and indeed with his anti-mentalist and anti-Cartesian epistemology. Knowledge is 'a mode of interaction' (435); it is connected to social practice. There is even a correspondence with his view on certainty: For Dewey, science is a matter of 'making sure, not of grasping antecedently given sureties . . . what is accepted as truth . . . is held subject to use' (154). [Dewey sees science as art and art as practice, and questions the separations in thinking between practice and theory, and between art and science (358; see also 381). I will return to this argument at the end of this book.] And both Dewey and Wittgenstein give language a central place with regard to the making of meaning: Dewey says that 'meanings do not come into being without language' (299) and calls language, understood as 'the tool of tools,' 'the cherishing mother of all significance,' including the meaning of technologies, which he sometimes calls 'agencies:'

> For other instrumentalities and agencies, the things usually thought of as appliances, agencies and furnishings, can originate and develop only in social groups made possible by language. Things become tools ceremonially and institutionally.
>
> (Dewey 1929, 186)

Thus, both language and the social, which are intrinsically bound up, make possible tool use. Tools become what they are by means of action and interaction, the repetition of which leads to 'their becoming institutionalized as tools'—which also depends on communication (187) and hence language. Afterwards, language can become more independent, when it is 'liberated from the contingencies of its prior use' when then in turn it can be used to regulate human interaction (193), for instance in the cases of law and bureaucracy, or indeed in the cases of philosophy and logical systems. But the origin lies in social experience. Furthermore (and again interesting in view of the next chapter), influenced by Malinowski, Dewey argues that language has also what Austin will call 'perlocutionary' force: It is 'a mode of action used for the sake of influencing the conduct of others in connection with the speaker' (206). In other words, Dewey recognized that we use words to do things and to get others to do things. This fits Dewey's view of language as a social tool: It is used for interaction and connected to the formation of habits.

Applied to technology use, this view of language "turned around" means that we can see *technology* as a social instrument that plays a crucial role in human interactions and makes possible social institutions and the social life. Like language, technology can be understood as an instrument of social cooperation. Technologies enable cooperation and may create a sense of sharing. Things become tools within a social context. Repetition and ceremony lead to habits and institutions. Tools must thus be understood within a social context. Furthermore, because of their social role, technologies can also be used for the sake of influencing the conduct of others, to make others do things. Language and other tools may also be conceptualized in a way that crosses dualisms such as mind/body, inner/outer, culture/nature, and so on. Like language, technology does not only function as a medium but also has the capacity to transform the social. At the same time, it is emergent from that social whole; it is a kind of by-product of the social or part of the social—at least if the social is understood in an experiential, embodied, and (inter)activity-related way, rather than in terms of reified forms, such as law and logical orders, which are merely forms of use and experience that have become crystallized and independent from experience, alienated from their origins. Moreover, the meaning of technology does not come about by declaring its meaning; rather, it comes about in and through use and interaction. Meaning, then, comes into being through language and technology *in use.* Language is a special kind of tool, as it is always used in connection with other tools.

With regard to the latter claim, it is interesting to note that Dewey expresses a conception of language as a 'supplement' to technology—participation depends on 'the use of extra-organic conditions, which supplement structural agencies, namely, tools and other persons, by means of language spoken and recorded'. And next to general remarks about language, technology, and meaning, he also points to a particular way language and technology are connected, which stresses the cultural dimension of meaning and language use, and which supports the claim I discuss later in this book that not only humans but also technologies "speak" (see in particular my discussion of Latour in Chapter 6): Language is 'the tool of tools' that makes possible tool use, since uses of appliances and utensils 'are bound up with directions, suggestions and records made possible by speech' (168). Thus, tools and words go hand in hand—literally, as our use of tools is linked to our use of words, for instance in the form of directions. This is also why he can associate experimentation in craftsmanship with finding 'new modes of language' (363). And in his *Logic* (1938), Dewey writes, in line with what we may call, in analogy with my reading of Wittgenstein, his "cultural" approach (but not "cultural" as in "nonmaterial,") that tools or machines can "say" something:

> A tool or machine . . . is not simply a simple or complex physical object having its own physical properties and effects, but is also a mode of language. For it says something, to those who understand it, about operations of use and their consequences. To the members of a primitive community a loom operated by steam or electricity says nothing. . . . In the present cultural setting, these objects are so intimately bound up with interests, occupations and purposes that they have an eloquent voice.
>
> (Dewey 1938, 46)

These are very interesting remarks which support and are compatible with my arguments in the next chapters. But let me for now conclude this brief comparison by saying that Dewey can also be used to further discuss the relations between language and technology, and to support the "cultural" interpretation of technology I develop here with the help of Wittgenstein—with "cultural" not understood as being entirely different or entirely unconnected to the material and the technological, but as crucially including it in and through use.

Making this link to Dewey may be somewhat unexpected: While it is not uncommon to use Dewey or pragmatism in philosophy of technology [see especially Hickman's (1990) work on Dewey's view of technology as a boundary-crossing tool and his stress on the social and political], here I suggest that attending to Dewey's view of *language* may help us to better understand technologies and their relations to language. This excursion also contributes to arguments that ask for more emphasis on the *social and political* dimension of technology—a dimension which I will argue is largely missing in postphenomenology (see Chapter 6), in contrast to, for instance, the work of Winner or Feenberg. It is also interesting that, perhaps in contrast to Wittgenstein, Dewey's view does not seem to stand at the expense of downplaying the *material* object: Dewey's view encompasses both the cultural and the material, crosses over. Finally, later in this book, the idea that language and technology make possible the social will be interpreted as a *transcendental* argument.

More on the "Cultural" Interpretation of Wittgenstein: Winner

Many philosophers of technology may agree with the Wittgensteinian view of technology I started to develop (for instance, Ihde may agree), but usually in their work they do not make a link to, and good use of, Wittgenstein's view of language (or any other philosophy of language for that matter), and hence miss a potentially interesting connection and insight(s). An exception is Winner, who in 'Technologies as Forms of Life' (2014) explicitly refers to Wittgenstein's concept 'forms of life.' As said, Wittgenstein argued that the speaking of a language is part of an activity or a form of life; similarly, Winner suggests technologies 'become woven into the texture of everyday existence' and then 'shed their tool-like qualities to become part of our very humanity' (Winner 2014, 55). Based on my interpretation of Wittgenstein presented in the previous pages, one could rephrase this as saying that understanding the instrumentality of technologies leads to understanding their noninstrumentality, since it is through their *use*, in everyday contexts, that technologies shape humanity. In this sense, technology shapes culture. But as Winner rightly argues, the influence goes in both directions. On the one hand, technologies bring about 'significant alterations in patterns of human activity and human institutions'—technologies make 'new worlds' (54):

> We do indeed "use" telephones, automobiles, electric lights, and computers in the conventional sense of picking them up and putting them down. But our world soon becomes one in which telephony, automobility, electric lighting, and computing are forms of life in the most powerful sense: life would scarcely be thinkable without them.
>
> (Winner 2014, 54)

Yet, on the other hand, the forms of life in which new technologies enter and are embedded also shape our use, interactions, and expectations. Wittgenstein also said, in his *Philosophy of Psychology* (fragments published with the *Investigations*), that forms of life are the 'given,' 'what has to be accepted:' 'What has to be accepted, the given, is—one might say—forms of life' (§345, 238e). This "given" exerts its influence on the use of technology and the related interactions. Winner writes, 'most of the transformations that occur in the wake of technological innovation are actually variations of very old patterns' and gives the example of the humanization of computers:

> Forms of life that we mastered before the coming of the computer shape our expectations as we begin to use the instrument. One strategy of software design, therefore, tries to "humanize" the computers by having them say "Hello" when the user logs in or having them respond with witty remarks when a person makes an error. We carry with us highly structured anticipations about entities that appear to participate, if only minimally, in forms of life and associated language games that are parts of human culture.
>
> (Winner 2014, 56)

Again, there is little doubt that technologies are tools, instruments. But they are also more than tools and instruments (as we usually conceive of these), in the sense that they shape our activities and our thinking, and are in turn shaped by culture (broadly understood). Technologies-in-use are interwoven and interdependent on social and cultural patterns and roles that existed before the innovation took place—patterns that are altered in the wake of the innovation, when the technologies are used. With Winner, we can place technologies in the context of the 'vast multiplicity of cultural practices that comprise our common world' (57). They are not only physical instruments and processes but they are linked to world-making, to the psychological, social, and political conditions. Technologies are tools that are part of language games, including the language game we call thinking and the language game of talking about ourselves (58). Indeed, Winner argues (58) that the way people talk about themselves has been influenced by the computer, for instance when we compare the mind to a computer.

This example of how (our use of) the computer frames our language is highly interesting for our project of investigating the relation between technology and language: Apparently, these relations are more intimate than presumed. If my interpretations of Heidegger, Winner, and Wittgenstein make sense, then we do not only talk *about* technology (assuming that technology and thinking are not intrinsically connected); technology also shapes our very thinking, for instance through metaphor, that is, through language and through use. Moreover, Winner's use of Wittgenstein adds an interesting dimension to the "cultural" interpretation: power and politics. Technologies are also social and political. If technology is about our common world, then technology is political. If the use of technologies shapes our societies and our cultures, then that use itself is political: Which technologies we use and how we use them matter to our games and ultimately to our form of life. Later in this book, I will say more about power and technology. [Note that elsewhere Winner (1980) has written about the politics of technology, without, however, relating this to Wittgenstein.]

Conclusion

To conclude for now, with Wittgenstein (and supported by Dewey), we move seamlessly from the instrumentality of technology to its hermeneutic and social-political role. One could say that it is *by taking the instrumentality of language seriously* that Wittgenstein moves to his view that language shapes our thinking and our (common) world. Similarly, it is *by taking the instrumentality of technology seriously* that we can move towards its hermeneutic and social-political role. Just as language makes our world, technologies also make our world. This world-making is not only a matter of humans setting goals and technologies being used as means to reach these goals. Rather, technologies influence these goals as world-making happens in the interaction between our *use* of technologies and our cultural-technological practices and patterns— patterns which were already there before that use, and which may also change as a result of that use. World-making is a matter of processes which lead to certain outcomes, and these processes cannot entirely be described in terms of our goals (ends) and instruments (means). The "goals" (and values, etc.) we say we have are themselves dependent parts of these patterns, this whole, and these outcomes; our goals may themselves be shaped by our use of technologies. This does not only render technologies more-than-instrumental but also more-than-human. Of course, they are "human" in the sense that we use them; but to the extent that their influence exceeds our explicit goals and intentions, they are also "nonhuman" and as such shape the human. In this way we go beyond what Heidegger called the instrumental and anthropological definitions of technology. It turns out that technology also "speaks."

Later in this book, I will develop this view further and say more about Wittgenstein and Heidegger. For now, it is important to hang on to the thought that technologies are more than instruments, and to recognize that there is indeed an analogy between language and technology: Both are used by us as instruments, as tools, but as used they are also more than that. As we can learn from Heidegger, they also shape our thinking and our "world." Yet, influenced by Wittgenstein, we can nevertheless learn from the instrumental understanding of language and technology, since it attends us to *use*. We use languages and technologies as instruments, and next to that, but also *because* of that *use*-fullness, they are more than instruments, they do more than we may expect on the basis of the instrumental definition: By using Wittgenstein, I have shown that it is through their use that they can have the Heideggerian hermeneutic effects ascribed to them by contemporary philosophers of technology. Paradoxically, it is through their use that technologies are more than things to be used. And this may inspire us to a more-than-human conception of technology; paradoxically, it is through their use by humans that they also partly leave the sphere of the human, or at least the sphere of human control.

Thus, in this section, I have already started to criticize an instrumental understanding of language and technology. Even if we understand language and technology as instruments, that does not mean that either can be reduced to *mere* instruments, if that means that the medium does not influence the message, that the object does not touch the subject. Both language and technology, and indeed language *as* technology, also shape our message and our goals. In the next chapters I will elaborate this view. Let me first further comment on speaking *about* technology.

2.2. Speaking about Technology: Narratives and Discourses about Technology and Beyond

Introduction

We speak about technology in various ways. But this speaking does not just consist of words and sentences. Words and sentences are part of larger wholes. In our societies and cultures, there are narratives about technology, for instance, and there are discourses about technology—academic and others. Insofar as the narratives are salient, philosophers of technology can articulate them. They can also criticize them and intervene in discourses about technology. Thinking about technology – understood as a linguistic activity (an activity mediated by language) that is "about" something (technology) – participates in these kinds of speaking about technology. It describes, articulates, argues, intervenes, responds, and so on. It also tries to do this, or at least should try to do this, in a critical way. That is, it should take distance from existing narratives and discourses, make their assumptions explicit, and criticize them, for instance.

But is current philosophy of technology also *critical enough*? In particular, heeding the linguistic turn, is it critical enough of its own use of *language*? Contemporary philosophy of technology has masterly revealed many mediations by *technology*. It has shown how technology can let the world appear in different ways [this is its hermeneutical role; see especially the work of Ihde (e.g., 1990); or to put it in a Heideggerian way: technology reveals] and it has shown how it influences what we do, giving it a normative and even moral role [see, for instance, the work of Verbeek (2005)]. But often these approaches are not sufficiently aware of the many mediations by *language* (1) in these technological activities and practices, (2) in any kind of practice, and (3) in *their own philosophical practice*. A more mature philosophy of technology needs to attain this awareness, understand the relations between technology and language, and reflect on the implications for its own practice.

One way to remedy these shortcomings is to (further) engage with philosophy of language and some of the well-known champions of the linguistic turn in the 20th century. There are theories about the important roles language plays as 'the vehicle of thought' (to use Wittgenstein's words again), and it seems wise for philosophers of technology to at least be aware of these theories and, if possible, to take them into account when developing theory in their own field. In this book I hope to contribute to this project by using Wittgenstein, Heidegger, Searle, McLuhan, and Ricoeur. In this section I discuss what may be problematic with studying and criticizing discourses and narratives *about* technology, and continue opening up avenues for bringing language and technology closer to one another.

Discourse and Narratives about Technology and Beyond

One of the tasks of a critical philosophy of technology is to evaluate the discourses and narratives about technology. This happens, for instance, when philosophers respond to public debates about artificial intelligence, automation, and robotics. One can study the public imagination about AI, for instance, and comment that often these debates are polarized between the promises and threats of AI, between overly optimistic (and naïve) and overly pessimistic (and often equally naïve) pictures of AI and robotics. As a philosopher, one can show that these narratives and discourses are always interpretations, and are never neutral or "objective" (hence, the subject "colors" the object). One can show the polarization, and one can then intervene in these discussions—by using words, language—and defend a different position and sketch a different picture. But in order to develop to achieve sufficient critical distance, one should also study and critically evaluate the *language* used to talk about these technologies and their future. One should not only study the "message" (about technology) but also the medium: the *language* of the discourse and narrative itself, which shapes the interpretations of technology. For instance, one may pay attention to the *metaphors* used in a particular debate, for example the metaphor of the machine.

Moreover, discourses and narratives about technology are likely to be connected to larger or at least other discourses and narratives, which are not necessarily or not explicitly about technology as such, but influence and shape our discourses and narratives about contemporary technology. Here, a historical and intercultural perspective is mandatory; we need to connect discourses and narratives about technology to a larger form of life of which they are part, to what Foucault called an "episteme" (Foucault 1970; see also Chapter 7). For instance, ancient Greek tragedy and the concept of *hubris* are a part of Western culture and may influence our narratives about technology today when these narratives warn about the danger and arrogance of doing what should not be done by humans. This can and has been applied to technology by philosophers in the 20th century: We should not "play God" by using modern technology. And the myth of Prometheus is perhaps more directly connected with thinking about technology: Prometheus was punished by Zeus for giving fire to humans. Today, the myth is used to warn about the unlimited use of technology. Another example is the Frankenstein narrative, a more recent trope, which is itself connected to all kinds of narratives about monsters, and which is also used to warn about technology: It might get out of human control. And recently I have argued that romantic thinking still shapes our thinking about, and use of, technology (Coeckelbergh 2017). In other words, before we start thinking and speaking about technology, there is already a form of life, an episteme, an order, a grammar, a web of meaning, a hermeneutic ecology, in which our thinking and speaking about technology is embedded. This ecology is historically grown. When we start speaking, we immediately hear the echo of the cultures and life forms of the past and present. This also applies when we speak about technology. A sufficiently critical eye can only be gained if one is also aware of the rich cultural histories in which current speaking about technology is embedded. Furthermore, considering Wittgensteinian arguments for a cultural approach, we have to attend to differences between forms of life in which discourses and narratives are embedded; we may study different technological cultures in order to better understand what *games* are being played with technology and what the meaning is of the activities that involve use of technologies.

However, it is not enough to study, reflect on, and evaluate these discourses and narratives *in* theory, to study different cultures and the meaning of technology within these cultures from a distance, or to develop theories about language and "technology" or, for instance, about AI in general. First, if the Wittgensteinian view I started to develop in the previous section makes sense, then it is necessary to look at language use (and technology use) in practice, including technological practices. Second, it is highly problematic if one remains uncritical of *one's own use of language as a philosopher of technology*. Such philosophical responses are "special cases" of language use, and these should themselves be an object of critical scrutiny.

If we do this double-faced exercise, then, we may detect the following general problem: Philosophical and theoretical discourse tends to take a lot of distance from the everyday. This may be partly justified, since a certain distance needs to be gained in order to reflect. But, on the other hand, to the extent that the reflections need to be relevant to practice (and what other *use* would they have, *as reflections?*) and maybe even *guide* practice, this theoretical distance may *itself* be *part of the problem*. Perhaps we should take some distance as philosophers, but distance-taking could be part of a dialectical process of proximity and distance. Heeding the later Wittgenstein, we should make sure that we do not stay in the fly bottle, that our language does come home from holiday. Wittgenstein famously said that philosophical problems arise when language goes on holiday (§38, 23e). In philosophy of technology, we could interpret this as a warning that we should not lose sight of the *everyday speaking about technology*, understood as the everyday use of words when using technology.

If we turn to that everyday speaking, we see that words are used in a particular way in the context of a particular technological activity. Use of language and use of (other) technologies are connected in everyday contexts, in which particular words and tools belong to larger wholes. Influenced by Wittgenstein, we can interpret speaking about technology as belonging to particular activities and language games. For instance, if I speak about my computer or word-processing program when writing a text, then this speaking is connected to a particular use of technology, of course (use of the computer and the program), but also to an activity (writing), a particular language game—that is, a particular use of language (e.g., swearing, trying to cope with an unexpected problem, etc.). Moreover, all these uses and activities belong to, are shaped by, constitute, and define a particular form of life in which specific expressions and tools are used as appropriate ones in a specific situation, in which these games are meaningful. Another example: If I talk to my navigation device in the car, then this language use is linked to a particular use of technology (using a car), but also to an activity (driving, navigating), a particular language game (e.g., dialogue, question and answer, thanking, swearing, etc.), and a particular form of life and culture in which this use of language, cars, and devices is at home.

When a new technology enters our lifeworld, then our speaking about the technology (and perhaps also our speaking *with* or *to* the technology, as in the case of the navigation device in the car—these are cases when "technology speaks" in a very literal sense) will be guided by the language games that are already available in our social interactions and cultures. For instance, a robot pet will be used and responded to in a way that belongs to the activities and language games "treating a pet" and "taking care of a pet," or more precisely, "asking a dog to do something," "cuddling a cat and saying sweet words to it," etc.—language games, which themselves imitate or even emulate some language games that are normally at home in human-human interactions and relations ("saying sweet words to a baby"), but in the course of cultural history have been extended to certain animals. Again, these examples show that for understanding these technologies, it is not sufficient to understand them as mere instruments; they are used to reach a goal, of course, but their use is embedded in a wider relational-linguistic whole, a form of life. These games and this form of life have a kind of normative presence as "givens" that shape our use of language and use of technologies here and now: They tell us how to use words and things. Our use of both language and technology has a history and a future, and belongs to a particular cultural understanding and cultural whole.

Acknowledging this "givenness" of a linguistic-technological form of life and accepting what we may call this linguistic-technological *holism* does not imply, however, the view that language and cultural contexts are entirely given and unchangeable. Humans, but also the technologies themselves, co-shape language. We are vehicles of language, but as vehicles (indeed, as instruments, as technologies) we are not merely being used by language; as users of language we also shape language and keep it alive and going. This happens not so much on an individual level, perhaps—some people contribute a lot more to changing language than others—but certainly on societal and cultural levels. For instance, language use while driving and way-finding, may change, as nearly everyone uses a navigation device. But it is not "technology" in general and in an abstract sense, as a concept, that changes language. And it is not the idea (or name) of a particular technology that changes language. It is not theory, in science or in philosophy, that changes language. It is the *use* of particular artefacts in particular contexts, activities, games, etc. that, ultimately, may change language. Language use and technology use go hand in hand. There are linguistic-technological activities and practices; language and technology are connected in use and experience, and it is only in theory (that is, afterwards) that we separate them.

Furthermore, we also can learn from Wittgenstein that there is variety, both within a culture and between cultures: Both language and technology can be used in various ways. Just as words or other signs are not married to their meaning and use, and can be understood and used in different ways, the meaning and use of technologies is not fixed. Technologies are hermeneutically unstable. (See also again Ihde on "multistability;" I will return to Ihde in Chapter 6.) Thus, if we understand technology in terms of (human) use, there is neither linguistic determinism nor technological determinism: We play our roles as users and there is variety in use. Nevertheless, it needs to be acknowledged that both language and technology are "given" to some extent. There are already linguistic, technological, and other activity patterns given in our practices, societies, and cultures—activities and patterns that were there before we were born and before our use started. When there is a new technology, or rather when it is *used*, it enters a form of life. While that form of life is not entirely fossilized, there is also not unlimited room for change. We give life to words and technologies; but there is also a sense in which they give life *to us*, in the sense that they make possible and shape the lives we live.

Let me return to the robot-pet example. Someone could say to the owner of a robot pet who cuddles the robot and says nice, "caring" things to it: "But this is a mere machine! You should not talk to it as if it were an animal or a human; this is wrong!" This remark (or more precisely: exclamation) is a normative-linguistic, pragmatic intervention, a case of doing things with words (to use Austin's phrase): One uses words to express and make a normative judgment, and to try to make the person do things in a different way. But to what extent *can* the pet owner change her behavior, given that, by its design, the robot pet encourages a particular use, a particular playing of a particular language game and practice? To what extent is there room for change if the technology is firmly "inscribed" in that game and practice—language/technology games and linguistic-technological practices that are in turn engrained in our culture? It is possible, of course, to explore different practices and different language games. It is possible, in principle, to use the robot in a different way. And perhaps there are alternative ways of dealing with (this) technology in other cultures (if these cultures are still "other" at all today, when all cultures are shaped by modernization processes). But such an alternative use and such a hermeneutic exercise needs work, needs a special effort on the part of the user, and ultimately will also have to refer to, and respond to, language games and practices that are already there, to the given. There is no such thing as an "invention" from zero; there is also no such thing as a private language (as Wittgenstein argued) or what we may call a "private technology." Technology is always part of technology games and of a form of life, which are in principle publicly accessible.

However, there is also an influence in the opposite direction: Our language-thinking about technology (and other things) is also influenced by the technology itself. New technologies can change our language. Here, Winner's example of the computer comes to mind again. Because of our decades-long *use* of computers (rather than referring to the "fact" that computers are around as "objects," as a dead "sign;" the sign only lives in practice), we start thinking-speaking about our activities, our mind, and our being by using computer language. Language does not only shape our interactions with and relation to technology; technology also shapes language. This may also happen to our use of language in relation to robots, particularly if we get more *used* to them. But there may not be *one* meaning that emerges from particular uses and games; some variety is possible, both intra-culturally and inter-culturally.

Heidegger's Language Machines

The insights of Heidegger and McLuhan can help again to elaborate this point. Let us take the example of the computer. The computer is more than a mere tool: It opens up a new kind of revealing, a new way of seeing reality; it opens up new worlds. But the meanings it brings are not determined; variety is possible. The computer can show the world as a standing reserve of data, but perhaps it can also disclose different worlds. Heidegger emphasized one meaning of technology, the enframing (*Gestell*), and this limitation can and has rightly been criticized; but his view that technology is a way of revealing remains helpful, and it is worth looking at what he said about information technology.

For this purpose, Heim's reading of Heidegger is useful. In his paper 'The Computer as Component' (1992), Heim quotes Heidegger, saying that, 'Maybe history and tradition will fit smoothly into the information retrieval systems which will serve as resource for the inevitable planning needs of a cybernetically organized mankind. The question is whether thinking too will end in the business of information processing' [Heidegger in *Parmenides* (1982); passage quoted and translated in Heim (1992), 311]. Clearly Heidegger highlights only one specific aspect of computing (information and information retrieval) here and is pessimistic about its influence on our thinking—indeed, he fears that thinking itself may end. But what makes this passage so interesting is *the very idea* of computers shaping our *thinking and speaking*. We can then add to this insight that the computer may lead to *various kinds of discourses and languages*.

This goes beyond analyzing Heidegger in terms of what kind of discourse he writes in, as Heim does in his paper. Heim suggests Heidegger's view fits in discourse or picture about the computer as opponent. This discourse is long-standing (for instance, it found expression in the writings of Dreyfus) and ongoing. For instance, it is intensified when there are contests where humans and machines meet one another as opponents in a game.[2] Heim instead proposes to see the computer as interwoven into the fabric of everyday life, as a component rather than opponent of humanity; he writes that we 'collaborate with technology' (317). According to him, technology is 'a mode of existence' and the computer a 'cultural phenomenon' (309). I am sympathetic to that discourse, but I do not think this view of the computer as a cultural phenomenon—and indeed, of any technology as a cultural phenomenon—is really on the same level as the "opponent" discourse. It is or should be, rather, a kind of meta-position or critical meta-discourse (a philosophical discourse), which has room for all kinds of discourses and narratives about the computer, including Heideggerian ones and others. Thus, Heidegger's discourse is more than a particular discourse: It refers to the philosophical position that technologies shape our language, thinking, and culture. Yet Heim's view is interesting, since it becomes clear that this particular "cultural phenomenon" interpretation and narrative could only emerge once computers became part of everyday life; only then, people started talking about computers in different ways.[3] Similarly, *if* robots became more part of the everyday, in other words, if they were *used* everywhere, then it is likely that our language and our thinking about them would also change. Perhaps this is already happening; perhaps our language and our thinking is already changing. But, as I read Heidegger, the point is that whatever the precise "content," whatever computers or robots may do to our language and thinking, it is clear *that* they have this kind of influence; or, to use McLuhan's phrase: It is clear that the medium shapes that content. The computer is a tool, but *as* a tool, it also does much more than being instrumental. It also has hermeneutic consequences. Using a different technology makes us use language in a different way and ultimately changes our language, and the language games and culture it is part of and constitutes.

Once we accept this role of the medium, we can have a discussion about the content, about what kind of changes to our language and thinking are made possible by the technology and *should* be encouraged by the technology. For instance, we can discuss about what computers do to our thinking. Heidegger played both games: He has argued *that* technology shapes our thinking, and at the same time he has expressed specific views about *what* technology does to our thinking. Let us now return to his *specific* views on information technology, given my interests in this book—in particular, its relation to *language*.

As Heim shows, the link between language and computers was directly addressed by Heidegger when he referred to the computer as 'the language machine' in his essay on Hebel (Heidegger 1957). Heidegger writes: 'The language machine regulates and adjusts in advance the mode of our possible use of language' (Heidegger 1957 in Heim 1992, 309). He then articulates his particular view on what this regulation and adjustment consists of: He thinks that it is one way modern technology controls the world. He writes that we are not the masters of the language machine; instead, 'the language machine takes language into its management, and thus masters the essence of the human being' (309). Heidegger thus already foresaw that what we now call word processing, a writing technology and "language machine" that did not even exist at the time he wrote, would influence our way of thinking, and would do so in a particular way: In his view, the (language) machine helps the modern management of everything. Moreover, commenting on the typewriter (which *was* a technology of this time), Heidegger writes that the mechanized script removes the hand from the realm of the word, which depersonalizes writing and 'degrades the word to a mere means for the traffic of communication' (a passage from Heidegger 1982 quoted and translated in Heim 1992, 311).

We could put this claim in the context of a *very* long-standing discussion about writing technology, beginning at least with Plato in ancient Greece and continuing to this day. (See also Ong again, who sees an evolution from oral to literate and then from chirographic to print technology/culture. And, indeed, according to Heidegger, the trouble starts with Plato.) And we could argue against this particular view, for instance by arguing—with Heidegger against Heidegger—that the new technologies also reveal new, interesting, and potentially positive possibilities. But regardless of Heidegger's particular view of what exactly the computer does to language, the most interesting point for my purposes here is the insight that technology *does something to language at all*, and that in this sense we can see the computer and other information technology as *language machines*. These technologies regulate our use of language, shape the conditions of possibility under which we use language—in other words, shape our language and thinking. Again, the medium shapes the message. Machines shape language. As Heim writes: 'Heidegger sensed the power of the machine as an agent for changing our relationship to the word' (311). Generalized, this means: Technologies do not only change the world, they also change the *word*. If we must use the language of agency at all, we can say that technology does not only function as a hermeneutic agent or hermeneutic mediator, as Ihde has shown, but also at the same time as a *linguistic* agent. Technology changes our (use of) language and (thereby) our thinking. Both Heidegger and McLuhan saw how new technologies, as 'the backdrop against which things appear' (313), defined and changed reality—also in a linguistic way. As language machines, then, information technologies shape our "doing things with words." This can not only be applied to the typewriter and word processing but to all kinds of devices we use today, including the Internet, smartphones and their apps and voice technologies, and robots, if they were to become part of everyday life.

Conclusion

To conclude, this view of technology no longer concerns speaking *about* technology. It is no longer merely about discourses and narratives about technology. According to this view, technology itself acquires linguistic significance, especially in the form of "language machines." Based on this Heideggerian insight concerning the influence technologies have on our language/thinking, philosophers of technology could analyze the precise ways in which this happens for various technologies, including the computer and other contemporary technologies such as robots or smartphones, which are also linguistic and hermeneutic "agents."

As said, currently the *linguistic* dimension is under-studied in contemporary philosophy of technology. Postphenomenology, for example, has focused on hermeneutic agency, but *language* has not received much attention. Philosophers of technology, such as Ihde and Verbeek, who turn to material artefacts to say more about the hermeneutic role of technology, do an excellent job at articulating and showing the way technologies, understood as artefacts, shape and mediate our understanding of the world. However, because of this nearly exclusive focus on the material artefact, the question regarding language and, in particular, the question regarding the precise relationship between language and technology, is not addressed, let alone that a framework is provided in which the mediations by *language* are integrated with those of technology—a framework, preferably, in which the relationship between language and technology is clarified. In order to bring in language, I have used Heidegger and Wittgenstein (and Dewey), whose views I reconfigured in order to serve as a tool to say something about technology. This also enabled me to start my reflection and essay on the relation between language and technology.

In the next chapters, I hope to further study this relation and explore various routes into integrating insights about language into philosophy of technology. I will sketch a few ways of how contemporary philosophy of technology (not only postphenomenology) *may be rewritten in order to account for the role of language*. This book is not so much a criticism of what is already going on in philosophy of technology; instead, it is meant to be a constructive effort to connect thinking about technology with thinking about language. It is a first attempt to systematically map, think through, and use various existing theories in philosophy of technology and philosophy of language in a way that fruitfully connects both fields. My engagement with Wittgenstein's and Heidegger's thinking about technology *and language* in the first two chapters has kicked off this project. The journey has begun.

Notes

1 Note that she remarks that for Wittgenstein, ethics, by contrast, remains private. If this is an adequate interpretation of Wittgenstein, then this is indeed unfortunate. Here, I will not go into this very interesting discussion about Wittgenstein's ethics: Did he have ethics at all, and, if so, what kind of ethics?

2 For instance, as I write this, a Google computer has won against the human Go world champion—see, for instance, www.geekwire.com/2016/ai-software-masters-the-game-of-go-historic-human-vs-machine-match-awaits.

3 See also work by Turkle and others who have described this change. Appropriate to the topic of this book, we can interpret Turkle has having studied the *language* used by computer users. For example, in *The Second Self* (1984), she described what children, computer programmers, etc. *say* about and to computers, and how it changed their thinking—indeed, in her 2004 introduction, she explains that she first studied computer programmers, but later computers became part of everyone's lives, and, as such, they shape our thinking. Interestingly, she observed how computers made people think of themselves as machines.

References

Coeckelbergh, Mark. 2017. *New Romantic Cyborgs: Romanticism, Information Technology, and the End of the Machine*. Cambridge, MA/London: MIT Press.

Dewey, John. 1929 (1925). *Experience and Nature*. London: George Allen & Unwin.

Dewey, John. 1938. *Logic: The Theory of Inquiry*. New York: Henry Holt and Company.

Foucault, Michel. 1994 (1966). *The Order of Things: An Archeology of the Human Sciences. A Translation of Les Mots et les choses*. New York: Vintage Books/ Random House.

Heidegger, Martin. 1977. "The Question Concerning Technology." In *The Question Concerning Technology and Other Essays*, edited by Martin Heidegger, translated by William Lovitt, 3–35. New York: Harper & Row.

Heidegger, Martin. 1982. *Parmenides*. Frankfurt: Vittorio Klostermann.

Heidegger, Martin. 1983 (1957). "Hebel—der Hausfreund. (Hebel—Friend of the House.)" In *Contemporary German Philosophy*, translated by Bruce Foltz and Michael Heim, 89–101. University Park, PA: Pennsylvania State University Press.

Heidegger, Martin. 1996 (1927). *Being and Time: A Translation of Sein und Zeit*. Translated by Joan Stambaugh. Albany, NY: State University of New York.

Heidegger, Martin and William McNeil, eds. 1998 (1967). *Pathmarks*. Cambridge: Cambridge University Press.

Heim, Michael. 1992. "The Computer as Component: Heidegger and McLuhan." *Philosophy and Literature* 16(2): 304–19.

Hekman, Susan. 2002. "The Moral Language Game." In *Feminist Interpretations of Ludwig Wittgenstein*, edited by Naomi Scheman and Peg O'Connor, 159–175. University Park, PA: Pennsylvania State University Press.

Hickman, Larry A. 1992 (1990). *John Dewey's Pragmatic Technology*. Bloomington/Indianapolis: Indiana University Press.

Ihde, Don. 1990. *Technology and the Lifeworld: From Garden to Earth*. Bloomington: Indiana University Press.

McLuhan, Marshall. 2001 (1964). *Understanding Media*. Abingdon, NY: Routledge.

Nordmann, Alfred. 2002. "Another New Wittgenstein: The Scientific and Engineering Background of the Tractatus." *Perspectives on Science* 10(3): 356–83.

Springs, Jason A. 2008. "What Cultural Theorists of Religion Have to Learn from Wittgenstein; Or, How to Read Geertz as a Practice Theorist." *Journal of the American Academy of Religion* 76(4): 934–69. doi:10.1093/jaarel/lfn087.

Sterrett, Susan. 2005. *Wittgenstein Flies a Kite*. New York: Pi Press.

Turkle, Sherry. 2004 (1984). *The Second Self: Computers and the Human Spirit*. Cambridge, MA: MIT Press.

Verbeek, Peter-Paul. 2005. *What Things Do*. University Park, PA: Pennsylvania State University Press.

Winch, Peter. 2003 (1958). *The Idea of a Social Science and Its Relation to Philosophy*, 2nd edition. Taylor & Francis e-Library.

Winner, Langdon. 1980. "Do Artifacts Have Politics?" *Daedalus* 109(1): 121–36.

Winner, Langdon. 2014. "Technologies as Forms of Life." In *Ethics and Emerging Technologies*, edited by Ronald Sandler, 48–60. Basingstoke/New York: Palgrave Macmillan.

Wittgenstein, Ludwig. 1969. *On Certainty*. Translated by Denis Paul and G.E.M. Anscombe. Oxford: Basil Blackwell.

Wittgenstein, Ludwig. 1993 (1967). "Remarks on Frazer's Golden Bough." In *Philosophical Occasions: 1912–1951*, edited by James Carl Klagge and Alfred Nordmann, 118–55. Indianapolis, IN: Hackett.

Wittgenstein, Ludwig. 2009 (1953). *Philosophische Untersuchungen/Philosophical Investigations*, Revised 4th edition. Translated by Elisabeth Anscombe, P.M.S. Hacker, and Joachim Schulte. Malden, MA/Oxford/Chichester: Wiley-Blackwell.

3 Giving Meaning to Technology

A Searlean Social Ontology of Technological Artefacts

3.1. Introduction

In the previous chapter, I already suggested several ways of going beyond the instrumental view of language, and explored what this means for thinking about technology. Wittgenstein's and Heidegger's views of language, however, are already quite radical in the sense that if the instrumental view of technology is on one end of a continuum, their view is at the very other end, entailing a very intimate connection between language and thinking, and—by the analogy I made—between technology and thinking. But there are many views that are situated in between these extremes. This chapter discusses and uses a view of language that is closer to the common instrumental view, but still takes significant distance from it: Searle's social ontology. Like Wittgenstein, Searle focuses on the use of *language*, and especially on the grammar of language; but, as we will see, Searle holds on to a view that clearly separates subject/language and object/physical reality. Like in Wittgenstein, the one-sided focus on language use is to the downplaying of the material object, which remains non-active and dead; the act is in human speech and human use. Yet although at first sight Searle's view concerns talking *about* technology, more is going on than in the merely conventional instrumentalist view of language: By talking about and to technology, we also give a particular meaning to technological artefacts. Hence, the use of language is no longer merely instrumental: It constitutes the meaning of objects. Subjects and objects start to approach one another; the subject "touches" the object now. For thinking about technology, this means that, in contrast to the instrumentalist view of language (and technology), what a technology or a technological artefact "is" can no longer be defined apart from the way we speak about (and to) it. Technology can no longer be understood without taking into account language—in particular, the way we give meaning to technology by using specific words and grammar.

Moreover, as in the previous discussion, but here more explicitly, there is—next to language and technology—a third term that plays a role in these accounts: the social. Heidegger's, Wittgenstein's, and Dewey's view also implies specific views of the social and its relation to language, and (sometimes inexplicit) views on technology and the social. For instance, Wittgenstein linked the use of language to language and games and forms of life, which, at least on the "cultural" interpretation of the later Wittgenstein, can be interpreted as constitutive of the social. And Dewey explicitly understood language and technology as making possible the social. Searle's view also contains presuppositions about the social. But, whereas Heidegger and Wittgenstein emphasize the "givenness" of the social (the givenness of modern society as enframing, the givenness of specific forms of life), Searle's social ontology stands more in the tradition of contractarianism and related views of the social, which see the social more as a human construct and, more specifically, as an outcome of agreement and intentionality. As we will see, Searle uses the term "collective intentionality." In other words, here, humans have more to say (literally) about what form the social takes. By using language, humans construct social institutions. Social-linguistic agency is emphasized. The agency, intention, and power of the language user or, more precisely, the collective of language users, is central. In contrast to Heidegger's view, here, humans remain the masters of language (and the masters of the social). And, in contrast to Wittgenstein's view, here, language does not "bewitch" us or lead us astray; instead, we use it as an instrument to collectively construct (social) reality. While language gets a more "active" role, humans are still in control: They speak, whereas technology and language are mute tools.

3.2. Searle's View of Language and Social Reality

Searle's view of language is part of a linguistic turn in philosophy, which no longer sees language as an entirely neutral instrument, but as a tool we use to create social reality. In this sense, words and sentences *do* things. They are active and *performative*. I will stress this performative aspect by interpreting Wittgenstein, but this is also present in Austin and Searle. (Note that Wittgenstein influenced Searle, but Searle rejected his antitheoretical stance.) Austin and Searle developed "speech-act" theory, which holds that communication is not just about getting across propositional content and *representing* the world. Instead, speech acts *create* social reality. In this sense, we 'do things with words,' as Austin put it (Austin 1962). Thus, here, the medium is no longer neutral, but shapes the message—at least to some extent. In Searle's view, there is still an unchanged physical-material reality, untouched by the subject. But the subject already shapes the meaning of the object in the sense that social meaning is created by speech acts.

This view of language is Searle's answer to what he calls the problem of social ontology. In *The Construction of Social Reality* (Searle 1995) and in 'Social Ontology' (Searle 2006), he asks how humans can 'create a reality of money, property, government, marriage' and so on (Searle 2006, 13). His answer is that humans do this by using language. In contrast to physical facts, Searle argues, 'social facts' are not observer independent but 'observer relative' (13), since they depend on humans, and, in particular, humans as language users. Language, according to Searle, constitutes social reality itself.

How does this work? Searle argues that we collectively accept to give a certain status to a person or object, which enables this person or object to perform a (social) function. He uses the term 'collective intentionality:' 'collective intentionality assigns a certain status to [a] person or object and that status enables the person or object to perform a function which could not be performed without the collective acceptance of that status' (Searle 2006, 17). He gives the example of paper money: A piece of paper performs a social function, but it does so 'not in virtue of its *physical* structure but in virtue of *collective attitudes*' (17; Searle's emphasis). Thus, money can perform its function only because we collectively *give* it that function. We accept that the piece of paper has the status of money.

This works by means of a particular use of language, and more specifically a particular grammatical form. Searle focuses on declaration: we *declare* that a particular object or person has a specific status. Searle calls this declaration a 'status function' (Searle 2006, 18), which has the logical form of 'X counts as Y in context C' (18). For instance, we agree and declare that 'this piece of paper counts as money in the context of trade and exchange.' In this way, social institutions take shape. They are not only made by physical-material means but also by language. Language does not represent something here; instead, the use of language is itself doing something and it is constitutive of the social. Moreover, this use of language does not only construct social reality; it also has normative implications. Status functions give us what Searle calls 'deontic powers:' If I have a parking ticket for this parking lot, for instance, I have the right to park here. Rights and obligations come with the (new) status. Status functions thus create a social and moral reality. Language transforms objects into meaningful and powerful social things. What the social "is" is constituted by our use of language.

In Searle's view, this ascription of status can only happen as a result of social agreement: The social institution with its meanings and deontic dimensions only comes into being if we *agree* that X counts as Y in context C. Moreover, this agreement and collective intentionality is *sufficient* for the social institutions to be created. In the end, for Searle, language *alone*—in the specific form of a status function with collective intentionality—can do the work of status ascription. We can have status functions even without physical objects (Searle 2006, 22). Consider the example of money again: electronic money does not really have a physical, material form; yet, if we agree that it counts as money in specific contexts, then it *is* money. Language itself is powerful enough, it seems, as long as there is collective intentionality and agreement. It is sufficient that there is an object—real or virtual—that is receptive to our meaning-giving by using language.

On this view, then, the social can be seen as the outcome of a kind of social contract agreed to in an 'original position,' in which we have collectively agreed that particular objects X count as Y in particular contexts C. Social institutions are the outcome of, created by, and perhaps also *justified* on the basis of, these status functions—that is, these uses of language in an (imaginary) original position. Furthermore, physical reality, physical and material objects, can have a role in the social if they are what I would call "dressed" with a particular status by us. Language, in the form of status declarations, then indeed *does things*: it gives meaning and normativity to what otherwise would be entirely nonsocial objects. Status declarations turn physical things—and even nonphysical things such as numbers—into perhaps not social agents, but at least things, tools, that help humans to construct social institutions and social reality. Physical things are thus dressed with a social layer and become part of what we may call the social performance. They are taken from, perhaps liberated from, the nonhuman world of physical facts and enter the human world. However, they remain entirely physical things; they are not themselves social nor cultural. Their social meaning and normativity does not touch their core ontological status. The social remains a shell or cloud around the physical core. Indeed, Searle is keen to keep social facts and physical facts separate.

The same goes for human agency and non-agent things: Ultimately, humans, as language users, have all the agency and intentionality, in particular collective agency and intentionality. Humans use language, in particular status declarations, and they use physical objects (but the latter are not really necessary) to construct social reality. But we could say that the grammatical form itself does not speak, is dead. It is the use by the human user that gives it life, and, in Searle's view, human users retain full control of that use. This is different from my interpretation of Wittgenstein's view, in which language games and a form of life take a more active role as "givens," and thus there is a sense in which language itself speaks. To conclude, while, according to Searle's view, clearly language is no longer a neutral instrument since it gives meaning to objects, humans—as collective—remain in control of language. Neither language nor technology, understood as artefacts, speak (yet). In terms of meaning, they are "dead;" they are given life by human language users. Humans speak (collectively), objects get meaning, and, hence, subject and object approach one another; but they are still neatly separated.

Let me further reflect on the implications of Searle's views for thinking about technology.

3.3. Implications for Thinking about Technology: Towards a Social Ontology of Technological Artefacts

What does Searle's view mean for thinking about technology? A first and perhaps most obvious application of Searle's view is to regard technological artefacts as potentially socially and normatively significant *if* they are given a particular status by humans as language users, if they are given meaning and "life." A second way of using Searle's view is to see language as a technology, and to then to apply his view of language to *technology* and to explore analogies between language and other technologies in order to say something about technology in general—an exercise that I also did in the previous chapters.

Let me start with the first use of Searle. The Searlean problem in philosophy of technology could be stated as follows: We see that technological artefacts play a role in the social life, even often a constitutive role, but it is mysterious how they can play that role, given that they are (supposed to be) mere physical objects. For instance, how is it possible that a robot can become "social" given that it is a mere thing, a machine? How is it even possible that a fork is used for eating, rather than, say, for scratching one's back? And, indeed, how is it possible that a medium and technology like money means so much to people and has all the functions it has, given that it is often a mere piece of paper or even just a number on the screen? Think also about "objects" in virtual worlds, such as virtual money, virtual weapons, or virtual houses in games: How is it possible that they have social meanings and all kinds of uses, in spite of their apparent ontological status as mere "virtual," not real, or at least *less real* or *not very real* objects? There seems to be a gap between, on the one hand, what these technological objects "are," in virtue of their physical and material or even immaterial properties, and on the other hand what they mean to people and are used for. Indeed, the mere physical-material structure and ontology of a technological artefact does not *determine* its meaning and use. That meaning and use at least *also* depends on the meaning and use we *give* it.

Now, in order to capture how all this is possible, how it works, and maybe how it can be justified, one could use Searle's theory: What explains and perhaps justifies the meaning and use of these technological objects is collective agreement and collective intentionality, and, in particular, status declarations. A Searlean philosopher of technology may argue as follows. It may be true that a robot "is" a mere thing, a machine, and so on, in virtue of its physical properties. But, if we agree that in specific contexts it is also a companion by declaring it as such, then socially speaking it also "is" a companion. It may be true that one can do all kinds of things with a fork, but if we collectively agree that in the context of eating it has the function of an eating tool, and embed it in, and make it co-constitute, the social institution of eating together, then this is also what a fork "is," regardless of its physical properties and whatever else could be done with it. (The same can be said about chopsticks, of course.) It may be true that paper money "is" just paper, but if we collectively assign the status of money to it and agree that in the contexts of trade and exchange it is money, then that makes it money in these contexts. The piece of paper then becomes part of the social institution of money. And if a collective agrees and declares that a virtual object, such as a virtual axe in a game or a virtual piece of gold in a virtual world, counts as an axe or counts as gold in specific game or virtual-world contexts, then these "are" also axes and pieces of gold in these contexts. Through language use, it seems, technological objects are decoupled from, or, if you like, *liberated* from, their ontological status in virtue of their physical properties, and acquire new meanings and normative dimensions, which are added to their core physical-material ontological nature, like clothes that dress a human body. Artefacts are dressed up as social artefacts, and *we* dress them up as such, through collective status declarations. It is up to humans, and in particular human collectives, to agree upon and *give* them their social and moral status (and therefore use) in particular contexts. And while often and initially that status and use will be somehow related to their physical properties, this is not a necessary relation; there is no determination. To take an example from Searle (1995): A wall may first serve as a physical border, but when the wall decays and only a line is left, that is sufficient for the functioning as a border, as long as humans give it that status of a border. And to return to my example of a computer game: it can be imagined that in an "original position," gamers agree to assign specific powers to a virtual sword and its bearer. The sword is not even physical. It is entirely nonmaterial. Yet the image on the screen is sufficient, as long as there is collective agreement to assign this particular status to the sword and its bearer, which can be made explicit in the form of, for instance, this status declaration: 'This virtual sword S counts as having powers P in these game contexts C1 and C2.' In other words, language, used in this particular way as described by Searle, enables language users to create a particular social reality. And physical and material objects may or may not play a role in this; imaginary objects or virtual objects can also be endowed with meaning and normativity.

This application of Searle's theory thus seems particularly helpful to explain how "virtual" and "digital" technological objects can acquire meaning and normative powers. If there is no link between the status of such objects and their physical reality, this is not a problem for Searle's approach; what matters is collective intentionality and agreement. And, in general, Searle's theory seems to provide an adequate way to explain and theorize about the role of language in creating social reality. Humans use language to create that reality.

Indeed, it seems that, on the basis of Searle's theory, language itself could be seen as a technology. With its specific structures and related functions, called grammar, it enables language users to do more than simply describing or representing reality. While, at least in Searle's view, physical-material reality remains untouched by it, *social* reality can be created by means of language. Language thus turns out to be a technology for the construction of social reality. Words and grammar are tools for social construction.

This raises the question if *other* technologies could have a similar function. Could it be that, in contrast to Searle's view, not only language but *all* technologies and media shape the social? Perhaps collective intentionality does not only use language to perform and implement its agreements and status declarations; perhaps other technologies can also create social reality and perform social institutions. Indeed, if we take the example of marriage, we could say that not only words but things also play a role: The marriage ring plays a role in a marriage ceremony, for instance. The ritual and social institution depends on linguistic declarations, but also on artefacts. We could construe a revised (more specifically, expanded) Searlean view, in which there is not only a grammar of words but also *a grammar of things*. Artefacts (physical, human-made "facts") are also used to bring about *social* facts. They have a more "active" role in terms of meaning than Searle is prepared to give them. For instance, money is not only the object of linguistic status declarations; it is also itself an agent of social construction and social change, since, as an exchange medium, it establishes a particular kind of value relation and social relation. With Simmel and Dewey, we can say that it also shapes that relation. We can imagine an original position in which we collectively intend and declare that a piece of paper has a certain value. But we cannot imagine a similar linguistic act when it comes to the relations between people that money establishes. For instance, we never agreed and decided that relations mediated by money would be less personal (to borrow a claim from Simmel); if this is the case at all, it came about as a result of our use of the medium and technology of money. When social meaning changes, this cannot only be understood as a result of linguistic declarations and other acts/performances with words; it is also a result of acts and *performances with things*. Thus, in this expanded and revised Searlean view, both language and things are tools we use to construct the social. (Below I will suggest a way to develop this point by picking up again Searle's example of the wall.)

Moreover, returning to the problem of unintended use, we can ask if the shaping of the social also happens in ways that are not limited to what humans as language users do and *intend* with language or (other) technologies. This question leads us to consider a more radical view of technology and its relation to language, in which technology itself "speaks." The idea is then that the meaning and normative dimension of technology cannot be reduced to what is given to it by humans, but that technology, also *because* of its own properties and structure, its own *grammar*, co-shapes social reality and perhaps all reality. Could it be that technology *itself* also "gives" a specific meaning to things, and has deontic aspects, which are not necessarily under full control of individual or collective agency? Could it be that, through using specific technologies, things acquire a specific meaning that may differ from the meaning we have assigned to it? Could it be that the uses of specific technologies lead to specific deontic powers, such as rights and obligations, which go beyond those we agreed upon? I have already suggested a more radical view in the previous chapters, and, in Part III, I will further develop the view that "technology speaks." But let me first further clarify the implications of the Searlean view sketched here for thinking about technology, and further substantiate and critically discuss it by using Wittgenstein, Heidegger, and Austin.

3.4. Conclusions for Thinking about Technology and Critical Discussion: Beyond Searle with Wittgenstein and Heidegger

Conclusions for Thinking About Technology

Let me summarize the previous sections. Like Wittgenstein and Austin, Searle focuses on the use of language. We do things with words. One of the things we do with words is giving social meaning and status to physical objects. In that sense, language does not merely represent or express social meaning and social reality; it also creates it. This gives a larger role to language than the common, instrumental view. Yet, as I argued, Searle's social ontology enables us not only to get beyond a merely instrumental and neutralist view of *language*; it also suggests at least three ways in which we can move beyond a merely instrumental and neutralist view of *technology*. First, Searle's view explains how gaps between physical properties of technological objects, and their social and normative meaning, can be bridged and are bridged: Humans, as language users, can collectively assign a particular status to things by means of a status declaration, status function. In particular, Searle's theory seems very useful for thinking about the meaning of "virtual" and "digital" technological objects. Second, we can interpret Searle's view as saying that language itself is a technology, which is instrumental in creating social reality. Third, this interpretation also raised the following question: If language, as a technology, creates social reality, what about other technologies? Perhaps they are also used to create social reality. Perhaps the creation of social reality is not only a linguistic affair as such but happens as a result of linguistic *and* material workings and mediations. Perhaps technology *also* has a "grammar" and *also* "speaks."

This view takes us beyond Searle. Influenced by Heidegger and Wittgenstein (and perhaps also Dewey), we must question (1) Searle's focus on human (linguistic) agency and his emphasis on agreement (the contractarian view of social reality as the result of agreement), and we must criticize (2) his assumption that we must strictly distinguish between an entirely non-social and nonlinguistic reality on the one hand, and a social and linguistic layer on the other hand. I will first question Searle's stress on human (linguistic) agency; then, when I turn to Wittgenstein (and Dewey), I will question social/physical dualism. (Note that I will also return to the questions regarding dualism and agency when I discuss Latour's nonmodern view in Chapter 6.)

Beyond Searle with Wittgenstein and Heidegger: Language Use, Technology Use, and the Given

In Searle's view, human agency reigns—also and especially linguistically: Even if language gets a more important role than in the instrumental view, it is the human that speaks. But, as we will see in the next parts of this book, there are other theories and approaches to the relation between language and technology that give a far more important role (and agency) to either language or technology. In these views, "what language does" is not limited to what humans collectively or individually agree it should do, and similarly "what technology does" (to use a phrase from Verbeek) is not limited to what humans collectively or individually agree it should do. As I already suggested with my interpretations of Heidegger, Wittgenstein, and Dewey, language itself may have more agency than we usually assume: In our daily use of language, we presuppose an instrumental view of language (language that *itself* makes possible this instrumental view—I return to this point when I introduce and take a transcendental turn in Chapters 5 and 6). Similarly, several philosophers of technology have pointed out that technology itself may "do" more than we intend. In Chapter 6 I will discuss Ihde (1990), Verbeek (2005), and Latour (1993, 2004). However, in contrast to Ihde, Verbeek, Latour, and others who turned to artefacts, I will emphasize the role of *language* in how technology unintentionally shapes the social, the human, etc., and attempt to (re)construct the position that technology also "speaks"—and more than we think—by taking into account the role of language. For now, I will focus on Wittgenstein and Heidegger, whose insights on language will help me to criticize Searle's view and to further develop the alternative view suggested in the previous chapters.

Like Wittgenstein, Searle focuses on language use. But Searle's view of linguistic agency is different from that of Wittgenstein and Heidegger, since it is rooted in a different view of language use. For Searle, what counts for meaning is individual or collective intentionality. In other words, there is one speaker: an individual or a collective. The focus is also on one word or one sentence, for instance, a sentence that takes the form of a declaration. That declaration is somehow self-sufficient and isolated from other declarations and other uses of language in other contexts. There are many social facts, just as there are many physical facts, but it seems that, in Searle's view, these social facts are not intrinsically related to one another. Moreover, as said, Searle assumes that there is a gap between the social and the physical. Physical facts are governed by natural laws, whereas social facts are a result of status functions, that is, of language use. But there is no "whole" in this view: There is little interest in how the sentences are related, how the facts are related, or how our declarations may be influenced by other or larger cultural patterns. In other words, it seems that we find ourselves in a kind of atomistic universe of words, sentences, and facts (social facts and physical facts).

Heidegger and Wittgenstein, by contrast, have a more holistic and relational view of language use in which particular speakers, words, and speech acts are connected to a larger whole: A particular use of language is linked to ways of thinking/speaking, such as modern thinking (Heidegger), or to activities and language games that are part of a form of life (Wittgenstein).

This means that language use and meaning cannot be reduced to the meaning and intention of one speaker, one word, etc. Speech acts, words, and their meanings are always connected to a larger whole, which also shapes how language is used and what is meant. (In the first chapter, I already referred to *semantic holism* in Wittgenstein, Davidson, and others.) Admittedly, Searle's view is relational in the sense that there is a relation between the speaker, the grammatical status function, and the object that is given meaning, but further relations are not visible in his ontology. In the Heideggerian-Wittgensteinian view I articulate here, by contrast, meaning is *all* about relations and is relational in a more holistic sense.

With a view toward enriching philosophy of technology with insights from philosophy of language, I propose to transfer this holistic approach from the domain of meaning and *language* to thinking about *technology*. Heidegger already connected the use of technology to larger patterns and wholes, for instance and most importantly, modern thinking (Heidegger 1977). Regardless, from Heidegger's particular claims about modern technology, we can infer from his *approach* the view that technological artefacts should not be understood as material things isolated from language and culture; instead, they shape that language and culture, and "make sense" only in relation to that larger whole. In this sense, technology is not just about artefacts; it is also about thinking. Moreover, as I already suggested in my introduction (and also keeping in mind Winner's view, referred to in the previous chapter), Wittgenstein can also be interpreted in this way. His holistic view of language can seamlessly be applied to technology: There are technological artefacts (similar to words) and there is the use of technology, which has a particular "grammar" and follows rules (a grammar or syntax of use), but these artefacts and rule-based uses of technologies must be connected to a whole, a form of life, within which those artefacts and those *uses* of technology and that grammar acquire meaning, which, in turn, is also influenced, made possible, shaped, and transformed by new technologies. (Similarly, Dewey sees language and technology as deeply connected to the social; in my interpretation of Dewey, language and technology make possible the social.)

The latter claim is a transcendental one, which I will elaborate on in Chapter 5; this will also lead us to a different meaning of "grammar." For now, let me further support this interpretation and application of Wittgenstein by revisiting the *Investigations*. Remember that, in the text, Wittgenstein constantly compares language to technology. Technology, in particular in the form of tools, is the metaphor Wittgenstein uses to make his central points about language: He compares words and their functions to 'tools in a toolbox' (§11, 9e). As I already pointed out, there is an unintended and salient claim about technology in this comparison, one which points us to *use* of technology. But Wittgenstein's intended meaning is also interesting for thinking about technology: He uses the toolbox metaphor to emphasize the variety of the use of words to *the variety of use of tools*: 'It is interesting to compare the diversity of the tools of language and of the ways they are used, the diversity of kinds of word and sentence, with what logicians have said about the structure of language. (This includes the author of the *Tractatus Logico-Philosophicus*.)' (§24, 15e). Thus, Wittgenstein argues here against an atomistic, non-relational understanding of language use, an understanding of language also that is disconnected from life. [This use of "life" language is not coincidental; the later Wittgenstein can also be interpreted as a "philosopher of life," a *Lebensphilosoph*, not only a philosopher of language. See Gier (1990, 285); see also my remarks in Coeckelbergh (2012, 135).] If our use of the tools of language is very much connected to all our activities, to the way *we* do things, to our historically grown habits and natural-social practices, to communal ways of doing, and, hence, a form of life, then the same could be said of other tools. Then technologies and their uses can also be connected to these activities, larger patterns, forms of life.

This "cultural," "life"-connected, and holistic view also has implications for the epistemology of technology and for learning technology. Just as it is true that, as Wittgenstein argues, we learn language not by first thinking and then learning the words, but rather *we learn thinking by using language* (see his remark about Augustine §32, 19e), it is equally true that we learn technology not by first thinking about artefacts and tools, and then applying these thoughts to them; instead, we learn a technology by using it. At the same time, through this learning-to-use, we also learn particular games and forms of life. We also learn a particular way of thinking, a culture, in this way. For instance, Heidegger would say that in modern times we learn "technological thinking" in this way: by *using* modern technologies. Or, to take up my example of gaming: by using the technology, we learn the game, but also participate in, and learn, a particular (game and gaming) culture. Just as a particular use of language is always related to a larger linguistic-semantic whole, a particular use of technology is always embedded within a larger technological-semantic whole, which is "cultural" and "material-technological" at the same time.

Thus, this is how learning works for everyday knowledge of technology. We learn technological ways of doing at the same time as we learn all kinds of activities. Acquiring this knowledge—or more precisely: this know-how, this tacit knowledge—goes together with learning to eat, drink, play, etc. in particular contexts, communities, societies, and cultures. There are already rules, patterns, forms of life. There is already a given, which is then, however, also shaped by our use and our interpretations.

If, on the other hand, we take technology *out of its everyday context and use*, and stare at it like a 'philosopher . . . staring at an object in front of him and repeating a name' (§32, 19e), as Wittgenstein put it, then indeed language/thinking goes on 'holiday.' Then we have theoretical knowledge of the technology. Now, creating this knowledge may well be part of what scientists and some philosophers of technology do; but this relation to technology is, literally, the exception rather than the rule. Just as Heidegger shows that the ready-to-hand or handiness (*zuhanden*) epistemic relation to technologies occurs in everyday use, and that the present-at-hand (*vorhanden*) mode is the exception (e.g., when a tool breaks down or when we theoretically contemplate the tool), this interpretation of Wittgenstein teaches us that a theoretical stance (in philosophy of technology and elsewhere) is a very specific relation to technology and the exception rather than the rule. (I admit that thinking/writing all this is itself already somewhat theoretical—it is nearly a theory of technology use; but perhaps it is a consolation that Wittgenstein seems to do the same for language use in the *Investigations*, against his own advice.)

Note also that Heidegger not only talks about the use of technology, about technology "at hand;" he also writes about the use of *language* in similar terms. In *Being and Time*, he warns for the tendency of language to appear as objectively present, as things, divorced from their sources of meaning in lived discourse. When writing about language and discourse, Heidegger claims that discourse has its worldly being in language as the totality of words and that words may be seen as "word-things":

> The way in which discourse gets expressed is language. This totality of words in which discourse has its own "worldly" being can thus be found as an innerworldly being, like something at hand. Language can be broken up into word-things objectively present.
>
> (Heidegger 1927, 151)

This appearance of objective presence is seen by Heidegger as problematic. As Livingston puts it:

> The worldly character of language is, here, not a matter of its actual or possible correlation to the totality of facts or situations in the world, but rather of its tendency to *appear* within the world as an objectively present totality of signs or of "word-things," abstracted and broken up with respect to the original sources of their meaning in the lived fluidity of discourse.
>
> (Livingston 2016, 227)

Yet, maybe seeing words as things is not so problematic in itself; rather, it is problematic when we consider them divorced from their *use*, when we no longer use them as tools, but stare at them as objectively present things. It is interesting in the citation above that Heidegger thinks words can be 'like something at hand.' Thus, like Wittgenstein, Heidegger points to the phenomenon of words as tools. Moreover, applying Heidegger's distinction between present-at-hand and ready-to-hand to language, one could say that sometimes language appears as word-things that have objective presence, while language is usually a tool, is ready to hand (*zuhanden*). When we use language, we do not notice it and do not think about it; we just use it. It is part of lived discourse and lived use. Similarly, we can say that, normally, in its everyday-use context, technology is not objectively present. It is at hand, but it does not even appear as tools or as things. It is not broken up. There is use and performance; "things" are part of that as tools, but do not usually appear as such.

Based on what Heidegger says about language, we can stress the holism of this view, and point to the strong analogy, if not proximity and similarity, between language and technology—not only in Wittgenstein but also in Heidegger. When he discusses references and signs in *Being and Time*, Heidegger makes an interesting comparison between the use of signs and the use of tools when he gives the example of the indicator of a car. He sees the indicator as a sign, and the sign as 'a useful thing which is at hand;' it has a "serviceability." It is, thus, like the hammer, which is also a 'useful thing' and which is also 'characterized by 'serviceability,' but is not a sign (Heidegger 1927, 73). Now, here, we are not interested in the precise difference between the tool and the sign, but in their similarity. Keeping in mind Wittgenstein, it is interesting to point to what they have in common, according to Heidegger: both are *used*. Moreover, Heidegger says that, as useful things, car indicators relate to a totality of 'useful things belonging to vehicles and traffic regulations' (73). Thus, as useful things, they are related to other useful things and to rules. The meaning of the car indicator lies in its use by the driver and by other participants in traffic. This is the meaning of the sign, a meaning that emerges from use. Again we encounter holism, a holism that is there in everyday use, when signs and tools do not appear as objectively present, but are at hand and are used, since they are part of activities and uses. Heidegger writes: 'A sign is not really "comprehended" when we stare at it . . . the sign is not really encountered' (74). Signs refer to behaviors and to activities, to 'what is actually "going on."' (74). This everyday context and these activities are what are relevant for us, what are meaningful for us. Our everyday experience of language and technology is holistic and is focused on activities and games, of which use of words and tools are part; the words and the tools are not objectively present when *in* use. (Note that Dewey would also agree with this stress on everyday experience and the link between meaning and activities/behaviors.)

Thus, just as there is a parallel in Wittgenstein between the use of tools and the use of language, there is a parallel in Heidegger between use of technology and use of language. Both thinkers seem to conceive of language as a tool (see also Dewey), and provide a more holistic view of words (we should always see words in the context of a language) and language (we should see language as the use of words in the context of our activities, rules, games, life), which by means of analogy and metaphor can be seamlessly applied to technology.

Seen from the point of view of this Wittgensteinian-Heideggerian approach, then, Searle's view can be criticized for seeing language as a totality of objectively present words, "word-things," consisting of linguistic objects, which then get arranged by grammar, understood as syntax (especially the grammar of declaration), and attached to physical things. Searle's focus is on use, but his language user resembles Wittgenstein's philosopher 'staring at an object in front of him and repeating a name' (§38, 23e): The individual or collective agent performs a speech act that baptizes the object and turns it into a social fact—by means of declaration. There is an original declaration, in the original contractarian position when we agreed, and then the name is repeated every time again, thus effectively continuing the life of the social institution. But this naming ritual is problematic, since it is too theoretical and lets words appear as objectively present. Wittgenstein writes in the *Investigations*:

> 'For philosophical problems arise when language *goes on holiday*. And *then* we may indeed imagine naming to be some remarkable mental act, as it were the baptism of an object. And we can also say the word "this" *to* the object, as it were *address* the object as "this"—a strange use of this word, which perhaps occurs only when philosophizing.'
>
> (§38, 23e)

Indeed, the Searlean language user "addresses" the object when one gives meaning to it, when one "names" it as money, for instance. In other words, from a Wittgensteinian perspective, one could frame Searle's approach as too *Augustinian* and too theoretical. (And from a Deweyan perspective: too nominalist and mentalist.) Searle's language user enters a world of naked physical facts and physical object, which one then tries to dress up with social meaning through the use of language. Moreover, the Searlean philosopher also has a distorted view of social reality: one enters one's social world as if one enters a foreign country and does not understand the social ontology of that country. The Searlean philosopher, as a player of the philosophical-social-ontology language game (a kind of meta-game), then tries to explain the social facts. But even this very talk of social "facts" is already an alienation from the social and reifies it into facts. According to a social, "cultural," and constructivist interpretation of Wittgenstein, by contrast, *we already know the social through our language use and through our (technological) practices*. There are no naked physical facts, and we cannot separate them from social facts. There is already meaning before we speak (before we use language, individually or collectively) and before we question meaning and particular meanings (e.g., as philosophers). What needs to be explained, if anything, are not social institutions, but the strange and alienating act of taking distance from the social. What needs to be explained, if anything, is the holiday.

In a similar way, we can think of the use of *technology* as only partly involving (individual or collective) agency and intentionality; instead, the social and the physical are already linked in everyday use and performance with technology, in using and living with technology, since these uses are connected to an already existing larger structure and whole. We can think of technology as being embedded in ways of doing and understanding that exceed what we do as language and technology users at any given moment in a particular situation (for instance, by means of declaration of the meaning of artefacts) or what we do with language and (other) technology as abstract contractarian individuals or collectives. Instead of focusing on, and presupposing, an abstract individual or collective contractarian user of technology, who on the basis of agreement on use then sets out to use a technological artefact in a particular way and gives meaning to it in a particular way, then, we should put the use of specific technologies in the context of the larger social-technological (and social-physical) whole of which they are part *and in context of the everyday activities and lives of which they are part— which are always already social and institutional.* The social (understood as social-technological) and institutional is already given; it does not need to be explained by means of a Searlean account. There are rituals and habits, but when it comes to both language and technology, the key ritual and habit is not naming, but *using and performing.* Naming is a very particular kind of use, habit, and performance, part of a very specific language/technology game and performative context. It is an exception rather than the rule, and it presupposes an already existing form of life, in which these particular performances make sense and get a sense, and in which material artefacts play a more active role. Thus, we need an approach to meaning that is more holistic, is more appreciative of how meaning giving works in everyday (performative) contexts, and at the same time brings back the object, recognizes the role of artefacts in the making of meaning. (In this sense, perhaps one could say that Searle's and Wittgenstein's approach is right when it rejects Augustinian nominalism, but is *not Augustinian enough* when it does not give the object its due.)

The closest Searle comes to recognizing the role of artefacts is when, in *The Construction of Social Reality* (1995), he offers a narrative of 'a primitive tribe' that builds a wall around its territory: First, people use the wall to mark their territory, but then 'the wall gradually evolves from being a physical barrier to being a symbolic barrier' and, when the wall decays, 'the only thing left is a line of stones,' which, however, is recognized as the border of the territory (Searle 1995, 39). Searle gives this example to stress that the meaning of the boundary is independent from the physical properties of the wall. For him, this is a matter of collective intentionality; Searle's emphasis is on the symbolic. But we can also (re)interpret this example as saying something about the social having a performative and holistic character, in which material objects play a key role as they are *used.* While it makes sense to say that, in his example, the people of the tribe assigned the status of boundary marker to the stones, it is equally meaningful to say that, in the game of marking one's territory and in the repeated performance (ritual) of confirming those boundaries, material artefacts—here, wall and the stones—play a key role in the game and the performance. Moreover, they do so not only as passive receivers of meaning. It is not only the case that humans speak but there is also a sense in which the wall and the stones "speak" and perform. They co-perform with the humans in the rituals of confirming the boundaries of the territory. Moreover, the collective intentionality Searle talks about, the recognition of the line of stones 'as marking the boundary of the territory in such a way that it affects their behavior,' is made possible by a form of life in which the boundary game (already) makes sense—a form of life and a game that, as the story suggests, came about not so much, or certainly not only, by means of a symbolic speech act, but at least also by means of the use and "performance" of a material artefact and through making that artefact. It is through the technological practice of wall building, and not only through speech acts, that a particular form of life has emerged. It may well be that in our culture(s) there is an evolution towards immateriality and towards the symbolic (as Simmel also recognized in his thinking about money), but without an initial role of the material artefact, this would not be possible. Moreover, to *sustain* that form of life and to sustain the boundary game, the wall and later the stones are *used* and play a part in the *repeated* performances of the tribe that confirm the boundary, in the relevant habits and rituals. When people no longer *use* the wall and the stones as a boundary, the moment that these artefacts no longer perform their function of border and no longer play a part in border performances, they *are* no longer a boundary. But if there is a form of life in which boundary games are crucial, new borders will emerge. There is a material-cultural background that is given, and new artefacts (next to natural entities) can play a role in the emergence of new borders. Collective intentionality and language use more generally is only *part* of this process, and if there is a status assignment at all (as Searle argues), then this status assignment is part of, or rather emerges from, this larger whole and this process of use and performance, in which material artefacts play a role. The use of words is not unimportant, but it is—literally—not the whole story; we have to add the use of material artefacts as equally significant in the making of meaning.

This point can also be made in response to Searle's example of primates using a stick to get bananas. Searle interprets this phenomenon as teaching us that 'animals can impose functions on natural phenomena,' and sees a 'radical break with other forms of life' when humans, 'through collective intentionality, impose functions on phenomena' when cooperation is needed, and create institutions; hence, he sees a radical break between prelinguistic and linguistic forms of life (Searle 1995, 40). But why see this as a radical break? We (still) use words *and* things to create and uphold human institutions, and primates also utter sounds. With Dewey, we could admit that there is a difference between what we do with sounds and what nonhuman animals do with sounds, but question that there is a radical break. Instead, the linguistic and nonlinguistic are part of the same social uses, processes, and activities, and there is no *radical* break with what animals do, in terms of social cooperation (consider also Dewey's naturalization and socialization of tool use). Searle limits the institutional and cultural to the use of words, but this need not be our limit. We can see 'the human capacity for tool using' (40), described by anthropologists, as equally important in the making of the social. The meaning of artefacts and the emergence of institutions is then made possible by the use of words, but also by the use of tools, and the function of things in the social is not only 'assigned' but also comes about and is maintained through use of the things and through performance with the things. It may be true, as Searle puts it, that there is a 'continuous line that goes from molecules and mountains to screwdrivers, levers, and beautiful sunsets, and then to legislatures, money, and nation-states' (41), but the 'central span on the bridge' is not only (linguistically mediated) collective intentionality; there is what one could call *collective use* and *collective performance*, which involves not only the use of words but also the use of artefacts. It may well be that modern complex institutions need words to function (41); but, simultaneously, they cannot do without material artefacts and structures. To take one of Searle's favorite examples: Even "immaterial" money such as an amount on a bank account or a digital currency such as Bitcoin needs the Internet, but also material infrastructure and devices, such as computers, servers, and wires (or, indeed, material equipment to make possible wireless communication), to work and function. Moreover, complex institutions often use the technology of writing and related technologies and artefacts, such as paper (think of bureaucracy, which is often referred to as "paperwork" and, even if the papers are in digital form, a material infrastructure is needed) and, again, computers and the Internet, which themselves are not only a matter of code that does things (software) but also material artefacts (hardware) and material infrastructures.

Shifting the emphasis to collective use and performance, and to the role of material artefacts, also implies, again, that the social is not only a matter of agreement. The use of language was already not entirely a matter of agreement—this alone already questions Searle's approach—but this is even more so, perhaps, with use of technologies. Generally, neither individuals nor collectives agree or can meaningfully be assumed to have agreed on using a technology and performing with a technology. While there is regulation of technological practices and policies about technology, it is striking how our use of technologies, understood in a broad hermeneutical and Wittgensteinian-Heideggerian sense, is *not* the outcome of agreement and explicit or implicit recognition of artefacts having a certain function. What Wittgenstein argued about language can be applied to technology: Learning how to use a technology is a matter of obtaining know-how, understood as implicit knowledge. And more often than not, as Heidegger already suggested, we do not think about technology when we *use* it, and new technologies tend to be used and change the social in ways that are very much outside of individual or collective control. For instance, it is difficult to uphold the idea that our use of computers or smartphones is a matter of collective agreement and that all their functions have been intended. Rather, through the use of words and things, a new technological culture has emerged, which can only uneasily be described only by using Searle's language of declaration and assignment. The meaning of our smartphones, for instance, emerged from the use (in response to functions created by designers and programmers) and is still developing in ways that are really hard to describe as giving a status to the phone. Its status emerged and emerges from use within a larger context in which certain social-communication games and a certain form of life already existed; the phone may change that form of life, but human intentionality and assigning functions is, at most, *part* of this process. Words do *less* and things do *more* than we commonly assume.

Thus, we can read use, performance, and materiality into Searle, taking his examples, but interpreting them in a way that emphasizes the use of material artefacts and performance with material artefacts. Moreover, based on Wittgenstein (and Heidegger and Dewey), we can say that there are already games and patterns. There are already forms of life that we mastered before the new technology entered our world. The emphasis is not on giving meaning to technology, but on how technology is *already* meaningful. However, the "givenness" of forms of life does not mean that forms of life or "worlds" are entirely unchangeable. As Winner rightly argued, technologies also change worlds as they bring about changes in patterns of human activity:

> significant alterations in patters of human activity and human institutions are already taking place. New worlds are being made. . . . The construction of a technical system that involves human beings as operating parts brings a reconstruction of social roles and relationships. . . . We do indeed "use" telephones, automobiles, electric lights, and computers in the conventional sense of picking them up and putting them down. But our world soon becomes one in which telephony, automobility, electric lighting, and computing are forms of life in the most powerful sense: life would scarcely be thinkable without them.
>
> (Winner 2014, 54)

Thus, we have to acknowledge that the given can always be, and is always being, changed. Here, this means: Technologies also change forms of life. The tools of Wittgenstein's toolbox need to be given a history (and, hence, also a future). There is variety in use, as Wittgenstein noticed, and, hence, there are cultural variations, variations in *place*. But it is important to add that both the tools themselves and these variations change over *time*, and therefore one could say that our forms of life also change over time. These changes may not be as radical as one might expect, want, or fear, but there are variations and changes in time. If language use, the use of technology, and the larger wholes they are part of and co-constitute are closely intertwined, then this time aspect implies that together they constitute a linguistic-technological-cultural *history*. We start using different technologies and use technologies in different ways, we speak in different ways (use new words or use old words in new ways), and we end up living in a slightly different world than we were born in. (Perhaps we also end up being different human beings than our parents and grandparents; however, here I will not further discuss the philosophical-anthropological implications.) I say a "slightly" different world, since, again, the emphasis in a Wittgensteinian-Heideggerian approach must remain on the persistence and perhaps "resistance" of the given: There are already older patterns, and like in improvisation in music (e.g., jazz, blues), change starts with new variations on these old patterns. It does not involve creation ex nihilo. In time, there are variations in use, and, in the end, there are variations in what I have called "technology games" (see also the previous chapters) and language games. Moreover, technology games and language games are connected and mutually influence one another. We *can* change the game by using different words and different tools. If this happens, the result is that the entire form of life changes a little. But, again, the change in terms of games and form of life is limited, not radical. Usually the game and certainly the form of life stay more or less unaltered; particular uses change and vary, but stay within the same game and form of life. In spite of ongoing local and incidental innovations in language and technology use, we usually continue to live within the same linguistic-technological grammar, understood as a form of life. A particular linguistic and/or technological syntax and a particular game my change a little; the form of life usually remains the same.

This view, inspired by Heidegger and Wittgenstein, will be further developed in the next chapters when I engage with other views according to which "language speaks" and "technology speaks." In the next part, I start with "language speaks." In Chapter 4, I will argue that, while theory about the social construction of artefacts that focuses on how the social and political shape technological development, and in particular the development of technological artefacts, recognizes that technology shapes society and acknowledges the historical character of technology-society relations, it unduly neglects the role of *language* in these social-technological processes. I will argue that language plays a crucial role in what is usually called "the social construction of artefacts," taking into account both Searle's and Wittgenstein's insights. Then, in Chapter 5, I will discuss radicalizations of the role of language: the view that *language*, not humans, speaks. In the following chapters (six and seven), I will further discuss the role of material artefacts and develop my turn to performance and process. However, let me first continue and elaborate my criticism of Searle and my interpretations of Wittgenstein by reading Austin and Wittgenstein's *On Certainty* and *The Blue Book*.

3.5. Critical Discussion Continued: Austin and Wittgenstein on Performance

Introduction

The previous section criticized Searle's focus on human linguistic agency, his nonrelational and nonholistic view, and his dualist ontology and contractarian view of the social. It also already jumped to a Wittgensteinian and Heideggerian view to show an alternative approach. Yet, *even within classic analytic philosophy of language*, Searle's social ontology can and has been criticized for focusing on only one kind of use of language and even one use of what is usually regarded as performative language: declaration/status function. As Wittgenstein also shows, there is much more diversity and variety of use of language; there are not only status functions. Searle's focus on declaration, at least in this later work, is an unnecessary limitation, especially when we want to make philosophy of language useful to philosophy of technology (see also Franssen and Koller 2016). Wittgenstein's comparison between words and tools in a toolbox was meant to show that the same word can be used in many ways to do/perform different things. If this is true, then the relation between words and things has much more variation than Searle suggests. Declaration is only one kind of relation. We can do many kinds of things with words, and hence, by analogy, with things: We can do more than ascribing a status or bringing into being a social fact or institution. Moreover, even within what is usually regarded as performative use of language, there is more variety. In this section, I turn to Austin, who stressed the performative use of language and distinguished between various 'speech acts' (for instance, direction and expression), and explore what this theory could imply for thinking about technology. In addition, I continue my engagement with Wittgenstein's texts, in order to arrive at a better view of using and performing with technology, including a better understanding of what kind of knowledge is involved.

Austin on Performatives and Variety of Speech Acts

Let us take a closer look at Austin's view, which influenced Searle's. Responding to a tradition obsessed with uses of language that were meant to say something about reality (e.g., propositions that express beliefs about the world, which can be true or false), Austin drew attention to utterances that do not describe or report something at all, but do something themselves, shape reality; instead of describing or reporting, their uttering 'is, or is a part of, the doing of an action' (Austin 1962, 5). His examples include marriage ('I . . . take this woman to be my lawful wedded wife') and naming a ship ('I name this ship the *Queen Elizabeth*') (5), which show that in such cases language is at the center of the performance: 'the uttering of the words is, indeed, usually a, or even *the*, leading incident in the performance of the act' (8). The words must be uttered in the appropriate circumstances and are connected with other actions (e.g., in the case of the ship, the smashing of a bottle). They are also supposed to involve the appropriate feelings, thoughts, and intentions (40); for example, when we make a promise, we should really intend to keep it, and the uttering must follow the right procedure (e.g., in a marriage ceremony). One could say, therefore, that the words uttered here are not so much *about* the performance (have the performance as their object), but *are* the performance; they are at least an essential part of it. In other words, in these cases 'to *say* something is to *do* something; or . . . *by* saying or *in* saying something we are doing something' (12). The utterances are called 'performatives' or 'performative utterances;' they cannot be true or false, but are meant to do something.

Austin's focus is thus on uses of language that are at the same time acts: Austin uses the term 'speech acts' (for instance, on page 146) to emphasize the link between speech and action. He focuses especially on 'illocutionary' acts. By this term, he means 'the performance of an act *in* saying something' (99) or, indeed, *by* saying something. Their meaning is related to their 'illocutionary force' (100), such as when I give an order or promise something. Thus, in these cases, when we say something, we do something (illocutionary). Moreover, sometimes we want others to do something (perlocutionary): 'perlocutionary acts' are about achieving something by saying something (108). The meaning of these utterances is not about what they are supposed to represent or say about the world (which can be true or false), but about what they are supposed to *do*. Illocutionary and perlocutionary acts are, thus, classes of *speech acts*. Austin then distinguishes between various types of illocutionary utterances, for instance 'verdictives' and 'commissives' (150 and following). For instance, by the latter, he means a speech act that is aimed at committing the speaker to a course of action, such as promising or proposing to someone (156–157). To conclude, for Austin, speech acts are utterances that have a *performative* function: In this case, saying something is already an act; when saying something, one also does something (e.g., naming or promising or ordering) or wants others to do something. In Austin's words: 'the performative should be doing something as opposed to just saying something' (Austin 1962, 132).

In his early work, Searle (1975), influenced by Austin, further classified illocutionary speech acts. He distinguished between representatives (e.g., saying what one believes), directives (e.g., requests, commands, advice), commissives (e.g., promises), expressives (e.g., congratulations, excuses), and declarations (e.g., baptisms, pronouncing husband and wife). The later Searle focuses on declarations in his social ontology, but with the early Searle, we must recognize that there are also these other kinds of speech acts, which *also* have all kinds of social functions. For instance, a promise creates an expectation on the part of the addressee that the person who utters the promise will do what (s)he promises. With Wittgenstein, we may say that the speech act of promising is part of a *language game*, which is always also a social game, involving rules/norms and expectations.

Applied to thinking about technology, recognizing this variety of speech acts implies that we can not only name and declare a technological artefact to be such and such; we can do much more with words and things. A lot of different kinds of performances are possible, in which words and artefacts play a crucial role. All depends on the relevant 'activities' and 'language games,' to use Wittgenstein's terms again. For example, technology and language can be used to ask someone to do something (it can be used in a directive way), or it can be used in art to express a feeling or a concept (it can be used in an expressive way). Analogous to the use of language, the use of technology can have various performative functions. For instance, if I promise something, then I can use (only) words, but I (or the person I promise to) can also use memory and writing technology. If I congratulate, I will use words, but maybe also add a present, that is, an artefact. Hence, words and objects are used in various ways, depending on the speech act (Austin) and what we may call the *technology-act*. Both acts can be combined, and are part of activities, language games (Wittgenstein), and a form of life. Use of technologies (tools) and use of language (words) can thus be coupled in various ways, depending on the kind of performance and social function/ meaning. Naming an artefact is only one particular, and probably rather exceptional use of language and technology. Usually, in everyday use and practice, we use words and things at hand to do things (and get others to do things), as part of social games, which belong to a culture and which already existed before we were born. Just as there is ordinary language (use), there is also ordinary technology (use); they are "ordinary" in the sense that they are used in daily life *and* (taking into account the etymology of the word) that they belong to a larger social-cultural order, which is always, at the same time, a linguistic and technological order. And within this order, within this form of life, words and tools are meaningful and useful.

Beyond the Baptism of the Object: Wittgenstein's Towel and Performances with Technology

Of course, everyone who thinks about technology may accept that technology can be used for various aims by individuals. But this is not exactly the point, and still retains a purely instrumental understanding of technology (or perhaps we should say *does not take its instrumentality seriously enough*) and a distorted view of use. It also has a misguided conception of "aim."

According to the view I am developing here, the "aim" is not taken as given (and as defined entirely by the individual agent), but is seen as deeply influenced, shaped, and, indeed, *made possible* (see Chapters 5 and 6) by language and technology, understood as words, sentences, tools, and artefacts that belong to larger wholes and patterns: games, structures, "grammars," and forms of life that are given prior to the speech act and the technology performance, and which are not just up to the individual (or even collective) speech or tech agent. For instance, if I use language and technology to make an excuse, I am not the first one who makes an excuse and not the first one using a particular writing technology; my use of words and tools will be embedded in already existing patterns and structures of language and technology that preexist my particular performative act. Thus, the point about variety of use is one that is directly related to a view of variety within a social-cultural whole, which exceeds and shapes individual use. In this sense, my intention and the corresponding "mental" content and "mind" are not private.

Thus, it does not suffice to refer to the "aim" of a technology and then define the technology as a "means" to reach that aim, assuming that language belongs to the part of defining the aim, whereas technology belongs to the means. Instead, the "aim" is already in the use of technology, *and is bound up with use of language*. There are not separate linguistic and technological performances, but rather what we may call *linguistic-technological* performances, which themselves also shape the "aim". For instance, if I write an e-mail to ask someone to do something, then I do not merely use the technology as a means to reach my aim. The technology may encourage particular speech-technology acts, and the use of e-mail is already related to "aims" that are embedded in particular games, which are in their turn related to other games in various ways. Perhaps I make a promise through e-mail, but give it less weight than a promise made face-to-face. Or particular social media may encourage particular uses of performative language, such as abusive language. I do not say that this is necessarily the case; this is merely an example of possible influence of the technology on the speech act, or rather an example of how technology act and speech act, materiality, and culture, are entangled in the concrete use and performance. The meanings that emerge in these uses and performances exceed what the technology is supposed to do or what the individual intends with it; the meaning depends on the use and performance itself, which is always embedded and shaped by larger patterns and structures. The "aim" or "end" is therefore not only or not so much in the "mind" of the individual user; the "aim" or "end" is rather an outcome of what happens with words and things, as these words and things contain their own "aims" and "ends," and as they are linked to a larger social whole, to language/technology games and a form of life in which certain "aims" and "ends" are already given and encouraged. (Note that this reasoning is also compatible with Dewey's questioning of the usual way we think of the relation between means and ends, also in his moral philosophy; however, I will not further discuss this here.)

This noninstrumental view of technology takes us once again beyond Searle (and Austin), since now the technology is no longer a passive recipient of the speech act (an initially meaningless object that I give meaning to by using words), but understood as "technology-act" that co-constitutes the speech act, is connected to the speech act, or even replaces it. This reconceptualization enables us to move from "doing things with words" to "doing things with words and things." For instance, I may ask someone to open the door by giving my keys. This is a technology-act. Of course I may do the same with a speech act, by *asking* to open the door, but this may not even be necessary in a particular context and given a particular history. Often the artefact suffices to do things or to get things done. The technology speaks, I let the technology speak (see also my discussion of Latour later in this book). It is not the case here that the performance has a material aspect which itself has no hermeneutic or social role, an object then worked upon by the subject; instead, the material artefact also influences the speech act and the meaning. Like Austin's words that are more active and performative than we usually think, here, artefacts are more active and performative than we usually think. We do things with words and we do things with things. But the latter cannot be sufficiently described by saying that we use things as a means to reach certain ends. In our use of the technology, the so-called "means" are more active and co-perform. Subject and object are entangled in the *technology act*/speech act. Whatever we do and whatever we want to do not only depend on words but also on things. Our intention is not only linguistic and "in our mind;" it is also technological and "in our things." (I will say more about this in the third part of the book, for instance when discussing Latour.)

Moreover, again the agent (as language user and as technology user) does not have full control. Here we can reconnect once more to Wittgenstein and Heidegger. There is already something that is given. Words and grammars are given. Just as speech acts are governed by a particular grammar belonging to that type of speech act, technology acts are also governed by a particular "grammar," since technology use is connected to technology games. In this sense, technology already has its "words" and "grammar." We know how to use it because there is already a know-how given, and it is given with the technology and with the words. Moreover, there is also the "grammar" of the larger form of life in which these activities and games are embedded.

Let me say more about the "small" or "local" grammar of use (I will elaborate the idea of grammar as a form of life later, when I add a transcendental argument). Technology has its grammar in a metaphorical sense, since, in their use, words and tools link to given meaning and structure, indeed, to given know-how, but it is also even literally the case: With a technological artefact comes a word (and a grammar). We could use Wittgenstein's example of holding the towel in *On Certainty*. Wittgenstein writes:

> If I say "Of course I know that that's a towel" I am making an utterance.' I have no thought of a verification. For me it is an immediate utterance. I don't think of past or future. (And of course it's the same for Moore, too.) It is just like directly taking hold of something, as I take hold of my towel without having doubts. And yet this direct taking-hold

corresponds to a sureness, not to a knowing. But don't I take hold of a thing's name like that, too?

(Wittgenstein 1969, §510–511)

Thus, when I take hold of a towel, I take hold of the thing/artefact, but also of the word/name! When I use an artefact, I also take hold of its name, which is related to knowing-how-to-use. In other words, this example shows that *my use of technology and my use of language are entangled.* In use, both come together, the meaning spans both subject and object. This is not theoretical knowledge, but an implicit know-how. We trust, without having the thought of verification. There are no doubts. I take hold, because I know how, because I know this language/technology game; I know how to perform this speech/technology act. Another example: If I take and handle a hammer, I do not only take an object; I also "take" and "handle" a meaning and a function. I take and handle an "aim" and an "end." I take and handle meaning. Because I *know how.* (If I use new technology, for instance a robot, by contrast, I am less sure. Perhaps I do not yet know how to use it. Use and meaning are still more open in such a situation. What language game can be played with it? What meanings could arise? And if the function is still unclear, there is more room for new, different uses.)

This approach gives us a way in which language and technology go together, in a much more "intimate" or "close" way than in the Searlean view, and in a way that further elaborates the Wittgensteinian-Heideggerian view I am developing. The technology game is also a language game, and the language game is also a technology game—not in the superficial sense that a technology or artefact is used by us in order to do our speech-act performance, as a *means* to some external end, but rather in the deeper sense that the technology and material artefact shape what we do with words. We do things and we "do" words with things and with words. The structure of language shapes the use of technology, and the use of technology (what we do with things) can only be fully understood by referring to what we do with words. This point can be supported by using Austin and Searle, but it can be radicalized by interpreting Heidegger and Wittgenstein, or even Dewey, who suggested that we cannot *think* an outside to language. Here we could conclude that we cannot think outside *technology* and language. Language and technology are entangled and mutually shape one another.

Supporting this view requires a different, more creative reading of Heidegger and Wittgenstein, which more strongly links language and technology. In this chapter, I already mentioned Heidegger's *vorhanden/zuhanden* distinction, which can be applied to both language and technology, and interpreted his car-indicator example. I also repeated my interpretation of Wittgenstein's analogy between language use and technology use. Now the towel example has enabled us to take a further step. If we take these elements in Heidegger and Wittgenstein seriously (and give a "performative" twist to them), then we can conclude that in cases such as that of the towel, and indeed perhaps in all cases of language/technology use, there is no speech act in itself apart from the object and there is no object apart from the speech act. There is *one* speech-technological performance. This view goes again beyond Searle's view, which still keeps both ontologically separate. And, again, it questions the emphasis on human agency and intention in Searle's and Austin's view.

The result of this operation is that it is no longer entirely clear *who or what is speaking* and who or what is *performing*. We humans do things with words and things, but at the same time words and things do things with us. Whereas in Searle's view, the humans remain in the driver's seat as they *use* language and technology ("use" in the sense that there is still a gap between instrument users and the instruments, ends and means, subjects and objects), in the Wittgensteinian-Heideggerian view, I also start from "use," but develop and interpret this in such a way that Searlean dualist thinking is overcome. Language *and* technology are the vehicles of performance, and the vehicles have self-driving capacities: Here, "use" involves a dance and entanglement of subject and object, in which the instrument also uses the goal and the one who sets the goal—maybe even in which the borders between humans, things, and words are blurred. In this less modern view of performance, we are no longer the only agents, but become co-performers.

I will further develop this more "radical" or integrative view concerning entanglement and performance later. For now, let me finish my criticism of Searle and further support the direction of thought I have taken here by what Wittgenstein says in *The Blue and Brown Books* (1958) about language use and thinking. If performances with words and things are very common, how "technological" is *thinking itself*?

Whereas the common view is that mental processes are necessary to, for instance, interpret signs and that 'the signs of our language seem dead without these mental processes' (3), Wittgenstein argues in *The Blue Book* that the signs become alive through use: 'if we had to name anything which is the life of the sign, we should have to say that it was its *use*' (4). (We can presume that Dewey would agree with this.) Moreover, Wittgenstein expresses his holistic view of meaning:

> The sign (the sentence) gets its significance from the system of signs, from the language to which it belongs. Roughly: understanding a sentence means understanding a language. As a part of the system of language, one may say, the sentence has life.
>
> (Wittgenstein 1958, 5)

But Wittgenstein stresses that this use and this holistic meaning is not a "mental" matter. Thinking, he argues, is not a 'mental activity' (6). Instead, for Wittgenstein, thinking is 'the activity of operating with signs' (6), and we use our hands and various tools for this operation, such as writing, the mouth, and pictures. He stresses this "medium" or "technology" side of thinking so much that he questions if there is such a thing as an agent or mind that thinks:

> The activity [of thinking] is performed by the hand, when we think by writing; by the mouth and larynx, when we think by speaking; and if we think by imagining signs or pictures, I can give you no agent that thinks.
>
> (Wittgenstein 1958, 6)

If there is an agent that does the thinking, Wittgenstein suggests, it is the hand, the mouth, etc. Instead of the thought (the message), he emphasizes the '(bodily) activities' (7) and the tools (the medium).

If we apply this view about language use and thinking to *technology* use and thinking, it implies that (1) technological artefacts are neither dead objects nor mere signs of a meaning that is to be found "elsewhere," but instead acquire and "give" meaning in and through use, that they are "alive" in and through use. It also means that (2) the question of agency needs to be rethought. Wittgenstein says: 'I can give you no agent that thinks' (6), and writes that if the mind is an agent, it is 'an agent in a different sense from that in which the hand can be said to be the agent in writing' (7). We could interpret this as saying that the instruments we use for using language and for thinking are at least *also* agents. If thinking itself is a performance, then it is a performance in which both human and nonhuman elements take part, including, for instance, words, hands, pens, etc. It is a performance where agency is no longer on the part of the human (mind) alone. Agency is distributed; it is also, and if Wittgenstein is right perhaps even mainly, in our activities and tools. With regard to the use of technology, one could say that agency is distributed among human and nonhuman elements. There is a human being, there are activities, and there are tools. But it is a lot less clear where the agency is. It seems to be distributed.

This is an interesting and less modern ontology (see also Latour later in this book), but here the point is not that we need a different ontology, an ontology of humans and nonhumans. Rather, I am exploring and experimenting with a different view of (technology) use and its implications for agency, a view of *language and technology in action*. (An emphasis that, again, not only Wittgenstein but also Dewey would approve of.) The emphasis is on the strong connection between use of language and use of technology, a connection which is not only metaphorical: Use of technology also involves use of language(-signs) and vice versa. The "dead" things and "dead" signs do not need Searle's declaration and other mental-linguistic acts to become alive (at least, if we interpret Searle's concept of intentionality in a "mentalist" way); they become, or are already, alive in use, in the language-technology performance. The lives of artefacts and the lives of signs are in use and performance.

Let me clarify this. Searle puts emphasis on intentionality. But, if we use the ideas of speech-act theory and combine them with Wittgenstein's view of language use in the ways proposed here, then there is no need for a strong emphasis on intentionality, at least not if Searle's concept of intentionality is interpreted as requiring a "mind" and "mental processes." The concept of *performance* enables us to shed a different light on thinking and language use (which, according to Wittgenstein, are basically the same) and on technology use. It suggests that there is no need for a kind of mental magic (individual or collective), which then gives meaning to technological artefacts; instead, the meaning of technological artefacts arises in and from their use, understood as performance, process, and operation—a performance, process, and operation in which language plays a key role and in which agency is distributed amongst humans, objects, and signs. In this view, technology becomes co-performer, cooperator, and perhaps also co-speaker.

In the third part of this book, I will give more substance to the idea that "technology speaks." For now, let me conclude that Wittgenstein helps us to articulate the view, against Searle, that there is more to human-technology relations than a quick application of Searle's social ontology to technology suggests. In particular, I have argued that there is more variety in language use and in technology use; there is not only what Wittgenstein called 'the baptism of an object' (Investigations §38, 23e), and that the "nominalist" view was rightly criticized by Dewey. Moreover, like Dewey, Wittgenstein also helps us to move towards a view of technology use in which subject-object dualism and the mentalist stress on intentionality in Searle's view are overcome. Now that this step has been taken, we can move towards a phenomenology and hermeneutics of technology use in which language plays a role, but in which that role is not reduced to declaration, and in which language is more than an instrument: Together with technology, language becomes co-performer, co-agent, or co-operator.

However, more needs to be said about the role of language. In the next part of the book, I will discuss and learn from postmodern and poststructuralist thinking. I will also add a *transcendental* argument to my toolbox, and revise/re-interpret the Wittgensteinian view developed so far in these terms. In the third part of the book, I will say more about technology. It will aim at bringing together the postphenomenological "empirical turn" emphasis on materiality with the insights from thinking about language (and technology) gained in the previous parts of the book, which will lead to a revision and expansion of the postphenomenological framework.

Furthermore, the view developed so far is still vulnerable in a number of ways. For instance, more needs to be said about the precise ways in which material artefacts are connected to speech acts and to larger contexts of use, language games, and, indeed, "technology games." Such an exercise may help to further develop the "entanglement" and "performance" views suggested here. In addition, it might be helpful to discuss more concrete examples and to take into account limits of the Wittgensteinian (and Heideggerian) approach. For example, the Wittgensteinian approach articulated so far does not pay much attention to the materiality of the artefact; this lacuna needs to be filled by learning from postphenomenology. It is also worth exploring the thought that perhaps not every meaning of technology can be reduced to use or performance—not even in the broad sense developed here—and that the social life is more than rule following (at least Winch's interpretation of Wittgenstein seems to suggest that it is all about rule following). In spite of these problems, however, and partly *because* of these problems, I believe there is enough critical mass here to warrant further development and critical discussion by engaging with different theories and approaches to language and technology. In the next chapter, I turn to theory concerning the social construction of artefacts.

References

Austin, John Langshaw. 1962. *How To Do Things with Words*. Oxford: Clarendon Press.

Coeckelbergh, Mark. 2012. *Growing Moral Relations: Critique of Moral Status Ascription*. Basingstoke/New York: Palgrave Macmillan.

Franssen, Maarten and Stefan Koller. 2016. "Philosophy of Technology as a Serious Branch of Philosophy: The Empirical Turn as a Starting Point." In *Philosophy of Technology after the Empirical Turn*, edited by Maarten Franssen, Pieter E. Vermaas, Peter Kroes, and Anthonie W.M. Meijers, 31–61. Springer International Publishing Switzerland.

Gier, Nicholas. 1990. "Wittgenstein's Phenomenology Revisited." *Philosophy Today* 34(3): 273–88.

Heidegger, Martin. 1977. "The Question Concerning Technology." In *The Question Concerning Technology and Other Essays*, edited by Martin Heidegger, translated by William Lovitt, 3–35. New York: Harper & Row.

Heidegger, Martin. 1996 (1927). *Being and Time: A Translation of Sein und Zeit*. Translated by Joan Stambaugh. Albany, NY: State University of New York.

Ihde, Don. 1990. *Technology and the Lifeworld: From Garden to Earth*. Bloomington: Indiana University Press.

Latour, Bruno. 1993. *We Have Never Been Modern*. Cambridge, MA: Harvard University Press.

Latour, Bruno. 2004. *Politics of Nature: How to Bring the Sciences into Democracy*. Cambridge, MA: Harvard University Press.

Livingston, Paul M. 2016. "Wittgenstein Reads Heidegger, Heidegger Reads Wittgenstein: Thinking Language Bounding World." In *Beyond the Analytic-Continental Divide: Pluralist Philosophy in the Twenty-First Century*, edited by Jeffrey A. Bell, Andrew Cutrofello, and Paul M. Livingston, 222–48. New York/London: Routledge.

Searle, John R. 1975. "A Taxonomy of Illocutionary Acts." In *Language, Mind, and Knowledge*, edited by Keith Gunderson, 344–69. Minnesota, MN: University of Minnesota Press.

Searle, John R. 1995. *The Construction of Social Reality*. New York: Free Press.

Searle, John R. 2006. "Social Ontology." *Anthropological Theory* 6(1): 12–29.

Verbeek, Peter-Paul. 2005. *What Things Do*. University Park, PA: Pennsylvania State University Press.

Winner, Langdon. 2014. "Technologies as Forms of Life." In *Ethics and Emerging Technologies*, edited by Ronald Sandler, 48–60. Basingstoke/New York: Palgrave Macmillan.Wittgenstein, Ludwig. 1965 (1958). *The Blue and Brown Books. Preliminary Studies for the 'Philosophical Investigations.'* New York: Harper & Row.

Wittgenstein, Ludwig. 1969. *On Certainty*. Translated by Denis Paul and Elisabeth Anscombe. Oxford: Basil Blackwell.

Wittgenstein, Ludwig. 2009 (1953). *Philosophische Untersuchungen/Philosophical Investigations*, Revised 4th edition. Translated by Elisabeth. Anscombe, P.M.S. Hacker, and Joachim Schulte. Malden, MA/Oxford/Chichester: Wiley-Blackwell.

Part II

Language Speaks
(Subjects Change Objects)

4 Language and the Social Construction of Artefacts

4.1. Introduction

The previous chapter articulated and critically discussed the Searlean view that what technological artefacts "are" and mean is shaped both linguistically and socially: A status is ascribed to an object by means of language and on the basis of social agreement. This view was then contrasted with a Wittgensteinian-Heideggerian approach, which offered a very different view of the relation between language and technology, in particular through thinking about the use of language and (hence) the use of technology. This part of the book takes again a step back and first discusses views that go beyond Searle, but are different from the view I am developing. Like in the case of Searle's social ontology, these views are constructed and used as tools and "dialogue partners" for further developing and sharpening my own view.

The first is the view that artefacts are socially constructed, as found in Social Studies of Science and Technology (STS), in particular, in the seminal work of Pinch and Bijker. In this chapter, it is argued that this approach goes further than Searle in bridging the gap between subject and object, language and technology. "Construction" is stronger than "declaration:" here, the ontology of the artefact becomes *more* social than in Searle, in the sense that it already becomes less easy to isolate the artefact from the social. While it is not denied that the artefact has a material and physical nature, the artefact itself is understood as the outcome of a social (and political) process. However, it is then argued that in theory concerning the social construction of artefacts, the role of *language* is undertheorized. An attempt is made to remedy this by interpreting the social construction of artefacts as crucially involving language: What artefacts are and how they are used is indeed the outcome of a social process (and in turn, these artefacts also constitute the social), but this is always a social-linguistic process in which communicative-linguistic activities (or dimensions of activities) and elements, such as discourse and narratives, play a key role, and which presupposes a given language, which may constrain the development and use of technology. Here Wittgenstein's towel returns: Expanding this idea, I will argue that to design or invent a technology is also to design or invent a world, to define what the artefact is, technologically and linguistically. Moreover, that linguistic-social construction is always embedded in a larger "grammar" and form of life, which constrains the construction. (This argument about constraint and what is presupposed will be later, in the next chapters, turned into a transcendental argument.) Thus, controversies about technology and the groups that take part in them are always part of larger social-cultural wholes—a relation that can be theorized using Wittgensteinian concepts such as grammar, games, and form of life. It again turns out that, in line with the Wittgensteinian-Heideggerian view that I am developing, the use of language is crucial to what technology means and does.

My own view is then further developed by (1) "jumping back" to the sociology of knowledge and philosophy of science, as influenced by Wittgenstein, which preceded Pinch's and Bijker's view, and (2) "jumping forward" to Latour's view, which was developed later than Pinch and Bijker's work, in the early 1990s, and which suggests a more radical way in which artefacts and language are intertwined.

Another view I construct (in the next chapter) does not underemphasize but overemphasizes the role of language. It can be interpreted as a radicalization of Heidegger, Wittgenstein, and McLuhan: Only language speaks, and humans and technologies are muted. In this view, technologies and technological artefacts are reduced to (textual, written) signs. There is only text. There is only the medium. In this sense, language itself becomes the only speaker. Humans are vehicles for language, and the materiality of artefacts disappears, since only the realm of signs and text is left.

This "extreme" view that only language speaks will be elaborated by using poststructuralist and postmodern French thinking: Derrida, Lyotard, and Baudrillard. (In addition, I will also comment on Virilio and Mersch.) Although I will argue that, in their obsession with language, these views are too one sided, it is also shown that there are some interesting elements in these works that could be used for thinking about technology, both in terms of content (the message: the stress is of course on the importance of language, but they also speak about *technology*) and in terms of approach (the medium, so to speak, including, for instance, a *transcendental* argument and attention to the singular). It will be argued that there is some truth in Heideggerian claims that language speaks, since it helps us to gain a more critical relation to speaking and thinking about technology. For instance, the very words we use to speak about technology may mislead us into thinking that technology is a mere "instrument." Yet, in its radical version, this language-centered approach renders material objects invisible or even dispensable; they are eclipsed by their linguistic existence as signs or symbols. (This dematerialization of the artefact also partly happens in Wittgenstein insofar as he does not pay sufficient attention to material artefacts, and it certainly happens when Searle suggests that a material object is not even necessary, that linguistic declaration suffices. Consider, for instance, electronic money.) This view is contrasted with the interpretation of Heidegger,

Wittgenstein, and McLuhan proposed in this book, which—like postphenomenology and other "empirical turn" approaches—tries to give the materiality of artefacts its due, but—in contrast to the empirical turn—does this now in a way that acknowledges and accounts for the key role of language in shaping our use and relation to technology and the world.

Let me begin with constructing and discussing the first view, the social construction of artefacts, which helpfully enables us to question Searle's dualistic assumptions (language versus technology, subject versus object, declaration versus physical reality, etc.) and makes a connection between the social and material, but underemphasizes and under-theorizes the role of language in what technologies mean, are, and become when we invent, use, and perform with them.

4.2. The Social Construction of Artefacts: Pinch and Bijker, and Beyond

The Social Construction of Artefacts and the Discourse and Imagination about Technology

In 1984, Pinch and Bijker published a seminal article on 'the social construction of facts and artefacts' (Pinch and Bijker 1984) in which they argued for a research program that understands technological artefacts as social constructs. Inspired by, and responding to, social-constructivist views in sociology of science, they argued that not only in science knowledge-claims are socially constructed (in the lab and in the wider community of scientists, where there are controversies about the "truth" that is being "discovered"); technological artefacts and technological innovation are subject to the same kinds of social processes and, indeed, in a sense are "made," "produced" by these processes. What does this mean?

In their work, the authors seek to explain the success of an artefact. They proceed by comparing processes of technological innovation and design to processes of scientific discovery. Just as in science, there is at first 'interpretative flexibility' (Pinch and Bijker 1984, 409)—initially, scientific findings are open to more than one interpretation—but this flexibility soon disappears through social mechanisms that terminate controversies and, hence, lead to 'closure' concerning the technology in a particular social-cultural milieu. Technological innovation, then, is also subject to various social mechanisms that reduce flexibility with regard to what the technology might become, and, hence, lead to closure. The authors propose to describe the variation and selection in the developmental process of a technological artefact, and show that in principle there are always other possibilities: The end result could have been a different (variation of the) technology. They give the example of the development of the bicycle: At the end of the 19th century, there were different variants that were rivals, but then some variants "survived" whereas others "died" (to use the pseudo-Darwinian language of the authors; 411). Similarly, some problems are seen as relevant, others not. The authors explain this by pointing to social factors; the artefacts and their meaning are socially constructed. In these processes of innovation, they argue, 'a crucial role is played by the social groups concerned with the artefact, and by the meanings which those groups give to the artefact: A problem is only defined as such, when there is a social group for which it constitutes a "problem"' (414). Members of a social group 'share the same set of meanings, attached to a specific artefact' (414). Again, Pinch and Bijker stress that various solutions and variations are possible, but that through these social processes, one form and one meaning become prevalent. Initially, the technological artefact can be interpreted and designed in different ways. There is not only 'flexibility in how people think of, or interpret, artefacts;" there is also 'flexibility in how artefacts are *designed*. There is not just one possible way, or one best way, of designing an artefact' (421). They emphasize that 'there is nothing "natural" or logically necessary' about a specific form of closure (428). This leaves room for social processes to shape the design and interpretation of technologies.

With some interpretative flexibility, one could say that with this approach, the Searlean more abstract approach discussed in the previous chapter becomes concretized, "sociologized," historicized, radicalized, and related to concrete innovation and design processes. Searle's contractarian imaginary agreement (original position) is replaced by a concrete process of how closure is reached. Meanings are once again attached to artefacts, but, here, the more socially concrete question is asked, *who* attaches the meaning, and, in particular, which *social* group attaches the meaning—or better *shapes* the meaning *as the artefact is designed*. It is also historicized, since the development and history of the artefact is considered and described. The time dimension is brought into play. Social ontology is replaced by social history. Moreover, this view is more radical with regard to conceptualizing the relation between technology and the social, since what the artefact "is" and what the "status" of the artefact is cannot easily be disentangled: If meanings are always given to artefacts by particular groups *through the design process understood as a social process*, then the meaning of the artefact cannot be defined in a socially neutral way. (In contrast to postmodern approaches, there is, however, still a sense in which the artefact is and remains material; it is not yet totally dematerialized. There are still traces of social-natural and social-material dualism.) Here, the focus is on how the artefact came into being in the first place: There is not only interpretative flexibility "afterwards," when there is already an artefact but also flexibility with regard to the *design* of the artefact. What matters is not only what people think and say about an artefact but also how it is designed. The design of the artefact is thus integrated with its social history: There is a socio-technological history of the artefact, in which technology and the social mutually shape one another. This is an advance compared to Searle's social ontology, in which such a history remains invisible and in which the social remains at an abstract, contractarian, "a priori" level.

However, what is missing in the social-constructivist account as described and interpreted so far is explicit attention to the role of *language*. Of course, "controversies" are mentioned and it is possible to read the text as implying that words and their meanings play a role in the process of innovation (e.g., the word bicycle, 'ordinary,' etc.). But the latter suggestion is already my interpretation and revision; an explicit discussion of the role of language in the social construction of artefacts is missing. The closest Pinch and Bijker come to acknowledging the role of language is when they use the term 'rhetorical closure' (425): Rhetorical arguments are used in science as a closure mechanism, and this happens also in technology development. The authors explain: 'The key point is whether the relevant social groups see the problem as being solved. When it comes to a new technology, advertising can play an important role in shaping the meaning which a social group gives to an artefact' (427). The authors give the example of the claim that bicycles were said to give you 'almost absolute safety,' which is indeed 'a rhetorical move' (427). Thus, here, implicitly language is seen as playing a role in the process of technology development, and we could interpret these claims as hinting at the performative aspect of language use Wittgenstein (and Austin and Searle) emphasized: Here, people "do things with words" in order to reach particular goals with regard to technology development. I propose to go a step further and reinterpret controversies about technology as communicative-linguistic processes, and interpret the 'wider sociopolitical milieu' (428) of these processes in Wittgensteinian terms (language games, forms of life). The discussions about technology development and the rhetorical moves related to it crucially involve uses of language, which are in turn related to language games and forms of life. The "moves" are mediated and made possible by language, and they are never entirely new: There are already language games available in our culture, which structure these discussions and rhetorical moves, for instance the game of expressing concerns about safety (and taking away these concerns). Moreover, using Wittgenstein, one could highlight how language *does* things, how there is variety of language use related to activities. For example, 'ignoring a problem' or 'moving on' (430) are varieties of language use, language games, and are related to activities of problem solving, for instance. In other words, they can be seen as specific speech acts that play a role in science and technology development, and that are embedded in the grammar of language use and the lifeworld.

This view goes significantly and substantially beyond Pinch and Bijker's view; if it still is social constructivism at all, it is a very specific interpretation of the social and its relation to artefacts, and it is much more aware of, and in recognition of, the crucial role language plays in technological innovation and in social-technological closure and acceptance of the technology. For example, it sees the social as a form of life which is more "given" than is suggested by Pinch and Bijker. Whereas, in their view, social groups have a relatively high degree of agency when it comes to deciding which artefact/design variant becomes successful; in a Wittgensteinian view, this agency is not absent, but is seen as fundamentally shaped by larger grammars and cultural-linguistic wholes. Speech acts by social groups are already embedded in, and constrained by, a given: in language games, a culture, and a form of life, which make possible speech acts by social groups and technology acts by designers. (See also my transcendental argument in the next chapter and later.)

This is not to say that explicit attention to language is *entirely* absent in contemporary STS, which is still influenced by Pinch and Bijker, but which also has seen further developments since then. Some people working in STS are influenced by the work of Bruno *Latour*, who integrates language in his view of the social, although the stress is on artefacts as actants (see later in this chapter and in Chapter 6). And some STS researchers use the method of *discourse analysis*, that is, they work on language *about* technology. They analyze texts related to science and technology, such as policy reports, political speeches, media texts, etc. This method can and has also been used by social constructivists: Here, social construction is understood to be manifest in, and to happen on the basis of, language as discourse. Many scholars using discourse analysis are influenced by Foucault and/or feminism and (other) critical theory, and, hence, emphasize issues concerning power, knowledge/truth, and their relation to language. Consider, for instance Roderick (2016), who uses a critical-discourse-studies approach, which understands technology and culture 'as being equally constitutive of the other' (1). He analyzes texts about technology in a way that assumes a close connection between discourse and technology—rather than seeing one of these as having an "impact" on the other—and calls attention to 'the ways in which power relations are invested in technology' (2). However, within discourse analysis there are many approaches. Furthermore and related to discourse analysis, it needs to be acknowledged that there is also attention to *metaphors* and *images* in STS research, for instance, in technology assessment. Bölker's et al. (2010) work on information and images of the human focuses on the role of metaphors in visions and scenarios about future technologies, especially when these scenarios are presented not only to experts but also to a wider public (11). Consider also work on the role of *imagination* in technological practices within STS, ethics of technology, responsible innovation, and related fields, for instance, Swierstra et al. (2009), van der Burg (2014), Coeckelbergh and Wackers (2007), Coeckelbergh and Mesman (2006), and Coeckelbergh (2006).

Yet, most these approaches (Latour is an exception) tend to assume a relation between language and technology that goes little further than the "about" relation I identified in the beginning of this book: They study language *about* technology. The discourses, the metaphors and images, the visions and imaginations, they all say something about technology and even play a role within technological practices. It is important to pay attention to these texts, metaphors, and visions in technology studies. But, usually in these approaches, language does not yet fully "touch" the technology; that is, as objects of discourse, the artefacts and the material processes remain somewhat separate from language. Subject and object remain separated. The social shapes technologies by means of discourse and imagination; but there is still a gap between the social-linguistic and the material-technological. With the exception of poststructuralist and postmodern versions (as we will see in the next section, the term "discourse" is also used in poststructuralism and postmodernism), a more "active" role of language is not on the menu.

Moreover, even if and when STS makes a closer connection between language and technology, subject and object, there was (certainly in the beginning) little attention to *use*. In contrast to the Wittgensteinian view suggested in the previous chapters, initially, the social-constructivist view focused a lot on the development of the artefact, next to the interpretation and acceptance of the artefact, and less on *use* of the artefact—let alone on how this use of the artefact relates to the use of language. The meaning of the artefact is not seen as related to its use, but as being constructed in the processes(es) of its development and acceptance, which are seen as socially shaped processes separate from use. Hence, the social is still seen as separate from use—the use of language and the use of technology. In Pinch and Bijker, the link to use is only indirect, for instance, when the authors mention the 'vibration problem' (422). The point of view is that of the engineer/innovator, not the user. If use is considered at all in this view, it is use *in* design, innovation, and development. But this use-in-design remains implicit and un-conceptualized.

Today, this has changed somewhat, perhaps; there is more work on use in STS. For instance, Oudshoorn and Pinch (2003) have put together a volume, which is more interested in 'how users consume, modify, domesticate, design, reconfigure, and resist technologies . . . in whatever users do with technology' (3), and includes feminist scholarship on power relations, semiotic approaches (see Akrich and Latour later in this book), and other approaches. But even here, the editors are keen to put their focus on use in the context of technology development. This is all about 'the capacity of users to shape technological development in all phases of technological innovation.' They assume that use is a phase in innovation, or another side of 'the innovation coin' (16). Moreover, use itself is insufficiently conceptualized; more work is needed in this area.

If most work in social constructivism typically focuses on the origins of the technology and on the dynamics of its development (and acceptance) rather than its use and (social) consequences *regardless of the innovation process*, this has to do with its origin. As Winner (1993) has shown, the approach has its origin in the sociology of science (see next section), which is about the construction of knowledge; this approach has then been applied to innovation. But we need to know more about the relation between meaning and use. Taking distance from social constructivism (especially in its initial version), then, we can go beyond questions regarding the design, development, and acceptance of a technology, and look at uses of technologies and at how these uses are embedded in social contexts. This could also be achieved by engaging with more recent literature in STS that is more focused on use (see again Oudshoorn and Pinch 2003, for instance). However, here I will focus on a philosophical source, which helps me to further develop my own conceptual framework about the relations between language and technology: Wittgenstein. Influenced by Wittgenstein, we can look at how use is more deeply connected with language and with the social. We can explore views that bring subjects and objects, language and artefacts, closer to one another.

Revising Social Constructivism in Order to Account for the Use of Language: To Design an Artefact Is to Design a Word and a World

From a Wittgensteinian angle, one could say that artefacts are *already* social and meaningful through their use and their embeddedness in a form of life, "before" there is a particular social-institutional process, with its explicit controversies about the meaning and design of the technologies and with its meaning-giving by particular groups. The whole does not determine the meaning; there is indeed interpretative flexibility in design and innovation processes. But this flexibility and this meaning is constrained by the use of the technologies (and in the case of new technology: constrained by the use of existing, but similar technologies), and by the related language games and cultural context that are already there before the innovation takes place and that shape both the innovation and the use. When people innovate, there is already a know-how and a meaning related to use of existing technologies. When groups and individuals do things with language in contexts of innovation—in particular, when they use rhetoric, that is, when they use language in a performative way to reach certain aims of giving meaning to the technology—this language use only *contributes* to the meaning of the artefact. There is also the use of the (other) technologies: old technologies and new technologies during development and after they have been developed. And there is already language use related to these old technologies, which will be relevant to language use related to the new technologies. For instance, the mobile-phone user still used the language of the old telephone; language only slowly changed (if at all) with the use of the smartphone. Interpretation and meaning-giving by "the social" should not be divorced from use, including both technology use and language use. Otherwise, one arrives at a view as in Searle and in social constructivism, where in the end a duality is maintained between a meaningless, naked object and a meaning-giving individual and social subject. In use and performance, by contrast, the subject is not alienated from the object, since meaning emerges in use and in the use context (activities, games, form of life). Use binds subject and object together.

Based on this approach, one could reinterpret and revise the social-constructivist view (as applied to technology) in a way that says more about the role of language and about the use of technologies *in* technology development. In particular, one could study *doing things with words and with things* in innovation and design processes. Given the general lack of attention to language in social constructivism (as identified previously, but acknowledging exceptions such as discourse analysis and semiotic approaches), one could especially focus on the relation between design choices and *language*. I already suggested analyzing what happens in terms of speech acts. This would rely on Searle's and/or Austin's concepts, for instance. But one could also further develop and apply the revised and more integrated Wittgensteinian speech-act/technology-act approach initiated in the previous chapter. In particular, taking up the example of Wittgenstein's towel, one could reinterpret what is happening, in terms of innovation as being at the same time technological *and linguistic*. Just as to take hold of a towel means to take hold of a name; to *design* an artefact is also to design a name. To take up Bijker's example of the bicycle: What the case of the history of the bicycle shows in this respect, is that to design and invent a technology is not only to invent an artefact; it is also at the same time to invent a concept, that is, a word and a name. To design a thing is not only designing a material thing; is also simultaneously to define what the thing "is." For understanding the social construction of the bicycle, for example, this means that the social does not only shape the construction of the bicycle as artefact, but also of the bicycle as concept/word/name. The meaning of "bicycle" is shaped by social, material, and *linguistic* processes, which interact or rather are aspects of the same social-material-linguistic process. Hence, we could say that design and innovation processes are not only more social but also more *linguistic* than usually presupposed. Innovation, as a social activity in a social-cultural context, is a matter of language innovation as much as it is a matter of engineering or so-called "technical" innovation.

Of course, social constructivism accepts and even emphasizes that the social context is important. But using Wittgenstein (and perhaps radicalizing discourse analysis and semiotic approaches within STS), we can now argue that this context is always also linguistic; it is also discourse. Keeping in mind the *social* studies of science's focus on groups, one could study the particular language used by particular social groups. This may already be done when discourse of particular groups is studied by STS researchers. But it is important to try to better conceptualize what happens in and with this language use: This use of language is not isolated and limited to a particular innovation practice (understood as use-context), but is in turn part of a wider context, which is a use-context and a cultural context simultaneously. Inspired by Wittgenstein, one could understand that language use as part of activities, language games, and a form of life. What these social groups do with words should not be taken as "black box" in itself; *opening the linguistic black box* means, then, to look at language use and relate that use to wider practical and cultural contexts. For example, the design of a humanoid robot may be influenced by what various social groups say about it; but the discourse of these groups is in turn imbedded in a cultural whole within which certain things are said about robots. For instance, in the west, it is said that robots are machines, that they don't have souls, that humans are not machines, and so on. Particular use of language in social robotics innovation can and must be placed in the context of these discourses and ultimately this form of life (here: Western modern form of life, which has a history, including, for example, Descartes, but also kinds of technological practices, such as automata performances). A Wittgensteinian focus on, and approach to, language use can thus helpfully guide, reformulate and redirect what STS researchers are already doing when they study discourse about technologies and/or use of technologies.

Moreover, it is important to note that this focus on language does not come at the expense of the material turn: In this approach, the social constructivist's emphasis on the material is not absent at all. In my application of the towel case, I focused on language, but, in use, the linguistic is always closely connected with the material-technological. As I said, there is doing things with words and doing things with things. Starting from this approach, we can recognize that in design processes mediated by technology (for a discussion about the term mediation, see later in this book) the material artefact is still visible. But it is not present as an object separated from subject, as a material thing separated from discourse. In use, that is in use of words and use of things, both subject and object come together. The "material" artefact is there, but it is not merely material; it is already imbued with meaning, a meaning that emerges in use. The "cultural" is there, but it is not merely linguistic; the words make sense only when connected to the material artefact in and through use and performance.

Furthermore, design and innovation also involve use. Design itself is a kind of use: use of other tools to create the artefact, and trying out *using* the artefact. Design and innovation broadly understood are not only happening on the drawing board (today: the computer); innovation is also about trying out the things and experimenting with them; that is, it is also about use. And the knowledge used by the designer and engineer is also based on know-how learned from use. For example, if one (re)designs a bicycle, knowledge of using a bicycle (an old bicycle) feeds into the design process, the design itself involves the use of tools and generates know-how on the basis of that use, and then one tries out the bicycle, which is again use and creates know-how. There are already uses, activities, and games related to cycling, before one starts to (re)design a new bicycle. The same can be said about robots and other supposedly more advanced technologies, which, for instance, involve the use of code, words, and related know-how. Hence, one could say that meaning of technological artefacts emerges (1) in the use of the technology in a general use context ("after" design) and (2) also specifically in the use of technologies in a design-and-innovation context, including at least the use of older technologies, the use of tools to design the artefact, and the "test" use of the artefact.

A better, more comprehensive approach to the social construction of technology could then integrate existing approaches in STS with these Wittgensteinian insights about language and use, and study the use of language *and* the use of technology by particular groups in design-and-innovation processes (which is also a "use context") *and* different contexts (usually called "use" contexts). This would include the various ways in which words and things get combined in social-technological processes that lead to what are later called "innovations" (and before, e.g., in a research proposal). It would also put these use experiences and use practices in the context of the history of relevant games and forms of life. In other words, sociotechnological change could be studied in a way that also looks at socio-*linguistic*-technological change, with a focus on use, understood as including use of language and use of artefacts, and understood in a Wittgensteinian holistic way. And this does not exclude room for looking at power and interests. On the contrary, use of language can be directly connected to such interests, as the point about rhetoric (made within STS, but usually unconnected to Wittgensteinian thinking) already shows. But the rhetoric and the use of power in particular contexts of innovation need to be connected to language games and technology games understood as power games, and to forms of power in a particular form of life.

Consider the example of the development of social robots. Particular groups say engineers and scientists who work on robots, but also robotics companies, may have an interest in developing "social robots"—not only the artefact but also the *name*. Perhaps the name "social robots" is chosen because "social robots" sounds sympathetic and may increase acceptance. More generally and radically, we could say that when engineers-in-context develop an artefact, then this development is not only the development of an artefact but also the design of a name. Moreover, with the name come particular activities and games, for instance, the activity of having a conversation with a pet or the game of asking someone to fetch something. These activities and games are, in turn, not culturally neutral, but involve all kinds of meanings available in a particular form of life, which shapes the use of language in relation to these robots (in development and later use) and, indeed, the use of these robots. For instance, in the West there are social meanings of competition and slavery available in the culture, which feed into the meanings that emerge in the development and use of robots, when there is a discussion about whether or not robots should be our "slaves." The invention of new technologies does not only introduce new words into our language (e.g., "social robot" or, previously, "robot"); it also introduces an entire "grammar"—that is, entire language games and other games—and an entire language. It introduces a "given" that is already there "before" the innovation process and "before" the use; at the same time, through use of language and use of technology, the linguistic-technological invention may also introduce a (small) change to our form of life.

Social scientists who study technological innovation, then, should not only focus on (what they see as) the social and the material, or on the discourse *about* artefacts; they should pay more attention to how discourse *itself* is changed in and through innovation processes, how new language games emerge together with the material artefacts-in-context. To the extent they are already doing this, for instance, by means of discourse analysis or semiotics, the Wittgensteinian framework I propose can help them to better understand and reconceptualize how technologies and language are related.

However, there is always the danger of slipping into the common instrumental view on the relation between language and technology, which maintains a large conceptual gap between the two. Let me repeat that if in such studies the role of language would only be seen in terms of language *about* technology, this would miss the point that language also *shapes* the technology. Instead of an external relation between "language" and "technology," there is what one could call the social-linguistic construction of artefacts—in various use contexts (including design and development, which also involves use of technology) and by various social groups. Technology and technological innovation are indeed socially defined here, but this defining is not only the defining of an artefact but also the defining of a word, grammar, discourse, and language. To take up the example of social robotics again: The very term "social robots" will also guide the design of the robot, by continuously reminding the developers that they have to make sure that the robot enables friendly and smooth interaction with the user, an interaction that resembles that of human-human interaction, that is safe, etc. Once there is the term, there is also the game. A new term, together with a new artefact, changes the game. If one is no longer creating a new "machine," but rather a "social robot," then this is a game change. From now on, once there is this new term and this new artefact, these computer scientists and engineers are playing a different game. With their uses of technology *and* their uses of language, they have contributed, and are contributing, to the making of that game—together, of course, with all kinds of *other* social actors and stakeholders. At the same time, these games are never entirely *new*. Usually they are variations of already-existing games, all of which are part of a larger whole. As said, in this case, there were already particular forms of interaction and use available before "social robots" entered the stage. There were already activities, language games and technology games. Consider, for instance, "meeting a friend," "helping someone," and "meeting someone you know." Moreover, in use, variety of use is possible. Different people may interpret and relate to the robot differently in their use: their use of words (e.g., how they address the robot or talk about it) and their use of the artefact. But again, these variations on use need to be placed in a "grammatical" and holistic framework. It is important to stress this variation and holism: In line with social constructivism's concept of interpretative flexibility, but also taking into account Wittgenstein's and Heidegger's approaches, it is important not to get stuck in thinking about this in terms of one-to-one relations between words and objects, but instead to stress the variety of meaning and the *interpretative*, hermeneutic aspects of language use and how this language use is holistically embedded in, and contributes to, language games and forms of life.

Moreover, this approach does not come at the expense of a historical perspective, which social constructivism and STS have adopted, but quite the contrary: It includes or even *is* a historic approach, in particular, one which reconceptualizes a particular innovation history in linguistic-material terms (as a history of words and a history of artefacts) on the basis of understanding it as a history of use(s). Now, we are neither only concerned with the history of discourse nor only with the history of artefacts and innovation, but with both at the same time, since words and artefacts are bound together in this new understanding of innovation processes and their history. To take up the social robotics example again: previously, the word "social" was not connected with "robot." But, in the course of innovation trajectories and socio-linguistic-technological change, that is, in the course of innovation history as a social, material, *and* linguistic process of use(s), meanings have been changed. Meanings have shifted and gamed in such a way that now "social robot" is increasingly meaningful, makes sense. And this change of meaning and, indeed, rise of particular *interpretations* may support the interests of particular groups (robotics researchers, robotics companies, etc.). This shows again how *making robots* and *making sense* go together, and how use of language (e.g., rhetoric and semantics) and use of technology (use of robots for social interaction) are coupled.

Furthermore, like any other historic approach, it recognizes the importance and weight of the past as shaping the present. A Wittgensteinian-Heideggerian perspective acknowledges the "givenness" and influence of existing games and forms of life on our current use. To pick up the example again: Now, this development of "social robots" is something "new;" but, at the same time, this "new" is made possible by meanings that were already there. It is not the case that entirely new meaning is created; instead, meanings that were already available in the social domain (e.g., activities and games that are linked to relations between pets and pet owners, relations between humans), are now transferred to human-robot relations. Keeping in mind Wittgenstein, it is important to see such cases of social-linguistic-technological development not as isolated use-and-development cases, but as uses and performances referring to, and being embedded in, patterns of language and technology use, activity, and ultimately forms of life that already existed before the particular social-linguistic-technological change happened. For instance, if a "care robot" is developed, then for this artefact and innovation to make sense, the meaning and use of this robot will have to refer to previous uses of robotic technology (e.g., in care contexts) and to human-human care practices. It will have to be connected to present and past experiences and practices, understood as including the use of words and the use of things.

Thus, while it is true that meaning emerges in use, this does not mean that new individual users of a particular technology entirely "decide" or "control" its meaning. Instead, their *making sense* of the technology in and through use is connected with the way others have used and made sense of the technology before the new application emerged, and with the way people have done things before and have used words within a specific practice. Care practices, for instance, already had a specific "grammar" before the care robot entered the scene. The use of the care robot may change the practice a little, but the "grammar of care" will still structure and shape its use, for instance, by means of (implicit or explicit) rules. It will also structure and shape the use of *words* alongside the use of artefacts. Again, there will be larger games and patterns of care which encourage a specific kind of use of words and a specific kind of use of technology. At the same time, of course, use of the technology and the care practice will also be influenced by the material artefact. But this influence should not be studied in isolation from existing and changing language people in the practice use, and existing and changing "grammars" of the practice. Thus, the "degrees of freedom" of semantic agency—in technology development and elsewhere—are limited. Semantic moves understood as uses and performances, and as involving linguistic and material moves, also "new" ones, always take place in a space constrained by existing games and forms of life, which make possible and shape specific uses, moves, and developments.

Note that the same relational and holistic argument can be applied to knowledge and framed in terms of knowledge (consider also again social constructivism's origins, which are all theories about knowledge). This was already implicit in the previous argument, but let me make it more explicit now. We could say that the kind of knowledge that is at work in use (use of language and use of technology) has the form of know-how. Use involves and requires knowing how to use words and knowing how to use things. As said, that use is influenced and guided (Wittgenstein might say: "ruled") by larger linguistic-social-technological structures and patterns, and ultimately by what Wittgenstein calls a 'form of life.' For knowledge, this means that the knowledge generated in innovation, design, and similar technology-development processes, is at the same time "new" and "old," since it is always also related to knowledge that is already there. Before new technologies (and new words) are being invented, we already use technologies and we already use language. This experience with use gives us know-how we can put to work in new uses, an experience and know-how which is in turn related to all kinds of activities and games. New technologies will thus also involve new uses and know-how, but these new uses and new know-how will only make sense, will only be meaningful, and, indeed, will only work if they refer to uses and know-how that is already there. Stronger (but this already uses a transcendental argument, which I will introduce in the next chapter): These old, existing uses, games, and forms of know-how *make possible* the new invention. There are creative variations, there is improvisation, there are original inventions. But all these variations, improvisations, and inventions are only possible because there are already uses and there is already a semantic-technological whole that scaffold these technological inventions. Compare to improvisation in music, which creates something "new," but also relies on patterns that are already given, e.g., in jazz culture. It also relies on previous experience and know-how about using a particular instrument (e.g., a piano) and playing together. And to stay closer to my previous example: Social robotics relies on existing technologies (e.g., robotic platforms or parts that are already there before the design of the robot), but also on existing social and linguistic meanings, for instance, about friendship or taking care of a pet, which concern relations that involve particular uses of words, next to uses of things, and which are part of our experience and know-how "before" we relate to the technology and "before" the technology is developed. Without this wider social-cultural context, including words and ways of saying things (grammars that are in turn related to activities and games), *it would not be possible* to invent a "social robot." Innovation is made possible by these existing meanings and grammars.

In order to further show the benefits of this Wittgensteinian approach for philosophy of technology, let us "go back in time" to the sociology and philosophy of science, which preceded social constructivism (including the sociology of knowledge), and then "move forward in time" to Latour's view, which emerged in the beginning of the 1990s and which will be discussed in further detail with regard to its materialism and its claims about modernity in the next part of the book.

4.3. Wittgenstein and the Philosophy and Sociology of Science

Wittgenstein's thinking was first brought into sociology by Peter Winch, who in *The Idea of A Social Science* (1958), emphasized Wittgenstein's link between our use of language and the background of our shared practices, "form of life." He also took inspiration from Wittgenstein's notion of following a rule. Rejecting the 'master-scientist' conception of the philosopher and positivist ways of doing social science, Winch emphasized the role of understanding (Weber: *Verstehen*) in the social life. Inspired by the *Philosophical Investigations*, he argued that language is 'based on a common life in which many individuals participate' (33)—indeed, on a social context, in which rule following and interpretation play a crucial role.

Later in the sociology of scientific knowledge (Bloor, Pickering, Lynch, Friedman, etc.), Wittgenstein's concept "forms of life" has guided people who study scientific communities as social practices. For instance, inspired by Winch, Collins has referred to forms of life in science; Friedman then criticized this translation of Wittgenstein into an empirical research program.

There has also been discussion about whether meaning is determined by a community's social practices, and how Wittgenstein's discussion of rule-following needs to be interpreted. (For an overview of such discussions, see Stern 2002.)

In spite of these disagreements, however, Winch's interpretation of Wittgenstein was and is appealing in philosophy of science. It does not only link language to the wider practical-cultural context but also opens up a different approach to science and scientific knowledge. If "forms of life" means specific social practices, then this can inspire study of such practices. Moreover, Wittgensteinians such as Winch provide an interesting anti-representational view that takes into account not only know-that but also and especially know-how. Hence, for instance, Collins has been interested in tacit knowledge and has shown how such knowledge plays a role in science. In an influential article, he argued that 'all types of knowledge, however pure, consist, in part, of tacit rules which may be impossible to formulate in principle' (Collins 1974, 167) and showed that such inarticulate knowledge plays a role in the building of a laser, as part of a specific local scientific culture.

This Winchean-Wittgensteinian approach is not only interesting for philosophers of science, but could also helpfully be applied to the study of technology, including the study of innovation processes and technological change. As I already argued previously, not only language but also technology has its roots and meaning in a common life in which we participate. New technologies arise in a given social context; when a new technology is designed and developed, there are already meaningful practices. While those practices do not *determine* the development, meaning, and use of the technology, they shape what the technology is (means) and becomes. Moreover, in these processes, tacit knowledge plays an important role: Sometimes there are explicit rules, but especially at the level of the practice and of society and culture, many rules are implicit. Thus, if we integrate Winch's interpretation of Wittgenstein with the insights about language use and technology use in the previous section, we arrive at a view according to which both the use of language and the use of technology in the development of new technologies is embedded within larger language games and technology games (and ultimately within a 'form of life'), which gives us rules and shape our *Verstehen*, our interpretation and understanding of technology—but this is always an interpretation-in-use. Thus, the approach I have been developing in the previous section and chapters can be further supported by starting from philosophy of science.

However, in STS, this Wittgensteinian route has not been taken because of the empirical turn. In the 1980s and 1990s, people like Bijker and Latour turned to the materiality and to the network character of scientific work. And as the "old-fashioned" philosophy and sociology of science retreated to the background, so did the interest in language, rules, interpretation, and other Wittgensteinian preoccupations. Science and society were now about artefacts and networks (and for some, also power and interests). Words, discourse, and language were, at best, secondary phenomena.

The same can be said about philosophy of technology, which has also taken an empirical turn, and this "empirical" orientation often did not take advantage of, or entirely neglected, insights into use of language or insights from (Wittgensteinian) philosophy and sociology of science. For example, postphenomenology's reading of Latour and Bijker has largely and success-fully filtered out the link to thinking about language. In the work of Ihde and Verbeek, in particular, the material artefact takes center stage. As we will see in Chapter 6, Ihde's hermeneutics becomes so material that the sym-bolic and the linguistic move to the background. But this is unnecessary and undesirable, and there are alternative possibilities. It is possible to retrieve and expand the linguistic dimension in STS and philosophy of technology, and it is possible to give this operation a Wittgensteinian touch. I already pointed to Winner's suggestions in philosophy of technology. With regard to science and technology studies, I now propose to turn to a particular phase in Latour's work, which shows a lineage that goes back to the Witt-gensteinian phase of social constructivism, and in which there is still a keen interest in language via semiotics. In the next section, I will explain this semiotic dimension and what one may call the "Wittgensteinian" aspect of Latour, and I will argue that retrieving and discussing these dimensions and aspects can help us to better conceptualize the relations between humans, language, and technology. This section thus further supports and develops the Wittgensteinian-Heideggerian approach, and contributes to the overall framework I aim for in this book. In the next part of the book, I will then further discuss what this approach means for the relation between language and technology, between meaning and materiality, and between words and things, by engaging with postphenomenology, Latour's "nonmodern" work (another, later phase in his work), and other theory and approaches inside and outside philosophy of technology.

4.4. Akrich's and Latour's Semiotic Approach: Performance and Process

Latour's influential 1991 book *We Have Never Been Modern* was directly inspired by Shapin and Schaffer's (1985) *Leviathan and the Air-Pump*, which itself stands in the Wittgensteinian tradition in social science and even makes a direct reference to Wittgenstein when outlining *its main thesis.* It is claimed that the experiments of the scientists were, after all, discussions about a "form of life:"

> 'We will show that the experimental production of matters of fact involved an immense amount of labour, that it rested upon the accep-tance of certain social and discursive conventions, and that it depended upon the production and protection of a special form of social organi-

zation. The experimental programme was, in Wittgenstein's phrases, a "language-game" and a "form of life." The acceptance or rejection of that programme amounted to the acceptance or rejection of the form of life that Boyle and his colleagues proposed.'

(Shapin and Schaffer 1985, 51)

I will return to *We Have Never Been Modern* later, in part III. But there are more links between Latour and Wittgenstein, such as an interest in performativity and a nonrepresentationalist view of language and knowledge. We can use Latour to further thinking about the relation between words and things, and language and technology. Let me start with constructing and interpreting Latour's view of *language* and its relation to technology.

Latour's (and postphenomenology's) turn to materiality can be interpreted as a reaction against postmodern and poststructuralist approaches, especially in French philosophy, which stressed the autonomy of language so much that it seemed that there was hardly anything or anyone else around with sufficient ontological status. (See also the next chapter.) At the same time, however, Latour is very much inspired by semiotics, which links his view of technology, nature, etc. to thinking about language. Akrich and Latour's semiotics can be interpreted as a kind of synthesis of language-oriented semiotics and the empirical turn to materiality and artefacts. It is a kind of "expanded semiotics" or "material semiotics:" one that is expanded from words to things. Let me explain this.

In their well-known paper 'A Summary of a Convenient Vocabulary for the Semiotics of Human and Nonhuman Assemblies' (1992), Akrich and Latour interpret semiotics, the study of meaning and how meaning is made, as not only concerned with textual and linguistic signs; instead, they stress the 'nontextual and nonlinguistic' dimension of meaning-making. In their view, semiotics 'may be applied to settings, machines, bodies, and programming languages as well as texts'—to 'assemblies of humans and nonhuman actants' (Akrich and Latour 1992, 259). In line with work in the sociology of knowledge that emphasizes know-how (see the previous section), and taking what we could call a *performative* view, Akrich and Latour focus on the distribution of 'competences and performances' in these assemblies. Their assumption is that not only actors but also nonhuman 'actants' have competences (and actors have this in addition to a character). The social scientist then analyzes what the various actors and actants are doing, also in the sense of which competences they develop. The analysis is guided by semiotic concepts that are normally reserved for working with text: script, description, inscription, and transcription. Interestingly, the authors suggest that script can be attached to artefacts, or rather, that a text is translated by a material form. If we accept that this is possible, we can then study these shifts between texts and (material) things. They give the example of hotel keys: The text "DO NOT FORGET TO BRING THE KEYS BACK" is "translated" by heavy weights attached to keys in order to encourage people to return the keys. Textual signs thus shift to a material form. For instance, the sign "FASTEN YOUR SEATBELT" shifts to an alarm (260). This example also shows the *performativity* of these translations and the role both language (words) and technology (here: artefacts) plays in it: It involves doing things with words and doing things with things. It also may involve doing things with words and things to get *others* to do things. Not only humans but also nonhumans are counted on to make people do things and to fulfill actions. This also implies that there is a normative aspect of these human/nonhuman performances. Devices can be prescriptive if they permit or forbid. Consider, for instance, again the sound(s) that prescribe "FASTEN YOUR SEATBELT." Tasks may be shifted to humans or nonhumans, which have their own competences or may delegate their competences to someone or something else. Ascription, understood as the attribution of the origin, that is, saying who the author is and who the designer is, is for Akrich and Latour a secondary process. Moreover, they call categories that sharply divide humans and nonhumans 'an artificial cutting point along association chains' (263). In the processes they study, there are associations and assemblies of actors and actants. Perhaps one could say that the ontology game is less important than the various associations and, indeed, *performances* themselves.

However, here my interest is not in the human/nonhuman discussion as such, but in the role of language and its relation to technology. For Akrich and Latour, as meaning moves across humans and technologies, it not necessarily and not always takes the form of language understood as (textual) signs. Meaning shifts between linguistic signs and material forms. This means that sometimes the Wittgensteinian human user of language is bypassed (which may or may not be seen as problematic). More generally, what is so interesting and innovative in this text is the close conceptual proximity between words and things, linguistic elements and artefacts. The idea of scripting artefacts—in the sense of artefacts that are given a script by us (delegation, etc.) and in the sense of artefacts giving a script to us, artefacts scripting *us*, humans—implies that words and things are not only semiotically linked but also that both language and technology do things.

Indeed, in terms of agency, scripts are not only executed by humans but also by machines and other technologies. More generally, Latour shares with philosophers of language such as Wittgenstein, Austine, and Searle an interest in performativity. But, here, this becomes not only the performativity of language understood in terms of human use (using language to do things; more precisely, doing things with *words* and making other people do things with words) but also the performativity of language as text without necessarily involving human agency, it seems, and the performativity of things, of artefacts (using *things* to do things and make other people do things). The ontological divide between "culture" (language) and technology (artefacts, machines, etc.) is crossed. It turns out that things can "speak." Later, I will say more about how to interpret this claim, but for now let me further compare Latour and Wittgensteinian, including what we may call "performative" thinking about language.

In an article that aims to contribute to bringing Wittgenstein's view of language and his epistemology to the social sciences (and, hence, stands in the Wittgensteinian sociology of science tradition described above), Barinaga (2009) has argued that most cultural research still assumes a representational view of meaning—a view that she contrasts with Wittgenstein's performative view. The representational view accepts what Latour (1993) calls the "big divide," since it places language on the side of culture, whereas objects words are referring to are placed on the side of nature. Wittgenstein, by contrast, has a performative view of language that sees words not as labels for things, but as *doing* all kinds of things. Meaning is in use. What matters is what words perform (and perhaps how they are performed). In Latour, Baringa finds an empirical implementation of this idea. Merging Latour and Wittgenstein, she argues that culture should not be studied as an objective and external reality; instead, one should study 'the chain of mirco-situations:' 'the local, concrete interaction,' and, indeed, the use people make of language to shape their reality—the way people make plans, justify their doings, and so on—words used here and now.

This is an interesting comparison, as it highlights again an important link between Latour and Wittgenstein: Both can be interpreted as holding a performative view of the social. However, a full implementation of Latour in the social sciences, which is inspired by but also moves *beyond* Wittgenstein, would not only need to look at the use of language but also at the use of artefacts. If "culture" is not transcendent, but must be studied in the chains of humans and nonhumans, the same is true for "technology." Latour is critical of the tendency to isolate language from the world, although he respects the autonomy of language (Hørstaker 2005). Therefore, a more radical integration of Latour and Wittgenstein would conceptually merge the use of language with the use of technology, giving *both* their due—as I try to do in this book.

Could semiotics help with this project? Let us further investigate Latour's (use of) semiotics. As Hørstaker (2005) has pointed out, concepts such as inscription, translation, shifting, etc. come from semiotics. Latour made 'semiotic concepts operational in social science research' (Hørstaker 2005, 22). In particular, he fuses semiotics with actor theory: He is influenced by *narrative* theories (this is also interesting for part III, when I will say more about narrative). Hørstaker compares Latour to Greimas's semiotics, and indeed Greimas's narrative semiotics looks at "actants" in narratives, where actants do not need to be humans. There are all kinds of helpers and traitors, not just humans. Moreover, for Latour in scientific texts, *performance* comes before competence. An object gets its competence after a series of performances. We delegate competences to it, but this must be understood as having the object *perform*, repeatedly. Furthermore, the "narrative program" in Greimas becomes in Latour "programs of action:" someone wants to achieve something, and then takes up an object that will enable them to achieve this goal (15). And, as we have seen, humans also inscribe a program of action in them for telling *others* to do something; others have to perform, and the technical object helps them with that. This interpretation thus seems to bring back the human user and human agency. But, at the same time, there is also performance and agency on the part of the nonhuman, technological artefact. Consider the speed bump, for instance: The bump translates the program of action of the authorities and prescribes low speed. Hence, the speed bump makes others perform. Moreover, it also performs itself. This requires competence. Wanting-to-do requires knowing-how-to-do. The competences are a presupposition (17). Now, we may ascribe a "competence" to the speed bump. However, there is a sense in which ascription only occurs "afterwards;" "first," there is the performance. First, the bump does a series of performances, and then we can ascribe this competence to it. Thus, technical mediation and translation can be understood as narration and as *performance*, which rely on competences, and which involve both humans and nonhumans. Latour's programs of action have a narrative structure; this is an interesting side of Latour's work, which is often neglected when, for instance, a central concept, such as actor network, is interpreted in a nonnarrative way, as if it was suggesting a static ontology rather than a *process*. And this narrative and performative dimension of technological mediation is an interesting insight, which could be added to postphenomenology and mediation theory (see Chapter 6).

But, one could ask (especially if one adheres to an ontological, or what Heidegger would call "ontic," way kind of thinking), what is the precise status of the artefact here, if compared to language? What is the precise relation between the linguistic and the nonlinguistic? According to Hørstaker, Latour thinks that there is 'no known or knowable reality that is a nonlinguistic reality' (14). This puts him in line with Heidegger and Wittgenstein, and especially with semiotics and postmodernism (see the next chapter). Yet, at the same time, for Latour, the collective is built by technical objects and humans; this seems to presuppose that there is a nonlinguistic reality. But why is this question so important? Why is it the right question in the first place? Perhaps the solution to this paradox is to consider also that Latour is influenced by process philosophy, and use that approach to answer and change the question. As when I referred to narrativity, the idea is to replace static ontology with a more dynamic view, to bring in the *time* dimension. If we focus on process, then perhaps we could see the object (and the human subject) as the *emergent* outcome of a *performative process*. In this process, actors and actants make a world, make their world. In this process, objects are given *a voice* and acquire a voice, "speak." Technological artefacts are not silent intermediaries. They are used in human narratives and action programs, but as they perform in these narratives and programs, they also acquire their competences *and* their own voices. Thus, this process approach, combined with the idea that things can have a voice, helps us to bridge what Latour has called "great divide." I will say more about things that "speak" when I discuss Latour's politics of nature in Chapter 6.

To conclude, semiotics can help us to conceptually link language and technology, but not every kind of semiotics. Using Latour and Wittgenstein, we can give a narrative, performative, and emergence/process twist to semiotics, and see meaning as emerging from a performance process, which has a narrative structure. This enables us to bring humans and things, subjects and objects, culture and materiality, text and artefacts, closer to one another. In Chapter 6 I will further support this interpretation by drawing on Latour's nonmodern thinking and on Pickering's "performative" reading of Latour's posthumanism (1995).

However, in Latour, ultimately language is still frequently used as a metaphor, in particular the metaphor of "text." His focus is on giving *artefacts* a voice. Script, narrative, translation, and so on belong to linguistic semiotics, but concrete and lived language itself (understood as use and performance) is not so prominently present in his thinking. One could say, also in line with Pickering (see later), that Latour transfers the symmetrical thinking of semiotics to the social construction of artefacts and to the social, more generally speaking, thus rethinking the relation between subjects and objects and harvesting semiotics' posthumanist potential. But the relation between language and artefacts remains undertheorized, in the sense that the language user remains out of view, for instance, but also in the sense that there might be more kinds of *mediations* done by artefacts. In the next part of the book, I will say more about language and about the relation between language and materiality. In Chapter 6, I will engage with postphenomenology and, in Chapter 7, I will further discuss the relation between narrativity and technology. But, first, I construct another "extreme" position, according to which only language "speaks." I zoom in on postmodernism and poststructuralism (which includes poststructuralist semiotics and a postmodern interpretation of Heidegger and Wittgenstein), which tend to place so much emphasis on the autonomy of language (e.g., text) that both humans (and their bodies) and concrete, material technologies are muted and disappear.

References

Akrich, Madeleine and Bruno Latour. 1992. "A Summary of a Convenient Vocabulary for the Semiotics of Human and Nonhuman Assemblies." In *Shaping Technology/Building Society: Studies in Sociotechnical Change*, edited by Wiebe E. Bijker and John Law, 259–64. Cambridge, MA: MIT Press.

Barinaga, Ester. 2009. "A Performative View of Language—Methodological Considerations and Consequences for the Study of Culture." *Forum: Qualitative Social Research/Forum Qualitative Sozialforschung* 10(1), Art. 24. http://nbn-resolving.de/urn:nbn:de:0114-fqs0901244

Bolker, Michael, Mathias Gutmann, and Wolfgang Hesse, eds. 2010. *Information und Menschenbild*. Heidelberg: Springer.

Burg, Simone van der. 2014. "On the Hermeneutic Need for Future Anticipation." *Journal of Responsible Innovation* 1(1): 99–102. doi:10.1080/23299460.2014. 882556.

Coeckelbergh, Mark. 2006. "Regulation or Responsibility? Autonomy, Moral Imagination, and Engineering." *Science, Technology, & Human Values* 31(3): 1–24. doi:jstor.org/stable/29733939.

Coeckelbergh, Mark and Jessica Mesman. 2006. "With Hope and Imagination: Imaginative Moral Decision-Making in Neonatal Intensive Care Units." *Ethical Theory and Moral Practice* 10(1): 3–21. doi:10.1007/s10677-006-9046-2.

Coeckelbergh, Mark and Ger Wackers. 2007. "Imagination, Distributed Responsibility, and Vulnerable Technological Systems: The Case of Snorre A." *Science & Engineering Ethics* 13(2): 235–48. doi:10.1007/s11948-007-9008-7.

Collins, Harold Maurice. 1974. "The TEA Set: Tacit Knowledge and Scientific Networks." *Science Studies* 4: 165–86.

Hørstaker, Roar. 2005. "Latour—Semiotics and Science Studies." *Science Studies* 18(2): 5–25.

Latour, Bruno. 1993 (1991). *We Have Never Been Modern.* Translated by Catherine Porter. Cambridge, MA: Harvard University Press.

Oudshoorn, Nelly and Trevor Pinch, eds. 2005 (2003). *How Users Matter: The Co-Construction of Users and Technology.* Cambridge, MA: MIT Press.

Pickering, Andrew. 1995. *The Mangle of Practice: Time, Agency, and Science.* Chicago, IL: University of Chicago Press.

Pinch, Trevor J. and Wiebe E. Bijker. 1984. "The Social Construction of Facts and Artefacts: Or How the Sociology of Science and the Sociology of Technology Might Benefit Each Other." *Social Studies of Science* 14(3): 399–441.

Roderick, Ian. 2016. *Critical Discourse Studies and Technology: A Multimodal Approach to Analyzing Technoculture.* London: Bloomsbury Academic.

Shapin, Steven and Simon Schaffer. 2011 (1985). *Leviathan and the Air-Pump: Hobbes, Boyle and the Experimental Life.* Princeton, NJ: Princeton University Press.

Stern, David. 2002. "Sociology of Science, Rule Following, and Forms of Life." In *History of Philosophy and Science*, edited by Michael Heidelberger and Friedrich Stadler, 347–67. Dordrecht: Kluwer.

Swierstra, Tsjalling, Dirk Stemerding, and Marianne Boenink. 2009. "Exploring Techno-Moral Change: The Case of the Obesity Pill." In *Evaluating New Technologies*, edited by Marcus Düwell and Paul Sollie, 119–38. The Netherlands: Springer.

Winch, Peter. 1990 (1958). *The Idea of a Social Science and its Relation to Philosophy*, 2nd edition. London: Routledge.

Winner, Langdon. 1993. "Upon Opening the Black Box and Finding It Empty: Social Constructivism and the Philosophy of Technology." *Science, Technology, & Human Values* 18(3): 362–78.

5   All about Language

Postmodern Interpretations,
or the Muting of Humans
and Technology

5.1. The Postmodern Interpretation of Heidegger, Wittgenstein, and McLuhan

In 20th-century thinking, there has been a poststructuralist and postmodern current that emphasized the autonomy of language to such an extent that it became difficult to give technology a place at all, or even the human. Like Wittgenstein and Searle, but in a much more extreme way, this current has focused on language at the expense of material objects, technology, and, indeed, the human. It is this problem, among others, which led many philosophers of technology and STS scholars to turn away from a "humanities" approach and to take a material and empirical turn, which was also a turn away from (thinking about) language. While I sympathize with this turn, this book also tries to recover some of the insights from philosophy of language and related fields—including the poststructuralist and postmodern current. Perhaps empirically and materially oriented philosophy of technology can learn something from these approaches.

But let me first construct the "extreme" position that only language speaks. A first way to do this is to entertain the thought that we live "in" language and that we are mere vehicles used by language. Heidegger (but also Wittgenstein and McLuhan) can be, and has been, interpreted in a way that "mutes" humans and technology, an interpretation that puts all the focus on the subject in a way that eclipses the object.

Heidegger

Heidegger famously said that 'language speaks' (Heidegger 1967, 57). Humans only speak 'insofar as they co-respond to language' (57). Language is given to us; there is already a spoken language and, when we speak, we respond to it. In that sense, there is not an outside of language. Language, Heidegger argued, is 'the house of being' (254), it is not just something we possess. We exist in it. To use a term from Wittgenstein: Language is a "form of life." Heidegger thus questioned the instrumental view of language. Language is not a mere tool, but a medium in the sense of a milieu. It is constitutive of experience and thinking; it actively shapes our understanding of the world. Language should not be objectified or instrumentalized; instead, it is what Heidegger calls a kind of "revealing:" it shapes what we know as it lets things appear in a specific way. But, to say that language speaks and humans merely co-respond to language, and especially to say that there is *nothing outside language*, radicalizes this view. Humans are reduced to mere vehicles of language. Language speaks. And technology and anything material remain out of sight.

Interestingly, Heidegger's view of *language* is very similar to Heidegger's view of *technology* (Heidegger 1977): Technology is not a mere instrument, but is also a way of revealing. Just as there is a revealing through language, there is a revealing through technology. Language, like technology, shapes how we experience the world; it shapes our possibilities of experience. And, again, this position can be radicalized by saying that there is nothing outside (modern) technology. Heidegger wrote about modern technology that there is a *Gestell* or (en)framing: Technology encompasses everything; everything that shows itself can do so only through the enframing by technology. Everything becomes a standing reserve. Technology performs a gathering together, which takes everything with it and in it. It creates an entire world. Again, there seems to be little room for human agency. We live under a kind of 'destining' (Heidegger 1977, 31), which Heidegger links to, 'That which, of himself, [man] can neither invent nor in any way make' (31). He also compares it to a 'constellation' (33). This can easily be interpreted as "technology speaks" and humans are muted; they are the mute vehicles of thought. Humans are the medium, the tool.

This view also has implications for thinking about the social. In contrast to Searle, Heidegger did not think that we use language to impose language onto the physical world. Instead, social ontology is already given in language. Language is the medium in which we think, also the way we think about the social. Or rather, language thinks through us, and hence language shapes what the social is. In that sense too, "there is nothing outside language." The social is unthinkable and unsociable outside language.

Wittgenstein and McLuhan

One could also reframe Wittgenstein and McLuhan by saying that "there is nothing outside language use" and that "there is nothing outside the medium." Indeed, Wittgenstein can also be interpreted in a more "radical" way: If language games and forms of life shape our use of language and give a grammar to our activities, then it seems that language uses us as much as we use language. A radicalized, language-focused version blends out the human user and gamer by overemphasizing language and *language* games. Furthermore, in my chapter on Searle, I pointed to how Wittgenstein approaches the question of agency in relation to thinking: Thinking is not "in" the human mind, but if it is "in" anywhere, it is in the hands, the mouth, etc. Similarly, one could say that when we speak, that is, when we use language, this is done by using all kinds of tools (words, grammar, voice, vocal folds, writing, text, etc.), but there is no longer a center of agency. Postmodern and poststructuralist thought have further decentered and, in this sense, *muted* the subject. For them, however, agency is not shared by things and bodily parts and organs, as in Wittgenstein; for them, agency— obsessed by writing technology and related terms such as text, they would say: authorship—is a matter of signs, symbols, and texts. When I am speaking, it is not me who is speaking; it is the text that speaks. In other words, it is *language* that takes over agency here; only language speaks.

Another way of putting this is to say that only the medium speaks. And, indeed, McLuhan can and has been interpreted in this way. If the medium is the message, then it seems that humans no longer have a voice. This is in line with the radical decentering proposed by postmodern theorists. Moreover, in postmodernism, this is coupled with a claim about a radical change in history. Kellner has interpreted McLuhan as influencing postmodern theorists like Baudrillard (Hammer and Kellner 2009, xxii) and even as anticipating postmodernity. McLuhan thought that we were entering a new stage in history in which a new electronic tribal culture would overcome alienation. Kellner reads him as 'a major anticipator of theories of a postmodern break, of a rupture with modernity' (Kellner 2010, 180) and points to 'his notion of a rupture with modernity and advent of a new postmodern era' (182). He writes:

> 'McLuhan not only provides brilliant insights into specific media . . . but has important general insights into the media, Western modernity, and the broad patterns of historical change in the modern era . . . Baudrillard would take up McLuhan's emphasis on the form of the media, his insight into the centrality of media in the contemporary era . . . as well as McLuhan's method of probes, explorations, fragments, and a mosaic shotgun approach that illustrate general theses and specific phenomenon under investigation.'
> (Kellner 2010, 184–185; see also Kellner 1989).

According to Kellner, postmodernity is characterized by 'a proliferation of signs, spectacle, information, and new media,' whereas modernity was 'centered on the production of things' (Kellner 2010, 187). In other words, in postmodernity, the medium speaks and language speaks. "Medium" is interpreted in a way that centers on signs and symbols. It is nonmaterial, "non-thingly." Things are banned to the industrial past with its commodities and products, its mechanization and, indeed, its material *technology*. As Deely has argued, postmodernism is about the revival of the sign and its noninstrumentality (Deely 2001). We should no longer occupy ourselves with things; things are passé. This is the thing-averse postmodernism the empirical and material turn in philosophy of technology turned away from.

In the next section, I will take a closer look at postmodernism. For instance, I will say more about Baudrillard and about Lyotard's postmodern interpretation of Wittgenstein. Let me first conclude this section: What does the postmodern interpretation of Heidegger, Wittgenstein, and McLuhan imply for thinking about technology (and the human)?

Conclusion

One the one hand, learning from these interpretations, we better acknowledge the important role language plays as a medium: as a medium of thinking and as a medium of other activities. This is also applicable to thinking about technology and technological practices, in philosophy of technology and elsewhere. First, when we think about technology, language thinks with us. For instance, the words "technology," "tools," "instruments," etc. themselves may suggest an instrumental interpretation of technology, which is not "wrong" (rather, it is 'uncannily correct,' as Heidegger put it), but it is a very limited and misleading conception of technology, given the noninstrumental functions and consequences of technology, and given that, as Heidegger argued, technology is also a mode of thinking and revealing. Second, as I have started to argue in the previous chapters, in philosophy of technology, we should acknowledge that not only technology but also *language* is a medium that shapes subjectivity and society. In particular, we should not only think about the use of technology but also the use of language: use of language *about* technology and use of language *while using* technology. Moreover, we can learn from Wittgenstein that this use should be interpreted holistically, and I have applied this not only to language use but also to technology use. Language and technology will always be part of a larger "grammar:" connected to activities, uses of language and technology are always part of language games (and "technology games"), and ultimately of a form of life. The ways we do things with words and do things with things are embedded within larger social and cultural patterns; technology, like language, is part of, and co-constitutes, that social bond and form of life. Finally, we can learn from McLuhan that media are never just media; they also shape and, indeed, transform our world. McLuhan shows the transformative power of new media.

On the other hand, the more radical interpretation that there is *only* language is highly problematic for philosophers of technology, especially if and insofar as it makes conceptualizing technology itself impossible. In contrast to the interpretations I am developing and sympathize with, which focus not only on language but also on technology, materiality, and the body, in the language and sign-oriented interpretations of Heidegger, Wittgenstein, and McLuhan, technology disappears. If the radical view means there is *only* language, then it seems that both the human and technology are *secondary* phenomena. In this interpretation, humans no longer speak, and technology does not have a chance to "speak." Here, humans cannot even speak *through* language; according to this radical view, it would be more correct to say that language speaks through them—if humans still have any (primary) reality at all.

Again, this is not the view I wish to defend; I will construct the view that humans, technology, and language speak. But it is worth constructing this "extreme" view for the purpose of constructing my own view. Moreover, while few people may hold this extreme view, it is clear that in the history of philosophy, currents of thought that have been labeled as "poststructuralist" and "postmodern" (Derrida, Foucault, Lyotard, Lacan, Deleuze, Baudrillard, etc.), mostly in French philosophy, have at least *tended* or *leaned* towards this extreme view. Given the project of this book to investigate the relations between humans, technology, and language, it would be unwise to simply neglect these writings, as current postphenomenology does. Therefore, in this chapter, I will read and interpret Derrida, Lyotard, and Baudrillard as contributing to the extreme view that only language speaks. I will also mention Virilio and Mersch—mainly as a contrast, since, interestingly, in *their* version of postmodern thinking, bodies and objects reappear.

That being said, I will also show that we can learn something from these poststructuralist and postmodern authors, even those who overemphasize language to the point of letting technology disappear. Thus, on the one hand, I read and (ab)use work by Derrida, Lyotard, and Baudrillard in order to construct the extreme, ideal-type view that "it is all about language" or that "language speaks." On the other hand, it is also my purpose to show that their writings are more ambiguous than one may presume, and that there are more elements or aspects in them that may help us with understanding technology than one might presume. I will show that this is true both in terms of the "message" or "content of their writings" (e.g., some explicit references to technology or technological practice) and in terms of the approaches (their attention to the role and limits of language, of course, but also a transcendental approach, and difference and alterity thinking).

5.2. Derrida, Lyotard, and Baudrillard; Virilio as Exception

Derrida

Whether or not it is justified to label them as "postmodern," Derrida's writings are famous for a focus on signs, reading and writing, (con)text and discourse, and, hence, language. His method of 'deconstruction' is a response to linguistics, in particular structuralist theories of language. When it comes to language, Derrida's focus is on writing as opposed to speech (in particular, oral speech, voice). Against what he called 'phonocentrism' (the belief that sounds and speech are primary, as opposed to written language), he argues that it is writing that opens up meaning. He is famous for claiming, in *Of Grammatology*, that 'there is nothing outside of the text' (Derrida 1974, 158). Thus, his focus is on language, and especially written language.

However, his writings do not entirely neglect technology or technological practice. For example, in his seminal text 'Structure, Sign, and Play in the Discourse of the Human Sciences' (1970), Derrida claimed, inspired by a term used by Lévi-Strauss, that 'discourse is bricoleur' (360). The bricoleur uses 'the instruments he finds at his disposition around him.' In other words, there is 'the necessity of borrowing one's concepts from the text of a heritage' (360). Derrida then contrasts the bricoleur with the engineer, who he—mistakenly—assumes to work from scratch rather than borrowing what is already there. But it is interesting that, here, technology is used as a metaphor to say something about what criticism and discourse does. Like Wittgenstein, Derrida compares language with tools. And like Heidegger and Wittgenstein, he emphasizes the givenness of language. But an important difference with Wittgenstein is that, here, the language user, the human language user, disappears from view, since 'discourse' *itself* becomes the bricoleur, not the language user.

This approach seems close to Heidegger's dictum that language speaks, but it is more radical. Here, language, in the form of discourse-text, completely takes over, to such an extent that the concrete material reality, of technology, for instance, seems to be gone—including any conception of the *human* being as bricoleur, on which his metaphor relies. In structuralism, material reality was still one of the three realities, next to language/structure and abstract ideas. Now, there is only text and discourse. Meaning seems to be divorced from anything material-technological or human, removed from the voice, also. This move was already prepared by structuralism, which viewed everything in terms of a deeper structure, thereby rendering concrete humans, voices, and material-technological realities epiphenomena. In this respect, Derrida's specific argument about decentering and the instability of meaning does not really change much, except that it even deletes the very idea of a "voice." Again, language appears here exclusively in the form of writing. No one is speaking, in the sense of "voicing" anything. The only technology left is writing. And, in contrast to Wittgenstein's and McLuhan's writings, this writing gets very immaterial, disembodied, and very much non-technological. Discourse plays around with signs. The technological and the human remain out of sight; they remain unwritten and certainly unvoiced.

This (over)emphasis on language as signs already started with Lévi-Strauss's structural anthropology (1963), which, inspired by the structural linguistics of de Saussure, abstracted from concrete material things and considered culture as a system of symbolic communication. Lévi-Strauss tried to understand the underlying "grammar" of cultures, the social structures or 'unconscious infrastructure' of cultures (Lévi-Strauss 1963, 33)—in the same way as Saussure argued that language is not about the individual speech act, but about grammar. For instance, he studied kinship in terms of "symbolic systems;" this focus on the symbolic helped him to integrate anthropology and linguistics (51). He argued that kinship terms are elements of meaning, which 'acquire meaning only if they are integrated into systems,' which, in turn, are 'built by the mind on the level of unconscious thought' (34). He was interested in 'the universal laws which make up the unconscious activity of the mind' (62), in how 'the human mind works (80), and even in 'the problem of the congruence between mind and universe' (90). This focus on the system and on the mind uses language as a model for its thinking, but de-emphasizes concrete humans and materiality. Lévi-Strauss's turn to linguistics was a turn away from the study of concrete beliefs and technologies; like in Derrida's key text, in Lévi-Strauss's main work, technology is only allowed as a metaphor (here, a comparison between a steel ax and a stone ax, 230). This linguistic turn in anthropology has then been radicalized in Derrida's poststructuralism.

Wittgenstein was also interested in rules and grammar, but, in his thinking, the structures are not "mental" and centered on mind (he is less dualist than Lévi-Strauss), and the use of tools and the concrete human being and body do not disappear. If we turned to Wittgenstein and to the structuralism Derrida responded to, but interpreted that structuralism in a less dualist and less symbol-oriented way, we could study not only the 'grammar' of the social but also of technology. We could study how we do things with words and how we do things with things. Influenced by Wittgenstein, we could study the tools of speaking: the hand, the mouth, the writing technologies, etc. But these tools would need to be related to their "grammar." We could study the linguistic-social in terms of the technological-social structures and infrastructures that make possible the individual speech act and perhaps also the individual technology act. We could study how people tinker (*bricoler*) with words and things—but always starting from the background of language games and what we may call "technology games." Emphasizing the structuralist and anthropological background against which Derrida thinks and plays, but rendering it more materialist, we could keep the holism of Wittgenstein and structuralism, but retrieve technology and let the material reenter the social.

To conclude, while Derrida's writings, like most writings in the tradition of poststructuralism and postmodernism, focus on language (in the form of written discourse), this does not mean that technology is entirely absent. Generally speaking, when it comes to thinking about technology, the writings of the poststructuralists and postmodernists may well be more interesting than may be expected, and can be interpreted and revised in a way that accounts for the role of technology. Derrida's use of the metaphor of the *bricoleur* opens up one route to do this. Let us now turn to Lyotard, who offers another, more directly Wittgensteinian path, which we can interpret for the purpose of better understanding technology.

Lyotard

Borrowing from Wittgenstein, Lyotard wrote about metanarratives and "language games" to emphasize that there are many systems and communities of meaning. In *The Postmodern Condition* (1979), he argues that, in the post-modern condition, there is an 'incredulity toward metanarratives' (xxiv). Narratives occupy the center of the work. For instance, he sees narrative in opposition to scientific knowledge (7). An alternative view would be to view scientific knowledge as a narrative. I will say more about narrativity soon. But let me start with Lyotard's view of language and his use of Wittgenstein. His very method is focused on language: 'emphasizing facts of language and, in particular, their pragmatic aspect' (8). He gives examples of deno-tative utterances, prescriptions, a promise, a narration, etc. He explicitly refers to Wittgenstein:

> Wittgenstein, taking up the study of language again from scratch, focuses his attention on the effects of different modes of discourse; he calls the various types of utterances he identifies along the way (a few of which I have listed) *language games*. What he means by this term is that each of the various categories of utterances can be defined in terms of rules specifying their properties and the uses to which they can be put—in exactly the same way as the game of chess is defined by a set of rules determining the properties of each of the pieces, in other words, the proper way to move them.
>
> (Lyotard 1979, 10)

Lyotard then emphasizes that there are many language games, and, related to that, the agonistic aspect of language use. There are not always rules that can be applied across discourses. Discourses may be incommensurable.

   This interpretation of language games is problematic if it is meant to be Wittgensteinian, since Wittgenstein seems to have understood language games as being part of the same language and form of life, and, hence, according to Wittgenstein's own view, there may be a lot less agonism and incommensurability than Lyotard suggests, at least if the discourses belong to the same form of life. When defining language games in terms of a set of rules, Lyotard also seems to neglect implicit knowledge. And with Burbules (2000), one may question Lyotard's claim that consensus necessarily does violence to heterogeneity: A new position can also open up new and *more* options. But, whether or not it is a "correct" interpretation of Wittgen-stein, it is interesting that Lyotard connects to a Wittgensteinian tradition that focuses on language and at least leaves room for, if not encourages, a social-cultural interpretation of this centrality of language: The social is to be understood in linguistic terms, and language is to be interpreted as social. This reminds us of Wittgenstein's holism and structuralism, and, indeed, of the "cultural" interpretation of Wittgenstein. But, interestingly, Lyotard highlights not only grammar but also another, related dimension of language—and, hence, of the social: narrativity. He defines 'the question of the social bond' as 'a language game' and claims that even before a child is born, it is 'already positioned as the referent in the story recounted by those around him, in relation to which he will inevitably chart his course' (15). So, in this sense, there is no outside of language. The social is a matter of language games, which, according to Lyotard, are varied and—again—have an agonistic character. As he puts it: 'The social bond is linguistic, but is not woven with a single thread' (40). (He then quotes Wittgenstein's comparison of language with an ancient city.) Moreover, the social also has a narrative structure, and who we are depends on the narratives in which we are embedded. This raises an interesting and stimulating question: What would this narrative understanding of the social mean for our thinking about technology? In Chapter 7 I will need to say more about this.

However, linking narrative with technology would require a revision of Lyotard's view. When he uses Wittgenstein, Lyotard neglects what we may call the "technological" and "bodily" Wittgenstein and, instead, focuses on the usual, "linguistic" Wittgenstein. Technology is in the background. Wittgenstein's toolbox, Wittgenstein's towel, and Wittgenstein's hands are not part of the picture. Yet, as I suggested with regard to Derrida, one could revise and expand Lyotard's view, and apply his ideas about language to *technology*. First, we should not only focus on utterances and speech acts (which are then placed in a larger context) but also on artefacts and what I have called "technology acts." Both kinds of acts are interconnected in processes of acting and performance in daily life. This brings back Wittgenstein's tools: external ones and those that are part of the body (hands). Second, the social may be a matter of language games; but these language games are always part of a larger whole. If this is true, then, applied to technology, it means that our use of technology is also part of a larger whole. As I have argued before when interpreting and using Wittgenstein, technology is linked to activities and, as such, it is part of games and part of a form of life. Third, as said, we could further reflect on the relation between technology and narrativity, which I will do in Chapter 7. Fourth, the idea of agonistic language games could be interesting for philosophy of technology. A particular discourse *about* technology, for instance, can have an antagonistic aspect. This idea can be connected to, for example, Hegelian thinking or to Foucaultian discourse analysis, which links forms of knowledge to ways of exercising power. I will say more about Foucault later, but let me already point out here that it is interesting that Foucault did talk about human subjects and discourse, but also about material architectures of power, such as the panopticon: a building designed by Bentham, which enables prisoners to be observed from a single point, without them being able to tell whether or not they are being observed. For Foucault, this shows how knowledge and power reinforce one another, in a way which makes, fabricates subjects (Foucault 1977). One could see such technological structures and infrastructures as being connected to various ways power is exercised, and generalize this to all technological uses and structures. Perhaps we could say that variety in use and structure, both in language and in technology, is linked to varieties of exercising power. To take up the prison example again: A different design would presumably support different ways of making knowledge, exercising power, and fabricating subjects. More generally, Lyotard can attend us to power issues in relation to both language and technology. Just as speech acts and discourse are connected to language games, which are, in turn, linked to power games, technology games, and a form of life, uses of technology are also linked to larger games, which also have a power aspect.

Let me also note that Lyotard makes an interesting link between narrative and knowledge, in particular know-how. He analyzes the nature of *narrative* knowledge. He argues that knowledge also includes know-how, knowing how to live, etc.: Knowledge is 'a question of competence' (18) and depends on knowledge and custom. Then, he argues that narratives define the criteria of competence, that the narrative form 'lends itself to a great variety of language games,' including evaluative statements, that their narration obeys certain rules (e.g., begins with a specific formula) and also involves a 'knowing how to hear,' and that narrative form follows a rhythm (20–21). He then contrasts this narrative knowledge with scientific knowledge, which involves only one language game (denotation) and is disconnected from other language games that form the social bond. Stronger, 'it is no longer a direct and shared component of the bond' (25). As I already suggested, this seems an inadequate view of science, which, if understood as a social practice, also involves narrative knowledge. We could develop this point by using Latour, for instance. However, the main criticism I wish to make here is that Lyotard analyzes these different forms of knowledge in *linguistic* terms. His thinking is part of the larger linguistic turn. And this means that technology and materiality, as such, are not part of the method. The same is true, to some extent, for humans. The main focus of Lyotard is language use, discourse, narratives, and language games. Real human beings become senders and receivers; in other words, they are only considered in their linguistic-communicative aspects.

However, as I already suggested, this need not be our limitation. We can apply these concepts, such as discourse, narrative, and language games, to technology. We can understand technologies as related to language games (and "technology games") and to the social bond. Competence is also an important concept. Perhaps it can teach us something about the kind of knowledge that is involved in technological innovation, which is about know-how. Furthermore, perhaps technology can also have a narrative form or function. I already remarked that there is a narrative dimension in Latour. Later in this book, in Chapter 7, I will use Ricoeur to introduce the concept of "narrative technologies." But it is interesting that we can also find a narrative aspect in Lyotard's postmodernism. Again, the focus on language should not deter us; Lyotard's view could also be further developed in a way that tells us something about technology, and I have made some suggestions for how to do this.

Moreover, it is fair to acknowledge that technology *does* play a role in Lyotard's book. He argues that technological transformations (especially cybernetics) have had an impact on knowledge: The transmission of information and the acquisition and circulation of learning changes with 'information-processing machines,' when 'learning is translated into quantities of information' (4). Thus, although technology is not really part of his approach, of his "medium," so to speak, in the very "message" of his postmodern meta-narrative, technology plays a role, even a very important one: Technology changes the mode of knowledge and thereby changes the social. Lyotard argues that 'along with the hegemony of computers comes a certain logic, and therefore a certain set of prescriptions determining which statements are accepted as "knowledge" statements' (4). Knowledge becomes a commodity. Lyotard even compares the circulation of learning with the circulation of money (6). Thus, here technology's role is not thought of in an instrumentalist way; instead, it is argued that the new technology changes the very nature of knowledge. This is in line with a Heideggerian approach to thinking about technology, which points to the epistemic and hermeneutic role of technology.

Furthermore, Lyotard is not only "pessimistic" about the new technology. Of course, he warns that 'computerization . . . could become the "dream" instrument for controlling and regulating the market system, extended to include knowledge itself;' but, at the same time, he writes that 'it could also aid groups discussing metaprescriptives by supplying them with the information they usually lack for making knowledgeable decisions' (67). He even puts forward the utopian idea of language games becoming 'games of perfect information at any given moment,' which would be made possible by giving the public 'free access to the memory and data banks'—which would be non-zero-sum games (67).

However, the point that new technology changes the nature of knowledge is more fundamental than Lyotard's opinion about particular new technologies or the pessimism/optimism or utopia/dystopia discussion. What is interesting here for philosophers of technology is the connection Lyotard makes between language games and new technologies that make certain games *possible*. This brings us to a transcendental argument, which as I will argue does *not* make it impossible to think the *materiality* of technology (see later in this chapter and Chapter 6).

To conclude, we can work on Lyotard to bring him closer to thinking about technology, but there remains a gap because of his focus on language. Lyotard shares this gap with other postmodern thinkers, most of whom also fail to do justice to materiality and technology. In Baudrillard's postmodernist view, for instance, materiality disappears completely as it is sucked up in a world of simulation—unless one revises his view and includes the material infrastructures that make possible the simulation. Let us turn to his work now.

Baudrillard

Another example of language-centered or sign-centered work in poststructuralism and postmodernism, which is nevertheless relevant to the present discussion about language and technology, is Baudrillard's view in *Simulacra and Simulation* (1981/1994). Although, in principle, this work *could* be interpreted and used as a criticism of focusing too much on symbols and signs, in Baudrillard's approach, the conceptual focus is entirely on signs, not on materiality or technology. It can more easily be interpreted as one of the most radically postmodern views, in the sense that there is *only* the (linguistic or sign) medium that speaks. Humans and technology are mute; they can only appear in simulation. Baudrillard (1981) claims that, today, simulation amounts to a model without an origin, the 'hyperreal' (1). There is only a map, no territory. There is no longer a difference between true and false, real and imaginary, reality and representation. The simulator produces true symptoms (3). The image or sign becomes a pure 'simulacrum,' which 'has no relation to any reality whatsoever' (6). We live in a kind of Disneyland. Medium and message are confused, and sender and receiver are also confused. Discourse circulates, a cycle 'that without distinction includes the positions of transmitter and receiver' (41). Thus, Baudrillard's view is not so much that we have to return to the real, but rather that we *cannot* grasp the real since the distinction between reality and representation has broken down. Nevertheless, we can read Baudrillard as being nostalgic about the real.

How does technology come into this picture, if at all? Let me suggest that, in Baudrillard's world, it could be given a dual role: as medium and as artefact. As a medium, technology becomes the message; there is no longer a message the medium transmits, since that would presuppose that they are separated. As artefact, technology belongs to the "real," which has been lost. Thus, any conception of technology as having something to do with the material is dissolved here. It might be there, but we do not know it and cannot know it. It is invisible.

Perhaps partly because he took distance from Marxism, Baudrillard has lost the attention to material things. He turned away from production, utility, and instrumental rationality to symbolic exchange, culture, and signs. There is a dialogue with French semiotics, which was all about signs. But where is technology? Of course technology appears in Baudrillard's "message:" he writes about television, computer cyberspace, and virtual reality. But, in his writings, these become rather immaterial technologies and spaces. Even labor becomes a sign. Baudrillard would accept the claim that computers, new media, etc.—that is, technologies—create hyperreality. But his focus is on the play of images, signs, simulacra—without connection to an external reality, without meaning—and the question as to how this hyperreality is produced is not asked. The medium is only seen as a milieu of signs; materiality remains invisible. More generally, insofar as poststructuralists and postmodernists have contributed to what they see as the death of the real, they have thereby also lost any conception of technology as artefact. Technology can only appear as medium (without message, or rather *as* message) or sign (without the reality the sign refers to). In the simulacrum, technology disappears.

One way to revise and expand Baudrillard's view in a way that nevertheless brings in technology and the material, would be to say that (material) technology *makes possible* the simulacrum. The blurred boundary between Disneyland and real cities, for instance, is made possible by concrete artefacts: in Disneyland and elsewhere, it is material artefacts that make up and make possible this mixed fantasy/real world. Just as in what Baudrillard calls the first order of simulacra, the image of the craftsman is made possible by (using) tools; in the second order of simulacra, industrial technology makes possible mass production and proliferation of copies, and the third order of simulacra is *also* made possible by technologies, in particular electric (Baudrillard's TV) and electronic, digital devices. Thus, with this revision, we can try to uncover and retrieve the materiality and technology that makes possible the simulacra.

Admittedly, later in Baudrillard's work the object returns (the object as opposed to the subject), but it is not a very material object. He presents a metaphysical narrative that alienates the object from anything material or real. The object is a hyperreal, simulated object. And if the object is real at all, our relation to it becomes rather mysterious. The object is a 'Sphinx' and it does not answer, except 'secretly to some enigma' (Baudrillard 1983, 205). Baudrillard imagines a universe where bodies travel at high speeds, but light would travel slowly, a universe in which we would be 'struck without ever seeing the obstacles approaching' (Baudrillard 1983, 194). The infrastructure only shows itself as a kind of invisible force, a trace of something that was once there: in 'seismic movements,' 'the requiem of the infrastructure' (195). In Baudrillard's universe, everything collapses and there has never been a ground. Energy is symbolic energy and art is merely simulation. Things 'only occur . . . through the magic of their effect' (201). And these effects are fatal.

Thus, according to this later view, it could be said that we can experience the effects of technology and the effects of technological artefacts. But it seems difficult, according to this view, to conceptualize other aspects of technology, in particular its material aspects. As said, one can try to make visible/conceptualize the material-technological structures and infrastructures that make possible the simulation. Perhaps one could also try to revise Baudrillard by using McLuhan, one of Baudrillard's forerunners, to bring back the technologies in and of the medium. But Baudrillard gives us very little material to work with in order to accomplish such revisions.

Finally, in this view, Wittgenstein's or Searle's language *users* are nearly absent. When both subjects and objects are sucked up in cycles of symbolic exchange and discourse, and get lost in the space of hyperreality in which material infrastructures can only appear as ghosts or traces, there is very little room for thinking about the use of language, let alone about the use of technology. There seems to be only one, symbolic order in which there are "events" and "spectacles," but there is no use and no concrete material practice, let alone embodied performance. In my Wittensteinian interpretation of use, use brings subject and object together, but it does not completely merge/dissolve them into a hyperreal universe. Moreover, in Lyotard, there were still language games—even if they became already alienated from concrete use of technology. And Derrida at least used the bricoleur (and the engineer) as metaphors. But, in Baudrillard, technology becomes medium, mysterious object, or ghostly infrastructure—if it appears at all. Within this kind of thinking, it seems impossible to think of technology as material artefact, use, and practice.

Virilio

An exception in the world of poststructuralism and postmodernism is the work of Paul Virilio, who can and has been interpreted as a (post)phenomenological thinker and who explicitly wrote about *technology*. I propose to read him as a philosopher of technology. In his works, he emphasizes the accident as always related to the invention of new technology, the distance from real space and time (like Baudrillard, Virilio was mainly thinking of TV here—this is the use and material practice that shaped his writing, but usually remains itself invisible), and the logic of speed associated with warfare and modern media and technologies. Like Baudrillard, Virilio also thought that war had become a media event. And like Baudrillard, his writings are rather dark and apocalyptic. But Virilio gives not only an important role to technology; in contrast to Baudrillard and others in this tradition, his writings are much more concrete and historical. He brings in the material and physical. He also brings in the body and the living.

In *Speed and Politics* (1977/2006), Virilio attends us to what makes war possible, the logistics of war: He writes about 'fuel, trucks and military vehicles' (98). He comments on historical events. He talks about war machines and prostheses (83). For example, he quotes the Italian futurist writer Marinetti, saying about the armored car that the human driving this car becomes inhuman, 'an animal body that disappears in the superpower of a metallic body able to annihilate time and space through its dynamic performances' (84). Such descriptions are closer to McLuhan or even to post-phenomenology (at least, if revised in terms of performance; see the next chapter), contemporary STS, and posthumanism than it is to the semiotics and postsemiotics of people like Derrida, Lyotard, and Baudrillard. This thinking is no longer only occupied with signs but also with the materiality of technology, with bodies, and with concrete human-technology relations. In Virilio's writings, we meet domesticated bodies (98), cloistered bodies put in a brothel and disciplined bodies (104), the 'vehicle-bodies of horses' (106), bodies of the 'living dead . . . possessed by wills other than their own' (106), and soul-less bodies 'assisted by technical prostheses' (128). There is 'biological manipulation' of Olympic champions and warriors (114). The engine is called 'a prosthesis of survival' and is put in the context of a historical shift from 'the metabolic vehicle to the technological vehicle, spilling its smoke like a last breath, a final symbolic manifestation of the motor-power of living bodies' (114). In the end, Virilio argues, the technological vehicle defeats the metabolic vehicle (115). Virilio narrates how slaves became movable commodities (125), claims that we should analyze the 'instruments and objects' brought over by ships after the war, and attends us to cars and 'gleaming kitchens' (128). In what he says about the 'animal domestication of the American citizen,' he even mentions household robots and bionic bodies, and puts them within a political and cultural (romantic) context:

> Significantly, the American government will not deem it necessary to establish a veritable welfare system on its own territory. It is convinced at the time that the promotion of a paternalistic and humanitarian *comfort* civilization will perfectly replace social aid through the *technical assistance of bodies*, from the household robot to the company psychiatrist or the latest model of car. Not unlike the way this country today nurtures a romantic taste for the revived bionic bodies of fascistic futurism, human bodies in which certain organs have been replaced by technological grafts, enabling these new heroes of surgical science to accomplish superhuman physical exploits.
>
> (Virilio 1977, 129)

Thus, in Virilio, technologies and bodies are not only present; they also become political. In his text, we find the bodies of proletarian soldiers and workers, the 'speed-body' of the fascist bomber (134), the mobilized bodies of the Nazis, and the unable bodies in the war and after the war. He talks about the suppression of *gestures*, about 'the physically and mentally handicapped' (142), and about missiles and 'war materials' (154). He argues that the military technologies impact time and space—and again: politics. For instance, a nuclear explosion leaves no time for political decisions (154). Wars become time wars. Speed is war (155) and war is speed. This means that computer programs have to make the decisions (155). Long before today's warnings about automated weapons (e.g., the discussion about "killer" drones), Virilio already raises the question regarding reduction of 'the time for human decision to intervene in the system' (156). There are 'automatisms' and procedures 'from which all political choice is absent' (163), and there is speed, 'the violence of speed' (167). Military technology, one may conclude, is therefore always political, as it reshapes bodies and time/space. Virilio's text thus provides interesting points of connection with contemporary philosophy of technology, including thinking about embodiment and technology (Ihde), and about robotics and military technologies.

To conclude, Virilio's thinking is no longer (mere) semiotics or postsemiotics. We are nearly ready for discussing the bodily turn and the material turn. In the next chapter, I will say more about Ihde's postphenomenology. But let me first finish my (de)tour of postmodernism and poststructuralism by indicating some developments and aspects in contemporary postmodern thinking that may be of interest to philosophers of technology, and that could helpfully be integrated in the framework I am developing in this book: thinking about objects and the real, transcendental arguments, and thinking about difference and encounters.

5.3. The Return of the Object and Mersch's Posthermeneutics

During the past decades, several authors in the poststructuralist and postmodern tradition have also followed the bodily turn and the material turn (of the 1980s and especially the 1990s) or have at least tried to combine the more semiotic, abstract dimension of this tradition with thinking about the body and about materiality. However, in their work, an uneasy tension remains between the abstract metaphysics and semiotics, on the one hand, and the body or the material, which are often turned into abstract terms like "the real," "the object," "the event," etc., on the other hand. Like in Baudrillard, this approach renders concrete artefacts rather mysterious.

For example, in the German speaking context, Dieter Mersch (2010) has proposed a 'Posthermeneutik' which tries to think what is *not* sign, *not* text, and *not* medium, what cannot (easily) be captured by discourse and cannot be (directly) communicated, but what needs to be presupposed in sign processes. Engaging with Wittgenstein's famous phrase in the *Tractatus*, Mersch (2010) points to the paradox of trying to say what cannot be said, of trying to use discourse against discourse (10; my translation). But what is outside text? According to Mersch, there is something that resists the symbolic. There are 'residues,' the 'shadows' (30) of hermeneutics: the materiality of things, the body in its '*Leiblichkeit*,' and signs of decay and finitude (13). They resist. They may show themselves, for example in art, in sport, in the erotic. In art, materiality shows itself; it cannot be (re)integrated in the sign process (35). Materiality resists. Like many postmodern writers, and remaining within the Heideggerian tradition of thinking (Heidegger also turned to poetry and art), Mersch plays the game of what I have called 'romantic epistemology' (Coeckelbergh 2017): He engages in what he calls 'a complex play of presence and absence' (14). His posthermeneutics tries to navigate between "unmediated" experience (which does not exist) and mysteriously remaining silent (15). One way he does this is using the concept of aura (92) and pointing to what happens, or rather to the happening (*Ereignis*, event) *before the question is raised as to what happens* (68). Moreover, poststructuralism and postmodernism remind us again of the importance of language as shaping our thinking (later I will say: as a transcendental condition for thinking, use, and performance). But Mersch tends to the extreme view that if and when we speak, it is language that answers, not the user of language. He interprets Heidegger's dictum "*Die Sprache Spricht*" (language speaks) as saying not only that we cannot fully control language but also that we have to submit to its order (79–80). Language has and takes priority. As language users, we are masters and slaves at the same time.

Of course, thinking about what can be said and what cannot be said is an important philosophical issue, which also found its expression in Wittgenstein's *Tractatus*, and which deserves attention. But, for philosophers of technology, this approach presents a problem: If the materiality of things can only appear as the residue of text and sign, then it seems that technology cannot say much about it, and that a philosophy of technology, understood as a talking about technology, becomes very difficult if not impossible. Yet, as we will see, a lot can and has been said about the materiality of things, even or also in hermeneutics—for example, in Ihde's hermeneutics or post-phenomenology. It would be a pity if we had to dismiss those discourses altogether, on the grounds that they do not capture "the real" as opposed to "the text," on the grounds that they always exclude something. These problems of poststructuralism and postmodernism need not be ours. If we do not start from the premise that everything is text and sign (or that everything is text except a residue, except some traces of the real, of the body, and of the material), then there is room for addressing other problems concerning technology.

That being said, we can learn from this tradition of thinking that every philosophy of technology is always also a discourse, a play with signs, and a language game. Following Mersch's advice, we can keep in mind that this will always mean that some things are left out, that philosophical discourse—also in the philosophy of technology—cannot and perhaps should not capture and colonize everything, that it should not be totalizing. Moreover, with both Heidegger and Mersch, we can keep in mind that if and when we speak, it is also language that speaks. We submit to its order. Philosophy of technology and everyday discourse about technology have their own order(s), which frame our thinking about technology. And with Lyotard and Wittgenstein, we can acknowledge that we generally follow the rules of the language games that are given to us and live within (and submit to) a form of life. Perhaps we can continue to learn from, and produce, philosophical discourse about technology, as long as we remain aware of these problems. We can try to be critical of the particular order of language we submit to when we think about technology. For instance, in everyday speaking about technology, we submit to an order in which it is only possible to conceive of technology as an instrument, a means to an end. Philosophy of technology can then understand itself as having as one of its tasks to reveal and criticize this order and develop alternative orders and discourses. Moreover, heeding Mersch and Wittgenstein (but, this time, the Wittgenstein from the *Tractatus*), some attention to what is *shown* about technology (in art for instance), rather than what is and can be said about technology, is helpful in philosophy of technology. We could explore a more posthermeneutic approach to philosophy of technology, or at least take into account its insights: Next to speaking about technology, we also need to attend to the shadows of philosophical discourse on technology—what cannot be said, but can only be shown, to use Wittgenstein's words from the *Tractatus*.

This implies that arts, including music, visual arts, and performance arts, can play a significant role in understanding technology: They can show things about technology that cannot be said (e.g., by philosophers or by scientists). So far, most philosophers of technology have paid little attention to the arts and to artistic research, and its methods and approaches. But if Mersch and (the early) Wittgenstein get it right, then this neglect is not justified. (I will return to this issue in my conclusion at the end of Chapter 8).

5.4. Redeeming Poststructuralism as Postphenomenology: Transcendental Arguments and Difference

Another route to redeem *some aspects* of postmodernism and poststructuralism is to interpret poststructuralist thinkers such as Derrida as *postphenomenology*, and then use this interpretation to develop a critique of contemporary, "mainstream" postphenomenology as it is known within philosophy of technology. This project has been proposed by Dominic Smith, who finds in Derrida, Foucault, and Deleuze a form of phenomenology that not only pays more attention to language (as I have already argued) but that also introduces a transcendental approach (which, according to me and Smith, is not incompatible with an empirical turn) and pays attention to difference next to sameness. Let me outline his argument, and link back to my criticisms of contemporary work in philosophy of technology as not sufficiently focused on language, to my own uses of the transcendental approach in moral philosophy and philosophy of technology, and to the Wittgensteinian and Heideggerian approach I am developing, which I will argue can helpfully be better understood and enriched with a transcendental approach. This will be a good preparation for the next chapters, which further critically engage with contemporary postphenomenological approaches in philosophy of technology.

Phenomenology and Transcendental Arguments

What is postphenomenology? Let us begin with phenomenology. It is true that traditional phenomenology was not very materialist at all; the focus was on the subject rather than the object, and this is one-sided. Smith (2015) writes that there is an idealist tendency in Husserl: The subject gives sense (*Sinngebung*). (The same can be said of Kant's notion of the transcendental. Consider also again my criticism of Searle earlier in this book, which *could* be reformulated as saying that Searle is too idealistic.) This subject-centeredness is, of course, also all present in postmodernism and poststructuralism, and has given rise to Ihde's and Verbeek's objections—mainly directed at Heidegger—to *any* transcendental approach. In response to what it takes to be old-fashioned phenomenology, contemporary postphenomenology sees subject and object as mutually constituting each other. In philosophy of technology, this becomes: technologies constitute subjects. Smith mentions Verbeek's example of the ultrasound, a technology that is understood not so much as representing, but as constituting the unborn (538). Describing what happens in terms of subject and objects mutually constituting one another is helpful (indeed, this will be part of the framework I propose in this book; I will say more about Ihde and Verbeek in the next chapter). But, as I read him, Smith is right to note at least three dangers in current postphenomenology, which can be linked to my own views and arguments in the following ways.

First, accounts such as those of Ihde and Verbeek indeed show how technologies are interpreted and embodied, and, hence, move the focus away from the subject. But, as I argued in this chapter and will also argue in the next chapters, these accounts also leave something out: the insights of the *linguistic turn* (538). For instance, while Ihde devoted some attention to language, following Merleau-Ponty against Heidegger he mainly focused on embodiment and material artefacts. But there is much more work to do. As I argued, one of the outstanding issues is the relationship between technology and language (Coeckelbergh 2015). Here, we can learn from poststructuralists, since they pay attention to words—even if they prioritize words over things as opposed to prioritizing things over words, as phenomenology does (539).

Second, for poststructuralists, language is not just a neutral tool we use to describe things. Instead, language is also part of the conditions of possibility of those descriptions, and should, as such, be analyzed and criticized. This point, which for Smith is a truly postphenomenological point, bring us to the *transcendental* approach. I agree with Smith that Ihde and Verbeek unduly (and I would add: unfairly) dismiss the transcendental approach, or are even 'hostile' to this approach, as Smith puts it (543). Inspired by Wittgenstein, Smith criticizes Ihde and Verbeek for what one could describe as *remaining on the surface* with their hermeneutic relations and translations, and argues for a critical philosophy of technology that, after Kant, moves 'towards the "transcendental" conditions under which thought and experience are possible at all' (544). Interestingly, in line with my own thinking and in contrast to Ihde and Verbeek, who did not only reject but also never considered engaging with this approach at all, Smith does *not* think the transcendental approach is incompatible with an empirical turn and does *not* think that it leads to a determinist, "autonomous" view of technology.

> Crucially, there is nothing here that commits us to reifying conditions for experience as if they were "autonomous," otherworldly, or unrelated to the empirical. To be sure, post-empirical turn wariness of the transcendental is justified insofar as certain philosophies do indulge a tendency to make conditions appear this way. . . . However, this wariness goes too far when it sanctions ignorance of the transcendental as

a form of argument: what it forfeits, in this case, is the potential for an approach to philosophy of technology that would be dynamically focused on the relation between the empirical and the transcendental.

(Smith 2015, 544)

And indeed, as I have argued in my book *Growing Moral Relations* (2012), *a transcendental approach does not necessarily exclude the empirical and the material*. On the contrary, I have argued, next to language, *technologies themselves* can be seen as constitutive of, or as being one of, the conditions of possibility of speaking about something—in that book: the moral standing of nonhumans. Now, this approach could also be applied to philosophy of technology: Our speaking about *technology* is made possible by conditions that are linguistic *and* material in nature. Both language and technology frame our way of speaking about technology. For instance, when we speak and think about robots, this speaking and thinking is made possible and shaped by a language in which the concept of "machines" plays a crucial role, and, at the same time, our use and engagement with actual technologies also shape our thinking. Earlier in this book, I already gave the example of computers, which frame our language to speak about humans. But, if robots were to become a more important part of everyday experience, they would also influence our thinking. Perhaps our thinking about ourselves would then also change: We might consider ourselves a kind of machine. To some extent, this already happens, for instance, when scientists (e.g., in psychology or cognitive science) use machines as models to think about humans.

Moreover, the transcendental approach can also be applied to understanding (use of) language and technology in general. In the previous pages I already could not resist inserting "made possible" and similar transcendental language in my arguments. Let me now clarify and begin to (further) develop this part of my view and framework by responding to Wittgenstein and Smith. Whereas Smith sees Wittgenstein as merely 'pointing' (544) in a postphenomenological direction, we can straightforwardly interpret Wittgenstein in a transcendental way (see also the next chapter). Wittgenstein clearly frames his project in the *Investigations* in terms of a transcendental approach. He writes:

> our investigation is directed not towards phenomena, but rather, as one might say, towards the 'possibilities' of phenomena. . . . Our inquiry is therefore a grammatical one.
>
> (Wittgenstein 1953/2009, §90, 47e)

This is not a mere "pointing;" it provides clear guidance for how we must read Wittgenstein. Taking this into account, we can rephrase his view of language as follows: The use of language is made possible by conditions that include what Wittgenstein calls "language games" and "forms of life."

The latter are conditions of possibility for, and therefore shape, our speaking and thinking. Taking a next step, we can then interpret these forms of life as including not only linguistic elements (words) but also technologies (artefacts, things, machines), which *also* condition our speaking and thinking. In Heideggerian terms: We should not remain at the "ontic" level, as Ihde and Verbeek tend to do when they (nearly) talk about "agents" constituting one another (see also Smith 2015, 538) and, hence, nearly present a *causal* kind of narrative about object and subject. Instead, we should *also* look at the ontological level, add "verticality" to the picture, so to speak: what makes possible specific human-technology relations, and what is it that makes possible this (speaking about) subject and object constituting one another? Asking this question will be important when I will further develop my framework in the next chapters—also, again in response to Ihde and Verbeek's postphenomenological approach—and adds an important dimension to the Wittgensteinian-Heideggerian approach, which should now also be explicitly taken to include or even constitute a transcendental argument.

It is important to stress that this "transcendental" turn does not render my proposed approach less empirical or material, let alone exclude it. As Smith rightly argues, a transcendental approach does not necessarily exclude an empirical orientation at all. He even proposes a 'transcendental empiricism' (546), which, in his case, is inspired by Deleuze. But one does not need Deleuze to recognize this point. It also can be shown *in* the field of philosophy of technology. Consider, for instance, my arguments about human-robot relations, in which I have relied on a transcendental approach, but at the same time kept an empirical focus, in the sense of relating it to concrete experience of and with technology—in this case, the phenomenology of interacting with robots. For example, in an article on the linguistic construction of artificial others (Coeckelbergh 2011), I have argued that robot ethics needs a description of how the robot and the human-robot relation appear to human consciousness, but that this phenomenology, this description of appearance-in-relation, is not enough: 'we need a more precise account of how this phenomenological process unfolds. How is the "social" appearance created? *What are the conditions under which a robot appears to us as a quasi-other?* How do we make sense of such human-robot relations?' (Coeckelbergh 2011, 62; my current emphasis). Now, asking under what conditions a robot appears to us as a quasi-other is a straightforward *transcendental* question. This kind of argument clearly does not exclude the empirical; on the contrary, answering it requires us to engage with the empirical, or better, with concrete experience and with the use of artefacts (robots) and of language (about robots). Looking at concrete experience and use is not new, but the novelty of the argument is that it links this experience with linguistic and technological conditions of possibility, which shape and structure that experience and use. For instance, one could explore if dualism about humans versus things is partly built-in in our language, in its words and grammar, and how this structures and shapes concrete experience, performances, and interactions with "machines," "robots," etc. A transcendental approach enables us to open up such routes for research. We could also point out that the development and use of particular material artefacts, such as computers, gave rise to a "technological grammar" that also conditions our thinking, for instance, our thinking about humans (e.g., the human as machine, or the negative of the machine—I have called the latter "negative anthropology"). To develop these points requires not less but *more* empirical work, including the study of the relevant (use of) artefacts and language.

Finally, as Smith suggests, a transcendental approach can also give us a chance to connect to a historical approach, ask questions concerning power. For instance, Foucault 'investigates the historical conditions for the language-games of power' (545). I suggest that, in thinking about technology, more effort could go toward investigating the *historical conditions* for the technology games of power. For example, how did our current discourse about humans and machines come about? What may be the power dimension of current language and technology games about the machine age, and what is its history? How do traditional power structures and historical discourses shape current discussions about sex robots, for instance?

To conclude, we can use the transcendental argument to say that not only language but also *technologies* are not mere tools we use, but are also conditions of possibility for our thinking and speaking, that is, make possible and condition our use of language (and, hence, thinking) and technologies. Earlier in this chapter, I already suggested that we can interpret and revise Lyotard in this direction; it turns out that this route also provides a way one can read Wittgenstein's view of language in general, and that the argument can be used in philosophy of technology. Furthermore, it became clear that using this argument does not necessarily exclude materiality. In addition, the transcendental argument applied to *language* can also be useful to *philosophy* of technology. As I already argued, it seems that the instrumental view of technology is made possible, encouraged, and shaped by our language, that is, by our daily, ordinary language (use), which makes it appear, through its words and grammar, that technologies are mere tools. The latter view can also be read as a "transcendental" interpretation of Wittgenstein, who alerted us to the language games and form of life that condition our speaking and thinking, also in philosophy. The instrumental view of technology and other views in philosophy of technology can hence be understood as themselves embedded in language games and a form of life. This explains why it is so difficult to move beyond the instrumental understanding of technology, why doing so may require even to invent a new philosophical language (and, indeed, this is how one may interpret much original work in philosophy of technology), but also why this is difficult and always partly fails, because our ordinary language "pushes" us towards the instrumental view. This is so, since our use of words as philosophers is part of our language, understood not only as words and sentences we use but also as a transcendental condition. The instrumental view is part of the ontological structure of our language, part of the current ontological constellation; it is a given. Perhaps it is part of our form of life. Or, as we may suggest in a Heideggerian fashion, it is our fate. In Heidegger, these two more general applications are even connected, since this problem does not only concern language but also, at the same time, technology: For Heidegger it is "technological thinking"—with modern technology in the role of a condition of possibility—that keeps us locked into an instrumental thinking about technology and, indeed, an instrumental thinking about all kinds of things and humans. Finally, one could give a historical twist to the transcendental argument about technologies and language, and investigate the historical conditions under which a particular kind of thinking about technologies unfolded and unfolds, including linguistic and technological conditions. One could look at how both language and technology itself shape our current language games and technology games, which are also always *power* games, and which have a history. Our discourses and activities are made possible and shaped by conditions that are linguistic-historical, technology-historical, and bring with it particular power structures that themselves have also historically evolved.

Thus, if postphenomenology and other empirically oriented contemporary philosophy of technology would take a transcendental turn, it would become more critical, and, hence, become at least *less* vulnerable to Smith's unfair, but nevertheless relevant, objection that it is 'a shallow and uncritical field, parasitically dependent upon developments in industry' (547). I do not think this is true, but it is clearly a danger for the field, and a transcendental approach seems to be part of the toolbox we need to avoid it. Furthermore, investigating the transcendental and historical conditions that make possible our current speaking about technology and our current use of technology, is not "anti-empirical," but instead may help us to further "deepen" and develop (post)phenomenological approaches to technological use and experience. Using Smith's work and linking back to Wittgenstein and Heidegger has helped me to further articulate these points.

Difference and Encounters

Third, philosophy of technology can also benefit from another point made by Smith, which highlights another way (empirically oriented) contemporary philosophy of technology can learn from postmodernism and poststructuralism. Foucault, Derrida, and Deleuze give us not only a transcendental approach that, perhaps in contrast to Winch's (1958) reading of Wittgenstein, points not only to rules but also to 'the wider conditions of their emergence—whether historical, economic, social, or ontological' (Smith 2015, 545); these writers also attend us to what is known in this philosophical tradition as *difference*. Indeed, philosophers usually make generalizations about technological experience. But it can also examine things (literally) in a different way. Inspired by Deleuze, Smith writes:

> One such way of re-examining presuppositions is to view technologies, not as instances of a general type (for example, an ICT, a weapon, or a vehicle), but rather in terms of singular and unrepeatable events or 'encounters' within a complex situation (for example, the ICT through which crushing or exalting news is learned, in a given place, at a given time, or the weapon or vehicle which one is forced to rely upon in a life or death situation.
>
> (Smith 2015, 547)

If it is part of a philosopher's task to examine presuppositions, then this is indeed one of the presuppositions that needs to be questioned. If Smith is right, then philosophers of technology should not be content with making general claims about technology and not even with making general claims about certain types of technology, but at least *also* attend to singular, particular events and *encounters* with technology. Another way of saying this is that they should pay more attention to difference as opposed to sameness. Smith suggests that there are not only different ways of relating to the same artefact, as philosophers of technology have sufficiently shown, but also that in cases where we normally talk about one artefact, there may be *different artefacts*, related to different faculties (548–549). I will not go into this point in detail (e.g., criticize Smith's faculty psychology presupposition or follow up his references to Deleuze), but this claim raises the question if current philosophers of technology pay enough attention to singular and particular events and artefacts, and if they have sufficiently conceptualized the various ways artefacts may appear to us. Postphenomenology, for instance, has rightly questioned Heidegger's focus on "technology" rather than technologies, and it discusses "cases" of concrete technologies. But, in practice, it is questionable if in the discussions of cases, sufficient attention is paid to the phenomenology of unique experiences and encounters, as opposed to merely drawing general lessons about human-technology relations and mediation—as opposed to doing and emphasizing the *same*. For instance, in concrete experiences and encounters with robots, more might be happening than what can be captured by Ihde's general schemes of human-technology-world relations (in the next chapter, I will discuss this theory). And in a concrete encounter, is there one robot or different robots, depending on how we perceive and experience it? Does Ihde's concept of multistability capture all there is to say about the technological object, or is more work needed?

We can also look at these issues concerning difference, uniqueness, and concrete encounters in the context of discussions about alterity. I already mentioned the robot as quasi-other (see also Ihde 1990) and, elsewhere, I have written about technology and alterity, in particular, machine alterity:

Critically responding to Levinas, I have focused on the concrete encounter with machines (Coeckelbergh 2016). I have asked if machines can appear as others and have shown that if we attend to the phenomenology of the encounter with an entity, we see that there is no full control of the epistemology and semantics of what goes on in this encounter. Some meanings of the technology 'may visit us uncalled, break in' and, indeed, 'surprise us' (Coeckelbergh 2016, 182). I have also asked how such a concrete encounter with an other or quasi-other, how such experience of alterity (here, mainly understood as alienness), is possible in the first place, given that we tend to categorize, generalize, familiarize, and domesticate the other and, in this way, reduce its otherness, shifting from difference to sameness (186–189). I have proposed an epistemology with an opening up and a hospitality that does not try to identify beforehand whether or not what or who we encounter is human or nonhuman, living or machine (191), and I have argued that language mediates our encounter with machines (192). Thus, here the question with regard to alterity and technology is approached in a way that attends us once again to difference, and, more precisely, to the unique, the particular, and the singular dimensions of concrete encounters with technology, instead of the general, the universal, and the regular. Moreover, my very question about what makes possible the experience of alterity is a transcendental question; this has helped me to make a distinct contribution to the discussion, which differs from, say, an analysis in terms of Ihde-type human-technology-world relations and mediations (see the next chapter). And the point I made about the role of language can be reformulated in a transcendental way: Language makes possible, and shapes, how the other appears to us and how technology appears to us. The transcendental approach has thus helped me to open up the discussion about alterity and technology in a way that makes a strong connection between language and technology in an epistemology and phenomenology of the encounter with the (quasi-)other.

To conclude, a transcendental and *broad*, revised postphenomenological approach (as, for instance, proposed by Smith and as already explored in my own work) could enrich the field and render it more critical—also, critical of its own *words*. Moreover, approaches focused on difference and encounters also tell us something about the social. They show that the social may not only be about rules, as the social-cultural interpretation of Wittgenstein suggests, and certainly if we follow Winch and others. Perhaps the social is at least *also* about encounters, about the singular and the unique, as opposed to only rules and the repetitive. Wittgenstein writes about use of language, not technology. But we could again apply this response to Wittgenstein's view of language to *technology*. If I say that there are "technology games," then this term may be understood as referring to games in which there is a lot of epistemic clarity with regard to social relations and relations with technology, and which can be fully described by describing the relevant rules. But this is not always the case, or perhaps there is always a dimension that is left out; as I just argued, there are also singular, unique encounters with technology, in which difference and alterity may emerge (or not). Thus, the concept of "games" does not describe and conceptualize all that is going on; it seems that we also need other concepts—if conceptual work is sufficient at all, since some things may only be shown. Or, one could conclude that concrete encounters with technology are always embedded in larger technology games, but that *in* these games there is a lot of room for epistemic confusion and surprises, for what can only be shown and not said (to paraphrase Wittgenstein's famous claim from the *Tractatus*), and, indeed, for unique encounters—also with machines. Finally, in the next chapters, I will further stress a process approach: In the encounter, both subject and object do not preexist, but rather emerge and are constituted. In that sense, there are "unique" objects (artefacts), as well as "unique" human subjects; their identity and difference emerge in encounters, understood as a process that is made possible by larger games, structures, and forms of life, but which we can neither fully control nor fully know—let alone fully describe by means of rules.

5.5. Conclusion

In the next chapters, we will go beyond poststructuralist and postmodernist views of the relation between language and technology, which generally place so much emphasis on signs and symbols that it becomes difficult to find an appropriate place for the materiality of technology and technological artefacts. If, as philosophers of technology, we read Wittgenstein and Heidegger in a very postmodern way that is entirely sign- and language-centered, we can no longer benefit from the material turn, and this is not desirable if we want a philosophy of technology that does justice to the material and artefactual dimension of technology.

Nevertheless, we should not throw away the postmodern child with the bathwater. I have shown in this chapter that Derrida, Lyotard, and Baudrillard offer some points of connection for thinking about technology, that contemporary postmodernist authors also provide some helpful approaches and insights, and, hence, that it is possible to arrive at a more nuanced view on what postmodernism and poststructuralism can contribute to philosophy of technology. First, it is striking that Derrida, like Wittgenstein, relies on technological (and artistic) practice for one of his key metaphors in his seminal text, *bricolage*. Second, Lyotard's use of Wittgenstein's concept of language games and his agonistic interpretation of these games may be used to further discuss and understand (thinking about) technology, and related games and forms of life, in a way that accounts for agonistic relations next to consensus. But, more importantly, we can use what Lyotard says about knowledge: that *narrative* defines the criteria of competence in a culture, and that there is there is a relation between (cybernetic) *technologies* and the form knowledge takes on in our society. The latter claim can be linked to Heideggerian thinking about technology, which also connected technology to epistemology. It may also be compatible with Foucault's work on knowledge and power, which also has a material dimension—even if that dimension in Foucault needs more attention (see Chapter 7). Third, Baudrillard's universe is also one of signs and symbols, signs that are no longer anchored in an external reality. Technology only appears as medium or perhaps mysterious object, if it appears at all. Yet, I have suggested that we can criticize and, if desirable, revise this view by acknowledging that, for hyperreality and its simulacra to occur, material technologies and infrastructures need to be in place. Fourth, like Heidegger and Wittgenstein, poststructuralist and postmodern thinkers attend us to the limits to linguistic agency. There is already a "given" symbolic order and, while we can contribute, edit, write, etc., as (linguistic) subjects we are also shaped by the larger whole. This is also a discussion that deserves a (more prominent) place in philosophy technology. Of course, we want to reject technological determinism. Similarly, we should reject a *linguistic* determinism. But this negative moment is not sufficient. Once we acknowledge that we are shaped by language, technology, and society, we need a more precise account of how exactly we are shaped by the larger whole, by the structures and the rules, by the grammar of language, and by the "grammar" of technology and of the social. What are the limits of technological agency? How and how much does technology shape us, even if and *as* we develop and use technology? Finally, whereas in Lyotard there is a utopian moment when it comes to new information technologies, most thinkers are pessimistic. Baudrillard is rather dystopian: The symbolic order he describes, the world of the hyperreal, is doomed. Simulation is fatal. And in all three thinkers, resistance is futile—to use a famous Star Trek quote. Virilio shares this dark view of technology, although at least he contributes to thinking about technology in a more direct way and raises interesting issues concerning speed and the role of war (technology) in technological development. We do not need to share this pessimism. But, beyond the optimism/pessimism discussion, it is important to better understand the relations between humans, technology, and language.

For this purpose, I have proposed adding a transcendental approach to the philosophy of technology toolbox—even and also to "empirical turn" approaches, and especially to those approaches that claim to be a kind of phenomenology. I have argued that we can interpret Wittgenstein *and* the interpretation of Wittgenstein I developed so far as making a transcendental argument understood in a broad way: Games and forms of life—which are not only "cultural" but also "material" and technological—make possible certain ways of speaking; that is, games and forms of life make possible and shape particular uses of language and, indeed, particular uses of technology. Our experience, use, and performance with technology is not only "mediated" by technology, as postphenomenology claims (see the next chapter); it is also *made possible* by technology and language as transcendental conditions. Hence, when doing phenomenology of technology, we must not restrict ourselves to descriptions of mediation relations understood at an "ontic" level (to use a Heideggerian distinction; see also Zwier et al. 2016), as if technology is only an "in between," mediating between humans and their world, but also interpret the mutual constitution of subject and object through technology in an "ontological" and transcendental way. According to this transcendental interpretation, both language and technology function not only as mediators but also as "underlying" conditions of possibilities, which make possible and shape both the subject-object relation (and the subject and object) *and the speaking about this relation*, for example, by philosophers of technology. Hence, the way contemporary postphenomenology describes the relation between subject and object is itself also subject to language and other conditions of possibility; gaining a critical relation to postphenomenology, or to any other approach in philosophy of technology for that matter, must include a study of its language. Both language and technology can thus be understood at the level of their use (here, we remain at a horizontal level, or rather, there is only one level; here, the methodological world is flat, so to speak) and at the level of conditions of possibility (vertical approach; a transcendental dimension is added). In the next chapter, I will return to, and continue, this transcendental argument as a response to Ihde's postphenomenology.

Furthermore, it is also necessary to see this transcendental conditioning in a dynamic, rather than only a static, way; it is necessary to add the time dimension to thinking about the relation between language and technology, and between subject and object. The mutual constitution of subject and object—now understood in transcendental terms—is not so much a state, but a process. Attending to conditions of possibility of use can also make us aware of the *historical*-cultural embeddedness of our use of language and technology, including different forms of knowledge and modes of power, in which our uses are always inscribed, and that also prescribe certain uses. Moreover, any theory about language and technology will necessarily *generalize* and *abstract* from concrete encounters with technology, in which alterity and difference may appear. It is important to attend to difference, uniqueness, and concreteness, and to reflect on the limits of what can be said about concrete uses, performances, and encounters.

However, more *can* be said about the games and a form of life that make possible and shape these uses, performances, appearances, and encounters. If language and technology play a role as conditions of possibility that make possible, shape, and constrain (but not determine) the use of, and encounter with, technology, then these games, forms of life, and conditions of possibility can be described on a general level by theory and philosophy, with a transcendental argument. But, to arrive at helpful and relevant views with regard to specific technology/language uses and performances, one also needs to study the "empirical" aspects of language and technology and their transcendental role. Indeed, in response to the empirical turn, it is important to stress again that this approach does not neglect the material dimension of technology. On the contrary, it is in their material form that technologies can play the role of condition of possibility. It is our very use of material arte-facts, and the games, cultures, and forms of life that emerge from that use, that constitute the conditions. Thus, this is not a proposal for a "cultural turn" or "linguistic turn," as opposed to a "material" or "technological" turn. There is no "culture" as opposed to, and separated from, "technol-ogy." Instead, both go together in use and in the ways these uses are shaped by, and themselves constitute and contribute to, larger linguistic-material wholes, such as games and forms of life. Studying this does not go against the empirical turn, but requires one. Studying games and forms of life may thus need to include for instance collaboration with the social sciences and with those who develop technologies. For example, developing this material in an empirical, but at the same time social, historical, holistic, and perhaps even transcendental and "difference-sensitive," approach may benefit from further critical engagement with STS, which can help to make underlying materialities visible and is also sensitive to the limits of individual agency in relation to the *social* whole. Of course, there are limitations. For instance, like postphenomenology, not all work and approaches in STS do sufficient justice to the *linguistic* aspects of the social and many authors in this field may reject a transcendental approach.

Finally, if we become aware of the transcendental role of language and technology, we should also recognize that, as philosophers of technology, our own thinking about technology is conditioned by the language and dis-course we use and by the technologies we use. For instance, we should be crit-ical of only using generalizations and remain aware that something always remains excluded when we speak about technology—also, in accounts of technology and the social (or better: technology in the social). Taking into account thinking about the limits of language and heeding Wittgenstein's and Heidegger's anti-theoretical stances, we should also explore ways of connecting with what is outside of theory, philosophy, and language. Let me say more about this and introduce the next part of the book.

Concluding This Part of the Book and Introducing the Next Part

In this part of the book, I have sketched two extreme views regarding the role of language vis-à-vis technology: One that tends to neglect language, and one that overemphasizes it. In the next part, I will again attempt to avoid these extremes, but, at the same time, still learn from them. This time, I will engage with "extreme," ideal-type views that let *technology* "speak." The view I would like to arrive at needs to take seriously the material dimen-sion of technology, as postphenomenology and STS do, but also aims to take on board the insights from philosophers of *language*—in particular, Wittgenstein, Heidegger, Searle, Ricoeur, and some postmodern thinkers. This includes not only learning from their messages (what they say, the con-tent of their performance) but also from their approaches (how they say it and how they move towards what they say).

I already started to work towards this integrated view in this chapter and in the previous chapters, but now a further effort is necessary to further develop it. In Chapter 6, I will discuss and use Ihde's postphenomenology and, again, Latour—this time, the Latour of *We Have Never Been Modern*—in order to try to move closer to a view that takes on board insights from thinking about *language* in thinking about *technology*. I will interpret Ihde and Latour as already having tried to reach a synthesis, or integrated view, but I will offer my own revisions, which I believe are more successful. Then, in Chapter 7, I will discuss Ricoeur's theory of narrativity for the same purpose.

The ideal-type view that will emerge from my constructions and discussions of these views, which I will use to develop my own approach, is the view that "technology speaks." With regard to this "speaking," I will also consider one particular kind of performative use of technology: Technology as doing prescription, giving order, and being moral; that is, technology that tells us what to do. Then, I will develop a more integrated view, taking on board the previous chapters and especially reconnecting with Heidegger and Wittgenstein. I will avoid a fatalist or deterministic view, but at the same time learn from Heideggerian, Wittgensteinian, and postmodern insights that it is difficult to change (some) orders—especially those orders that play the role of providing conditions of possibility for our dealing with, and our speaking about and with, technology.

I argue in this chapter that this critical, transcendental approach can and should also be applied to philosophy of technology. I argue for practicing philosophy of technology in a more critical way, a way that is critical in the sense that it includes not only making "critical" arguments and objections but also examining the *language* used by philosophers of technology (and, of course, by all users of technology) and studying other conditions of possibility that shape the field. Our very *thinking* about technology is made possible, structured, and constrained by the language and other technologies we use, such as a particular language (e.g., English), a particular discourse (e.g., about robots), and particular technologies (e.g., word processors, screens, smartphones, the Internet; these are at least some of the "hidden" technologies that form the context of this book). Perhaps the structures and processes of technology development and related funding structures also constitute such conditions and shape what we write and how we write. And philosophy of technology should also acknowledge that not everything can be captured by its discourse, that there will always be a dimension of technology and media use and experience that is left out, excluded, unwritten. There are always dimensions of concrete experiences and encounters that cannot be captured in words. These are insights based on a transcendental approach that is often neglected by thinkers of the empirical turn, and that can be learned from Heidegger, Wittgenstein, McLuhan, and, indeed, from the poststructuralist and postmodern tradition—regardless of the latter's shortcomings with regard to thinking about technology.

Of course, this conclusion also applies to my own writing and to the text of this book. As I am further developing my view in the next chapters, I shall try to keep in mind that it is also text and discourse, which is in its turn made possible and shaped by linguistic-cultural and material-technological conditions. Thus, the very project of this book should take a critical stance towards its own project as a language-mediated philosophical project, acknowledging that thinking about technology and language in the form of a writing practice and an academic profession has its own constraints and limitations, which have to do with the language and media/technologies it uses.

One way to take this insight into account is to turn our attention to other discourses and other practices, for instance, in the arts. For example, installation art, theatre, performance, and dance may offer practices and spaces in which things concerning technology can be shown, can appear, can show themselves, can happen, etc. Without that this revealing is and can always be entirely captured by use of language—oral speech, but also philosophical or academic text. And if it includes text and discourse at all, then art can remind us of the limitations of formalization (to put in the form of a rule, an algorithm, etc.) and generalization, which may blind us to difference, alterity, singularity, and unique encounters. At the end of the book, I will offer more reflections and meditations in this direction. Here, I wish to conclude with the suggestion that combining postphenomenology and posthermeneutics could help us attend to the conditions that make possible our speaking about technology and to the limits of that speaking: to the shadows of language and to the unique and singular in uses and performances with technology.

References

Akrich, Madeleine and Bruno Latour. 1992. "A Summary of a Convenient Vocabulary for the Semiotics of Human and Nonhuman Assemblies." In *Shaping Technology/Building Society: Studies in Sociotechnical Change*, edited by Wiebe E. Bijker, and John Law, 259–64. Cambridge, MA: MIT Press.

Baudrillard, Jean. 1988 (1983). "Fatal Strategies." In *Selected Writings*, edited by Jean Baudrillard and Mark Poster, 185–206. Stanford: Stanford University Press.

Baudrillard, Jean. 1994 (1981). *Simulacra and Simulation*. Translated by Sheila Faria Glaser. Ann Arbor: University of Michigan Press.

Burbules, Nicholas C. 2000. "Lyotard on Wittgenstein: The Differend, Language Games, and Education." In *Just Education*, edited by Paul Standish and Pradeep Dhillon, 36–53. London/New York: Routledge.

Coeckelbergh, Mark. 2011. "You, Robot: On the Linguistic Construction of Artificial Others." *AI & Society* 26: 61–9. doi:10.1007/s00146-010-0289-z.

Coeckelbergh, Mark. 2012. *Growing Moral Relations: Critique of Moral Status Ascription*. Basingstoke/New York: Palgrave Macmillan.

Coeckelbergh, Mark. 2015. "Language and Technology: Maps, Bridges, and Pathways." *AI & Society* (online first). doi:10.1007/s00146-015-0604-9.

Coeckelbergh, Mark. 2016. "Alterity ex Machina: The Encounter with Technology as an Epistemological-Ethical Drama." In *The Changing Face of Alterity: Communication, Technology and Other Subjects*, edited by David J. Gunkel, Ciro Marcondes Filho, and Dieter Mersch, 181–96. London/New York: Rowman & Littlefield.

Coeckelbergh, Mark. 2017. *New Romantic Cyborgs: Romanticism, Information Technology, and the End of the Machine*. Cambridge, MA/London: MIT Press.

Deely, John. 2001. *Four Ages of Understanding: The First Postmodern Survey of Philosophy from Ancient Times to the Turn of the Twenty-First Century*. Toronto: University of Toronto Press.

Derrida, Jacques. 1997 (1967). *Of Grammatology*, Corrected edition. Translated by Gayatri Chakravorty Spivak. Baltimore/London: Johns Hopkins University Press.

Derrida, Jacques. 2001 (1970). "Structure, Sign, and Play in the Discourse of the Human Sciences." In *Writing and Difference*, edited by Jacques Derrida, translated by Alan Bass, 351–70. London/New York: Routledge.

Foucault, Michel. 1977. *Discipline and Punish*. London: Tavistock.

Hammer, Rhonda and Douglas Kellner, eds. 2009. *Media/Cultural Studies: Critical Approaches*. New York: Peter Lang.Heidegger, Martin. 1977. "The Question Concerning Technology." In *The Question Concerning Technology and Other Essays*, edited by Martin Heidegger, translated by William Lovitt, 3–35. New York: Harper & Row.

Heidegger, Martin. 1998 (1967). *Pathmarks*. Edited by William McNeill. Cambridge, MA: Cambridge University Press.

Ihde, Don. 1990. *Technology and the Lifeworld: From Garden to Earth*. Bloomington: Indiana University Press.

Kellner, Douglas. 1989. "Resurrecting McLuhan? Jean Baudrillard and the Academy of Postmodernism." In *Communication for and Against Democracy*, edited by Marc Raboy and Peter A. Bruck, 131–46. Montreal/New York: Black Rose Books.

Kellner, Douglas. 2010. "Reflections on Modernity and Postmodernity in McLuhan and Baudrillard." In *Transforming McLuhan: Cultural, Critical, and Postmodern Perspectives*, edited by Paul Grosswiler, 179–202. New York: Peter Lang.

Lévi-Strauss, Claude. 1963. *Structural Anthropology*. New York: Basic Books.

Lyotard, Jean-François. 1984 (1979). *The Postmodern Condition: A Report on Knowledge*. Manchester: Manchester University Press.

Mersch, Dieter. 2010. *Posthermeneutik*. Berlin: Akademie Verlag.

Smith, Dominic. 2015. "Rewriting the Constitution: A Critique of 'Postphenomenology'." *Philosophy & Technology* 28(4): 533–51. doi:10.1007/s13347-014-0175-6.

Virilio, Paul. 2006 (1977). *Speed and Politics*. Translated by Marc Polizzotti. Los Angeles, CA: Semiotext(e).

Winch, Peter. 2003 (1958). *The Idea of a Social Science and Its Relation to Philosophy*, 2nd edition. Abingdon: Taylor & Francis e-Library.

Zwier, Jochem, Vincent Blok and Pieter Lemmens. 2016. "Phenomenology and the Empirical Turn: A Phenomenological Analysis of Postphenomenology." *Philosophy & Technology* 29(4): 313–33. doi:10.1007/s13347-016-0221-7.

Part III

Technology Speaks (Objects Change Subjects)

6 What Technology Tells Us (to Do) (Part 1)
Media, Artefacts, Networks

6.1. Introduction: "Technology Speaks?"

What would it mean to say that *"technology* speaks," and not only humans or language? In Wittgenstein and Searle, but also in work by poststructuralist and postmodern writers, the emphasis was on language. Here the extreme position was that only language speaks. Creative work was needed to bring back the material object and give technology its phenomenological and hermeneutic due. But there is another, more direct route to the material object. In this chapter, I develop and critically discuss the opposite extreme position that only technology "speaks" by engaging with theories of human-technology relations and the social that have taken what one may call a "material turn" in thinking about technology, in particular Ihde's postphenomenology and Latour's nonmodern view. I also link to McLuhan's view and media ecology, and to work by Akrich and Latour, which concern technology, but (still) pay more attention to language. However, in general, these by now well-established directions in contemporary thinking about technology, especially postphenomenology, lack systematic attention to the role of language and its relation to technology.

This lacuna is understandable given the historical development of postphenomenology and (its use of) Latour's thinking. Whereas in the 1970s and 1980s, "humanities" philosophers of technology and the poststructuralist and postmodern philosophers focus on "culture," signs, and language, in the 1990s and in the first decade of this century, philosophers such as Don Ihde in the U.S. and later Peter-Paul Verbeek and other Dutch philosophers of the empirical turn, shift the emphasis to the materiality of technology. They think about technological artefacts, about things. In France, Bruno Latour develops a nonmodern approach to social studies of science and technology in which humans and things become actors and "actants" in networks. And whereas in Latour's early work semiotics still plays a role, in Ihde's 'postphenomenology' and 'expanded hermeneutics' (Ihde 1998), language retreats to the background. It is in phenomenology and hermeneutics of things in which language is mostly used as a metaphor (in particular, the metaphor of text). Heidegger is rejected for being too nostalgic and too pessimistic, and with him, attention to language largely disappears. This has cleared the road for thinking about technology in terms of what *things* do (Verbeek 2005). Artefacts as mediators and actants are understood to have phenomenological-hermeneutic and even normative consequences. Humans and their language leave the front stage. Latour's politics gives a voice to things (Latour 2004) and Ihde's and Verbeek's views come closer to a more "extreme" position in which language no longer plays any significant role in understanding technology: It often appears that in their view, technological artefacts play their hermeneutic role and "do things" without mediation by language. Heidegger's, Wittgenstein's, and other insights about language are neglected, let alone that poststructuralist and postmodern views of language are taken into account. In this sense, language is muted, and humans, too—at least to the extent that the hermeneutic and normative agency of things is stressed. In addition, the social does not receive much attention. Ihde's and Verbeek's postphenomenology of human-technology relations takes place mainly between an individual, desocialized human subject and the world. In Latour, by contrast, the social is in the center, but is radically rethought in hybrid human/nonhuman terms—in other words, in a less modern way. But here a phenomenological and hermeneutic analysis is missing.

In spite of these shortcomings, however, we can try to interpret, modify, and develop these influential postphenomenological and nonmodern currents in thinking about technology in a direction that *does* give a more important role to language. I already started this project in the previous chapter. In this chapter, I present a more sustained and systematic attempt to develop a synthesis between the language-oriented position and the artefact-centered position. Initially, this project will stay "close" to Ihde and Latour. If one emphasized the hermeneutic side of Ihde and the semiotic side of Latour, then perhaps one *could* read these works as an attempt to provide such a synthesis. Yet, I will argue that this only partly succeeds, since there remains a gap with regard to thinking about *language* and humans, including a gap in conceptualizing the role of language in mediating our relation to technology. Therefore, I will argue that we need a better synthesis that takes seriously both the materiality of technology *and* its relations to language (use) and language users. I make suggestions for how this can be done, starting from some key texts by Ihde and Latour, but also significantly departing from them. In response to Ihde, I not only propose a revised and expanded mediation theory (staying "close" to Ihde's work) but also give a Wittgensteinian and transcendental twist to it. And after discussing Latour's nonmodern view, I try to read more language into Latour and stress its performative dimension. In addition, I will explore what these new directions mean for our understanding of the social. This exercise will be continued in the next chapters, when I will use Ricoeur and continue to use Heidegger and Wittgenstein in order to develop my own, more integrative view. However, let me start with McLuhan, who *does* combine thinking about technology/media with thinking about language in interesting ways.

I will also briefly revisit Akrich and Latour's work, which can be interpreted as another way of saying that "technology speaks."

6.2. What Technology Tells Us (To Do): McLuhan, Media Ecology, and Grammatical Inquiry

McLuhan: Language Is Mediated by Technology and Is Itself a Technology and Medium

A first way to conceptualize the idea that "technology speaks" is to interpret McLuhan's phrase that the medium is the message (McLuhan 1964). Technology as medium changes our perception of the world and how we see ourselves. Yet, in contrast to the thinkers of the empirical turn, McLuhan still pays attention to language. For McLuhan, (1) *language is mediated by writing technology* and (2) *language is technology*. He writes that the content of writing is speech (McLuhan 1964, 8); but speech is shaped by the medium, that is, it is shaped by writing as a technology—which then, in turn, is connected to other technologies, such as books, roads, 'electric' technologies and media, etc. He also discusses the effects of the phonetic alphabet as a technology (90–91) and contrasts literate with oral cultures. Thus, for McLuhan, language is not something abstract like "signs," but is connected to specific technologies and media, which then have their own effects on consciousness. In this sense, it is not only humans that speak; the medium or technology (e.g., writing, the alphabet, etc.) also "speak:" they have a wide-ranging effect on consciousness and culture. Furthermore, not only writing is a technology, as Ong (1982) also argued; even language itself is a technology, and thus, also has effects on consciousness and culture. Referring to Bergson, McLuhan claims that language is a technology that:

> has impaired and diminished the values of the collective unconscious. It is the extension of man in speech that enables the intellect to detach itself from the vastly wider reality. Without language, Bergson suggests, human intelligence would have remained totally involved in the objects of its attention. . . . Language extends and amplifies man but it also divides his faculties. His collective consciousness or intuitive awareness is diminished by his technical extension of consciousness that is speech.
> (McLuhan 1964, 86)

Thus, when we use language, we already use a technology, and this technology also "speaks" in the sense that it has effects on the way we relate to our environment and to others. McLuhan offers also a specific view of language, which follows a kind of "Fall and Salvation" narrative. First there is alienation, caused by speech acts. He seems to endorse Bergson's claim that 'speech acts . . . separate man from man, and mankind from the cosmic unconscious' (86). One could say that this alienation even gets worse when speech and language take a written form. For McLuhan, this "Tower of Babel" may come to an end, as he believes computers hold the promise to create a condition of 'universal understanding and unity' and, in the end, he thinks we might even 'by-pass languages in favour of a general cosmic consciousness,' speechless kind of consciousness, which leads to collective harmony and peace (87). Thus, in this view, technology speaks, until in the future we reach a condition in which neither humans nor technology speak. Then there is only a silent and harmonious medium/consciousness (assuming that medium and consciousness merge at that stage).

McLuhan can and has be interpreted in a technological-determinist way, which leaves little room for humans. It seems that technology has an autonomous development and that we have little or no freedom to change its course. Furthermore, when language is interpreted as a technology this *may* be interpreted as "muting language." But there are other, more interesting interpretations.

In media ecology, which is influenced by McLuhan, further attention is paid to the medium as an *environment*. In the so-called New York School (as opposed to the Toronto School), Postman focused on media environments and their structures and impact on people, which are understood as symbolic environments, like language. Again, the content is not important, but the medium is. We think that we use a technology or medium as an instrument, but actually the medium structures what we perceive and say/think, and specifies what we are permitted to do. For instance, the medium of writing fosters abstract thought (Postman 2000, 13). But what does this have to do with ecology? In 'The Humanism of Media Ecology' (2000), Postman claims that he uses the biological metaphor "ecology" in at least two ways:

> A medium is a technology within which a culture grows; that is to say, it gives form to a culture's politics, social organization, and habitual ways of thinking.
>
> (Postman 2000, 10)

And:

> We put the word "media" in the front of the word "ecology" to suggest that we were not simply interested in media, but in the ways in which the interaction between media and human beings give a culture its character and, one might say, help a culture to maintain symbolic balance.
>
> (Postman 2000, 11)

Thus, here, technology as medium also "speaks," in the sense that it is directly linked to culture and the symbolic. A culture is not something that is separate from technologies, but instead "grows" in it and is shaped by it. Postman argues that human beings live in a media environment, 'which consists of language, numbers, images, holograms, and all of the other symbols, techniques, and machinery that make us what we are' (11). According to this view, technology and language are not seen as separate, but are all part of one symbolic ecology. Here object and subject already come much closer. Both technology and language "speak."

Postman, however, remains (too) close to postmodernism in his emphasis on the symbolic. He also still makes a distinction between the natural environment and the cultural media environment; as we will see in Latour's nonmodern view, this split between nature and culture is questioned. Moreover, Postman is explicitly moralist (influenced by Rousseau, for instance) and is interested in moral growth (16), whereas McLuhan is not (11), or at least not primarily. Postman is also far more pessimistic than McLuhan about the possibilities of information technology. For Postman, the Internet turns us into information junkies and does not make it easier to see what information is significant and meaningful, let alone that it can solve our social problems. He writes: 'we are deluded into thinking that the serious social problems of our time would be solved if only we had more information, and still more information. But I hope I need not tell you that if children are starving in the world, and many are, it is not because we have insufficient information' (15). Postman's stress on the symbolic and his technopessimism is also clear in his book *Technopoly* (1992), which argues that "culture" surrenders to "technology." Thus, the symbolic and technological merge, in the sense that technology becomes dominant, but this remains a dualist way of thinking. By contrast, McLuhan, but also Latour and Ihde, tends to avoid speaking about technology as either good or bad, and instead enables us to analyze and understand our relation to technology. Many contemporary readers of these authors also move further towards posthumanism, rather than humanism. Postman, by contrast, is interested in the moral progress of humankind and, in contrast to, for instance, Verbeek's posthumanism, "moralizes technology"[1] in a dualist way, not recognizing that technology and the human, and technology and culture, have always been entangled. Hence, it is misleading to speak of a "surrender" of culture to technology, since that would imply that technology is external to human culture. Postman's narrative, a very familiar one in classic philosophy of technology, does not help us in forging a closer conceptual link between language/culture and technology. Thus, Postman's media ecology can be contrasted with posthumanisms that are less dualist, less moralist, and move beyond humanism. It is also still close to poststructuralism and postmodernism in its emphasis on the symbolic. Yet, in contrast to other humanists and in contrast to text-obsessed postmodernists, Postman is not interested in text as such, but more generally in media, which include texts, but also other things.

Indeed, it seems fair to say that, as Stephens has argued (Stephens 2014), posthumanists appreciate more fully 'the interrelations among humans, our cultures and technologies, and the more-than-human world' (Stephens 2014, abstract), since, for them, ecology and environment are not just used as a metaphor or an analogy—or, rather, one could say that they take seriously the metaphor. Stephens argues that McLuhan can and should be interpreted in a more genuinely ecological way than Postman did. For Stephens, 'the conceptual and the material cannot be neatly separated;' instead, there is a 'dense network of interaction between' the conceptual and the material (2031). According to Stephens, ecological thinking emphasizes relationships and process. It requires us to develop a non-anthropocentric perspective and a worldview that does justice to violence, dynamism, and conflict, and is not only focused on balance (2032). Based on McLuhan, he argues that new media do not bridge between humans and nature, but *are* nature. Today, 'the technological, biological, and social are merging in new ways' (2031). This sounds more like Latour's nonmodern ideas (and, of course, like other recent posthumanist thinking), to which we will turn later in this chapter.

Thus, both McLuhan and media ecology enable us to make sense of the thesis that technology "speaks." For media ecologists such as Postman, technology is not a mere instrument; instead, it is part of, and makes possible, a symbolic milieu. Like McLuhan, they believe that technology shapes culture to such an extent that, indeed, we can conclude that "technology speaks." But, in media ecology, technology is mainly interpreted in cultural terms. Postman mentions techniques and machinery, but the stress is more on the symbolic than on the material. The environment is linguistic, semantic, even informational. The biological and material are used as metaphor, but the metaphor is not really taken seriously. There are two worlds: the world of words and the world of nature. Perhaps posthumanist thinking can better account for the material aspects, while not necessarily downplaying the symbolic and linguistic dimension of human existence. Stephens is right when he suggests that a more posthumanist (and, indeed, substantive ecological or what we may call "strong ecological") interpretation of McLuhan can be given. Such an interpretation would enable us to stress both technology *and* language.

For the purpose of this book and to get closer to Stephen's interpretation and to posthumanism, then, we can read more *McLuhan* back into media ecology. This is possible if we interpret McLuhan in a way that lets both technologies (as media) and *language* speak. McLuhan understands language as both an environment and a technology—a very important one. In his view, language is hence deeply connected to technology (it even *is* technology), and even to biology. If there is an ecology, then, as media ecology claims from a McLuhanian point of view, this ecology is to be understood as one linguistic-biological-material milieu. This is also how I propose to read Stephens's point that language is 'part of who we are as collective and individual agents, bound together and constituted by culture and biology' and that it is always situated. For Stephens, there is just one, single ecology, a common ground, and in this ecology and these environments, transformations occur 'across linguistic-symbolic and material modes as processes that are both physical and metaphysical, mental and material' (Stephens 2014,

2034). There is not an "inner," "mental," and "subjective" realm of culture and language (words), as opposed to, or even isolated from, a physical, material, and "objective" world (things). Languages and organisms are part of the same whole (2035). There are not two worlds (or more), but one. Stephens rightly argues that Postman pays too little attention to 'the world of not words' (2039), and, indeed, holds a dualist view of two worlds in the first place, which should be overcome. By contrast, I am developing a less dualist and more integrated view, which has room for the speech of humans, language, and technologies/media—including their material aspects, their 'not words' dimension. In addition, Stephens invites us to acknowledge that 'nonhuman beings, neither more nor less than human beings, are part of an ontological field with metaphysical as well as physical dimensions' (2040). (His view also links McLuhan to postphenomenology and posthumanism, with its emphasis not on how media separate us from the world, but how they connect us to the world.)

McLuhan's Linguistic Turn: Laws of Media as a Grammatical Inquiry

However, if we consider other work by McLuhan, one may wonder if we really need Stephens's posthumanism or, indeed, media ecology to interpret McLuhan in a way that is less dualist and more ecological. In particular, if we take a closer look at his later, posthumously published, *Laws of Media* (1988), we find a more ecological and also an interesting "grammatical" approach, which further help us to conceptualize the relation between language and technology.

Indeed, my efforts to conceptualize a closer connection between language and technology can also be supported by McLuhan's own views. In his later work, McLuhan pays more attention to language and closely connects language and technology. Some even speak of McLuhan's 'linguistic turn' (Gordon 1997, 323), although McLuhan in fact *began* his academic path as a student of language, writing on grammar, logic, and rhetoric in his doctoral dissertation, for instance; in this sense, it is more appropriate to speak of a linguistic *return*. As Van Den Eede (2015) has helpfully pointed out, in the posthumously published book *Laws of Media*, Marshall McLuhan and his son Eric equate media/technology, metaphor, and language: All human artefacts are to be understood in a linguistic and metaphorical way. Now, we can read this text as offering not merely a repetition and development of earlier work by McLuhan, say, in *Understanding Media* (the medium is the message, new modes of perception, the alphabet and writing as abstraction of speech and other senses, the contemporary focus on the visual, and, more generally, the differences between writing and speech,[2] between oral culture and writing culture—see also, again, his student Ong), but as material that provides us with a further step in McLuhan's thinking by focusing on language, and that also supports further steps in this book to connect language and technology, in particular in transcendental and "grammatical" ways. In McLuhan's own words: The concern of the book is rhetoric and grammar; it aims 'to place the modern study of technology and artefacts on a humanistic and linguistic basis for the first time' (128). In a book that reveals McLuhan once again as a philosopher of technology (indeed he can and has been interpreted as such; see Van Den Eede 2012), we are asked to turn to 'the standard grammatical awareness of the correspondence of words and things' (221). Let me further unpack these claims about language and grammar by reading selected passages from the text; this will support the core thesis and framework of my book.

Already in the preface, the McLuhans say that they learn from the 'grammarians' Francis Bacon and Giambattista Vico, 'bringing the tools of literary training to bear on understanding the world and our part in it' (x). And this includes technological artefacts. After a quote from T.S. Eliot which clearly draws our attention to the critical study of language and introduces his comments on Bacon's 'Idols of the tribe' (see for instance, 88)—'our concern was speech, and speech impelled us/To purify the dialect of the tribe'—the authors summarize the main thoughts of their essay as follows: 'each of man's artefacts is in fact a kind of word, a metaphor that translates experience from one form into another' and all is language and all is artefacts, material and immaterial:

> It makes no difference whatever whether one considers as artefacts or as media things of a tangible "hardware" nature such as blows and clubs or forks and spoons, or tools and devices and engines, railways, spacecraft, radios, computers, and so on, or things of a "software" nature such as theories or laws of science, philosophical systems, remedies or even the diseases in medicine, forms or styles in painting or poetry or drama or music, and so on. All are equally artefacts, all equally human, all equally susceptible to analysis, all equally verbal in structure.
>
> (McLuhan and McLuhan 1988, 3)

Or to use a formulation later in the book:

> 'Our new dictionary includes all human artefacts as human speech, be they hardware or software, physical or mental or aesthetic entities, arts or sciences. Such former distinctions have no scientific relevance. As utterances, our artefacts are submissible to rhetorical (poetic) investigation; as words, they are susceptible to grammatical investigation.'
>
> (McLuhan and McLuhan 1988, 224)

The book then elaborates these claims in various ways, merging 'things and ideas,' 'physics and metaphysics' (3), and, recognizing 'the transforming power of language,' understanding all media 'and the whole cultural ground' as 'forms of language' (115), and understanding artefacts as 'speech' and 'metaphors:'

> all human artefacts are extensions of man, outerings or utterings of the human body or psyche, private or corporate. That is to say, they are speech, and they are translations of us, the users, from one form into another form: metaphors.
>
> (McLuhan and McLuhan 1988, 116)

Artefacts, thus, do not only extend us but also "translate" us. Technology 'becomes software, word' (224) and 'translates from one nature to another;" it is the user who is translated (118). Technologies are also 'metaphors of the body or its parts' (128):

> All human artefacts are human utterances, or outerings, and as such they are linguistic and rhetorical entities. At the same time, the etymology of all human technologies is to be found in the human body itself: they are, as it were, prosthetic devices, mutations, metaphors of the body or its parts.
>
> (McLuhan and McLuhan 1988, 128)

One point of departure, next to Bacon's Idol's, is the ancient Greek term *logos* and the ancient thought that *logos* and *cosmos* have the same structure (37–38): The book entertains the thought that all things are words, that everything is logos; at the same time, words are also artefacts. And both words and things are connected to deeds, to what we could call *performance*: 'words and deeds were related as were words and things,' since speech expresses what things are and words can be used to 'exercise rhetorical power of other men' (36), since when we consider 'the structure of experience of the utterer' the 'grammatical flips into rhetorical investigation' (116). The McLuhans explicitly refer to performance when they say that 'the structures of media dynamics are inseparable from performance' (116). They also return to the thesis in *Understanding Media* that technologies, media, and 'artefacts . . . are extensions of the physical human body or the mind' (93) and that they change us: 'Any new service environment, such as those created by the alphabet or railways or motor cars or telegraph or radio, deeply modifies the very nature and image of people who use it' (96). In particular, they are concerned that 'electric' technologies leave 'whole populations without personal or community values' (97) and give us a 'robot status' (see, for example, 98). (The artist's task is then to remedy this situation, by reconnecting biology and technology.) Another point of departure is Heidegger, whose text on technology (1977) is interpreted as an 'attempt to retrieve grammatical stress' (63) and reminds us of the 'the penetrative and configuring power of the new ground of Gutenberg technology' (65). McLuhan also responds to Aristotle's conception of metaphor.

It would take too much space to unpack and interpret this. For the purpose of this book, let us focus on the thesis that language and artefacts are the same. Whether or not one agrees with the McLuhans on this total equality and merging (one could summarize the position as: "everything is language, and language is everything") or on the specific effects they claim contemporary technologies have (destruction of human values), the similarity between *words and things* observed by the authors supports the synthetic project of this book to bring together thinking about things (technology) and thinking about words (language). Moreover, it is interesting that the McLuhans can only arrive at this position from a linguistic, in particular grammatical and rhetorical, point of view, and from considering the *use* and *performance* of words and logos (which is also a doing, there is the deed, what Austin called the speech act), which make it possible in the first place to compare what was previously incomparable: words and things. They discuss the structure and grammar of technologies. For instance, they write about the Stoic *logos spermatikos* as 'seeds embedded in things animate or inanimate that structure and inform them and provide the formal principles of their being and growth (becoming). This third logos is the root of grammar' (124). They say that their book is concerned with 'the structure of all human artefacts' (127). And when the McLuhans argue that the effects of technologies are not always visible, they refer to the linguist Ferdinand de Saussure (98), famous for his formalist and structuralist approach to language. McLuhans' 'laws' of media is a structural, grammatical, and, indeed, transcendental inquiry. Consider their first and most important law or question:

> What does the artefact enhance or intensify or make possible or accelerate? This can be asked concerning a waste basket, a painting, a steamroller, or a zipper, as well as about a proposition in Euclid or a law of physics. It can be asked about any word or phrase in any language.
> (McLuhan and McLuhan 1988, 98–99)

What matters in their analysis is not the content, but form, in particular change of form, history of form. Consider their third question:

> What recurrence or retrieval of earlier actions and services is brought into play simultaneously by the new form? What older, previously obsolesced ground is brought back and inheres in the new form?
> (McLuhan and McLuhan 1988, 99)

And:

> Any new technique or idea or tool, while enabling a new range of activities by the user, pushes aside the older ways of doing things.
> (McLuhan and McLuhan 1988, 99)

For instance, according to the McLuhans, the contraceptive pill has enhanced 'the programmable machine approach to the body' and has provided a 'base for promiscuity' (99). This can be interpreted as a claim about how new technology makes possible a change in the ways of doing and thinking, a change in the structure of relationships, and a change in a form of life. It is about what kind of life and existence the new technologies make possible. Technologies are put in context of a larger whole, a ground, a structure, a grammar. The method recommended by the authors is as follows, which, in a beautiful way, links the technological artefact to both language and the human body, and, in addition, attends us to the fact that a grammatical and transcendental inquiry does not necessarily mean that it is less *empirical*:

> The four aspects . . . require careful observation of the artefact in rela-
> tion to its ground, rather than consideration in the abstract. . . . In tet-
> rad form, the artefact is seen to be not neutral or passive, but an active
> logos or utterance of the human mind or body that transforms the user
> and his ground.
>
> (McLuhan and McLuhan 1988, 98)

I will return to these issues concerning body and transcendental argu-
ments later in this chapter and in the next chapter. My focus now is on the approach. It becomes clear that it is their "grammatical" question that enables the McLuhans to compare and equate technological artefact and languages, and to study transformations in meaning, perception, and exis-
tence made possible by the use of new technologies and media. [They even see their laws as 'an instrument for revealing and predicting the dynamics of situations and innovations' (105).] Their book is concerned with 'principal reversals of Western form wrought by electric information'—one of them being that 'commodities themselves assume more and more the character of information' (106). (Hence, it is no surprise, I may add, that one influential approach in philosophy of technology is philosophy of information.)

This grammatical-, use-, deed-, and performative-oriented starting point links McLuhan to Wittgenstein, whom I also propose to interpret in a transcendental way. It also links him to Ricoeur, in particular via his under-
standing of grammar as the 'interpretation' of texts and patterns (9) and his attention to the temporal (as the sequential) in his insistence that speech and writing have to be uttered (73). The book explicitly responds to Ricoeur's *The Rule of Metaphor* when it criticizes Ricoeur's conception of metaphor (121). And at the end of the book, the authors emphasize *flux* and—in congruence with Ricoeur's focus on hermeneutics—the need for continuous interpretation:

> As the information that constitutes the environment is perpetually in
> flux, so the need is not for fixed concepts but rather for the ancient skill
> of reading that book, for navigating through an ever uncharted and

uncharitable milieu. Else we will have no more control of this technology and environment than we have of the wind and the tides.

(McLuhan and McLuhan 1988, 239)

Thus, although McLuhan seems to suggest (in the beginning of the book and, perhaps, here at the end) that, in the end, there is no longer syntax, at least not in the sense of something fixed, his philosophical or scientific *starting* point is grammatical and, in the course of the book, he constantly discusses issues concerning (absence or presence of) *ground*. As the pill example and remark about the artefact in relation to its ground suggest, he also seems to share with Wittgenstein a holistic understanding of meaning. Consider also his remark, in line 'with the rhetoricians and grammarians,' that the formal cause of a work of art is 'the audience (the user) and the configuration of sensibilities in the culture at the time the artefact was produced' (89). In other words, when we speak or when we perceive an artefact, there is already a 'configuration,' there is already a grammar, which make possible the reception and perception, and, indeed, the *use*, of the artefact as meaningful. McLuhan shares also Heidegger's and Wittgenstein's insight that language is, in Cassirer's words quoted by McLuhan, 'a magic circle . . . from which there is no escape, save by stepping out of it into another'—except that, in contrast to Wittgenstein, McLuhan explicitly extends this idea to technologies, which are interpreted as words:

The ground that envelops the user of any new technological word completely massages and reshapes both user and culture. In this way too these words (extensions) have all the transforming power of the primal logos.

(McLuhan and McLuhan 1988, 226)

Thus, the McLuhans offer a very specific way of making sense of the phrase "technology speaks." Technologies are "words" that shape users and their (use) culture, by enhancing and making possible some things rather than other things. New technologies and media, then, have to be studied not only in terms of their use as such but also in terms of their grammar, which is first unknown: 'New media are new languages, their grammar and syntax yet unknown' (233). This seems to make it difficult to predict and study the unintended effects of new technologies; afterwards, their "grammar" may be clearer. Finally, the point about rhetoric and power attends us to the political dimension of language use—and, indeed, (*in deed*, as in the German: *in der Tat*) use of things.

Note that in McLuhan, this language-oriented inquiry does not imply that it is less material. "The object," to use the language of metaphysics, does not at all disappear. On the contrary, "the object" is very much present in this book, so present that Harman has proposed to extend what the McLuhans say about artefacts to all objects, in order to support his object-centered thinking (Harman 2009). Hence, McLuhan's linguistic turn is also at the same time a turn to the object (Van Den Eede 2017). In the next section, I will further discuss the relation between language and materiality by responding to Don Ihde's work. At the end of my book, I will also return to what McLuhan says about touch and art in the *Laws*. But let us first revisit Akrich and Latour, in order to say more about the performative function of things.

Revisiting Akrich and Latour's Semiotics of Things: Giving a Performative Function to Things

A second way to conceptualize the idea that "technology speaks," which pays more attention to the material than the postmoderns, but is—like McLuhan—guided by language as a metaphor and by the study of language, is to focus on the semiotic functions of the technological artefact. I already pointed out that Akrich and Latour's view (1992) is influenced by semiotics, and one could conclude from what we may call their "semiotics of things" that there is a sense in which artefacts are able to perform functions that are usually reserved for words and grammar. For instance, in their view, artefacts can "prescribe" if a particular prescription is delegated to them by humans. Another way to put this, which is closer to Akrich and Latour, is to say that *speech* is delegated from humans to artefacts. Now, the claim that artefacts can speak and prescribe can be understood literally (e.g., a computer or robot that speaks or a navigation device that prescribes what the human driver should do), but in Akrich and Latour it is mainly meant metaphorically. The human prescription is inscribed into the artefact, in the sense that the artefact takes over the performative function normally done by humans/language, which then has effects on the users of the artefact. Thus, in Akrich and Latour, language and technological artefacts are still very "close," since they have similar functions. As I have argued by using and expanding Wittgenstein and Austin, we use words *and things* to perform things—for instance, to prescribe. Both words and things, thus, have performative functions. Prefiguring Latour's later, nonmodern view, one could say that there is a theatre of humans and things, actors and actants, in which humans, technology, and language co-perform. (See also Chapter 8.)

That being said, Akrich and Latour's view has some shortcomings with regard to conceptually joining language and technology. Like in earlier humanisms and in (other) poststructuralisms and postmodernisms, there is a considerable emphasis on text and related technologies. Their metaphor is text and writing technology. The visual and the auditory, for instance, play a less important role; this is better in Ihde, who discusses visual and sound (e.g., Ihde 1998, 2007, and 2016). Furthermore, at first sight, it seems that in Akrich and Latour a more ecological perspective is missing; this is partly remedied in later work by Latour, to which we turn later in this chapter. But, apart from these problems, Latour and Akrich helpfully draw our attention to the *performativity* of *both* language and technological artefacts; this further strengthens the performative line of thinking I am developing. Moreover, they add not only the performative but also the *narrative* dimension of technology use. Acknowledging this performative and narrative dimension (see also the next chapter) can help to remedy the problem that, after the empirical turn in philosophy of technology, there was considerably less attention to language (and text). Simplified, one could say that, whereas in Postman and to some extent in Latour and Akrich's texts, there is too much emphasis on the symbolic and the linguistic, and, indeed, on writing and the textual, postphenomenology and other siblings of the material turn have overemphasized the material at the cost of neglecting language.

Let us first take a closer look at the postphenomenology of Don Ihde—not only to further construct the simplified or "extreme" view that only technology speaks but also to better understand his philosophical project, to learn from his attention to embodiment, to bring out the limitations of his emphasis on the material, and to show that core parts of his work (his scheme of human-technology-world relations and his related expanded hermeneutics) could be developed in a way that makes postphenomenology better account for the relations between humans, technology (also in its materiality), and *language*—understood as language use and as narrativity. I will read Wittgenstein into his thinking, highlight performance, and expand his scheme of human-technology-world relations to include language as mediator. Then I will propose a more substantial revision by using a transcendental argument and by shifting the focus more clearly to performance and the social.

6.3. What Technology Tells Us (to Do): Ihde's Postphenomenology and Expanded, Material Hermeneutics

Introduction

Don Ihde is one of the philosophers of technology who took an empirical, and, indeed, what we may call, a "material turn." Heidegger, Ellul, and other classic philosophers of technology were seen as 'retreating' into the linguistic terrain (Achterhuis 2001, 4–5), focusing too much on language and culture, and neglecting our relation to, and use of, concrete technologies and artefacts. In addition, they viewed Heidegger as too nostalgic and pessimistic (see, for instance, Verbeek 2005); they wanted a more constructive approach, which discusses the phenomenology of particular technologies. Less explicitly, they thereby also rejected and neglected the entire poststructuralist and postmodern tradition that followed in the wake of Heidegger. This meant, effectively, that language was no longer in the foreground of their analysis, if language appeared at all; text and discourse were banned to the background, or disappeared. The empirical turn was a turn from language to artefacts, from words to things.

However, interestingly, in Ihde's work, this does not mean that phenomenology and hermeneutics were abandoned; instead, these philosophical approaches are given a new direction. Instead of viewing experience and interpretation as mediated and shaped by *language*, Ihde (and, later, also Verbeek, Rosenberger, and others) proposed to study how *technological artefacts* mediate and shape our experience. Technology itself now played a role in phenomenology and hermeneutics, and not as a mere sign, symbol, or hermeneutically neutral instrument, but as a *medium* and world-making thing, as something that can play various hermeneutic and even normative roles in the relations we have to technology and to the world, as things that—to use Verbeek's words—*do* things (Verbeek 2005). (Note that, by contrast, relations with others, *social* relations, were not very much analyzed.) Thus, in the sense that Ihde's thinking is indeed an "expanded hermeneutics" (see below), it *could* be interpreted as an attempted synthesis between thinking about language and thinking about technology, albeit one that mainly borrowed the *method* from hermeneutics without borrowing its traditional object of study: It emphasized technology *at the expense of language*. Let me analyze Ihde's work in more detail in order to explain the novelty of this approach, to better understand the precise relation to "humanities," thinking in phenomenology and hermeneutics, which was focused on language, text, sign, symbol, etc., and to discuss its limitations with regard to thinking about language and technology. Then I will propose revisions to Ihde's framework.

Ihde's Material and Relational Phenomenology and Hermeneutics for Thinking about Technology

When he developed his approach, Ihde was influenced by Heidegger, Merleau-Ponty, Ricoeur, and other thinkers in the phenomenological and hermeneutical tradition. Hence, linguistic mediation was very familiar to him. Yet, he also turned away from that tradition in at least the following ways, which I will now articulate and discuss by drawing on two of Ihde's books: *Technology and the Lifeworld* (1990) and *Expanding Hermeneutics* (1998).

First, Ihde took the model of reading a text and applied it to reading an instrument, or, more generally: he took the hermeneutic model and applied it to science and *technology*. (In the next chapter, I will also make such a move by drawing directly on Ricoeur.) In Ihde's view, technology itself can have a hermeneutic role in the sense that it can be "read." For instance, we can literally read a thermometer, which enables us to interpret the world. But technology can also shape and mediate our experience and world in other ways (with "world" understood in a Heideggerian, phenomenological sense).

To fully understand this move, let us take a step back and look at Ihde's background and approach. In *Technology and the Lifeworld*, he starts from phenomenology, but considerably revises that approach. Ihde constructs his own form of phenomenology by turning away from the optimism/pessimism discourse about technology and from a romantic longing for a world without technology, from "Technology" with capital "T," which 'reifies' technologies (Ihde 1990, 26; Ihde thinks about Heidegger, Ellul, and others here), and from the transcendental perspective of phenomomenologists like Husserl and Heidegger, which he saw as too remote from materiality and the lifeworld (Ihde 2009, 22; see also De Preester 2010). Influenced by Merleau-Ponty, he stresses embodied perception and wants to understand the technological mediation of our experience of the world, which is, at the same time, very mundane (think of all the technologies we use on an everyday basis), but is also poorly understood. His starting point is an account of human-world relations (Ihde 1990, 23); in other words, he starts from *relationality* (a term which I think is more accurate than the term "relativistic," which Ihde also uses; note also the potential link with ecology here, also a relational view), in particular 'the relationality of the human experiencer to the field of experience' (25). He then outlines a number of relations, beginning with the simple relation I—relation—world (Ihde 1990, 23–24). Then Ihde adds a focus on technology, more precisely technology in its materiality, technology as artefact. He identifies and distinguishes between various human-technology-world relations, offering a 'phenomenology of human-technology relations' (Ihde 1990, 41) that has been very influential in establishing a 'postphenomenological' framework for thinking about technology. Let me summarize and simplify that scheme and these relations as follows:

- embodiment relations: (I—technology)—world
- hermeneutic relations: I—(technology—world)
- alterity relations: I—technology— (world)
- background relations

We have an **embodiment** relation to technology when we use the technology, but are not aware of technology as an object; in use, the technology becomes transparent, invisible. (Note that Ihde's use of the term "use" here is not accidental; I will return to this soon.) Ihde is inspired here by Heidegger's concept of ready-to-hand in *Being and Time*: There is a 'withdrawal of technology from within direct experience' (Ihde 1990, 32), the equipment withdraws in use (33). Consider the use of glasses: We do not see the glasses we are wearing; we look through them, they mediate our experience without us being aware of them. The wearer of the glasses embodies the technology: 'I—glasses—world' (73). Similarly, when I drive a car, I experience the road and the surroundings through driving the car (74). Ihde also gives the example of a window, which withdraws from vision: It is not usually the object of my vision (47). Earlier, Ihde also notes, following Merleau-Ponty's examples about the woman with a feather in her hat and the blind man's stick, that the lived, experienced body can be extended through artefacts (39). In all these cases, Ihde notes, the design of the artefact must fit the use (note, again, the term "use"); the technology adapts to the perception and actions of humans (74). (We may add that the opposite is also the case: We adapt our perceptions and actions to technology, to the car, for instance.) Ihde also argues that this transparency of technology is exactly want we want as a user: As a user, I wish to have the technology and its transformational effects, but I 'want it in such a way that I am basically unaware of its presence' (75). In other words, we want such technologies to be transparent instruments that have the intended effects. Or, we could add: We want to use them, but we do not want to be aware of us using them.

In a **hermeneutic** relation, on the other hand, the technology is visible and is interpreted ("read"). Think of a thermometer or a clock: What warmth or time *is*, is read through the instrument. This is a hermeneutic perception (63). But the instrument does not merely represent time; it changes our very perception and concept of time, already contributes to a particular interpretation of time. Here, Ihde already makes a reference to hermeneutics, which he further develops in his book *Expanded Hermeneutics* (see below). He focuses on *reading* (Ihde 1990, 80). Like text, technology may become the object of reading and, more generally, of perception; yet, at the same time, they also refer beyond themselves (82). What 'presents itself is the "world" of the text' (84). The world is made present. Similarly, Ihde argues, technologies can make present a world. The technology is present-at-hand, but not necessarily because it breaks down as in Heidegger, but rather it is present in a way that presents a world. It is also like text, since it interprets and requires interpretation, although, at this point, Ihde does not say much about interpretation, at least not in these terms. The emphasis is on the "reading." 'In a hermeneutic relation, the world is first transformed into text, which, in turn, is read' (92). Thus, here, Ihde maintains a close analogy with text and language. We could say that we have the experience of "reading" rather than "using."

In an **alterity** relation, by contrast, we have a relation *to the technology* (rather than to the technology-world), which appears as a 'quasi-other' (Ihde 1990, 98–100). Here, technology is no longer the means by which something else is made present, but rather it becomes this "something else" (an object that is present-at-hand and perhaps confronts us, in German: *Gegen-stand*) and sometimes even *someone* else (quasi-other). Think of a robot. Here, technology appears as more than an instrument we use and interpret; it becomes like an other (human) we can relate to. Ihde asks in what sense technologies can be other (97) and refers to Levinas's term 'alterity' (98).[3] Sometimes, when we interact with technology, we have the sense of 'interacting with something other than me' (100), for instance, in video games or when we interact with automata, robots, etc. Ihde stresses that this remains a quasi-other, and notes that like the dream of full transparency, the dream of the technological otherness (the wish for real otherness) is actually wishing for what technology is not. Thus, here, Ihde seems to maintain a naturalist and dualist understanding of reality, which makes a strict distinction between "what reality really is" (according to science) and "what we think reality is or what we perceive it to be;" he seems to refrain from going all the way to a phenomenological epistemology. But, in any case, here, technologies become 'focal entities' (107); they are, so to speak, in the foreground. I would add that here we no longer have the experience that we "use" a technology, since the technology is experienced as an other.

Finally, there are '**background** relations' (108): some technologies are designed to function in the background (like the thermostat) or are sheltering humans off from the rest of the environment. These relations are meant to be the opposite of the more focal relations discussed previously: We do not directly experience the technology and the relation, but direct experience is nevertheless structured by the background relation. Background relations are non-neutral and shape perception and experience, even if they stay in the background. For example, the thermostat shapes our experience of heat and cold. Here, we can add that, like in embodiment, we also do not experience that we use the technology; instead, we may experience the use of other technologies that are in the foreground.

Ihde also uses the terms 'magnification' and 'reduction' to argue that technologies are not neutral and 'do not simply replicate nontechnological situations' (Ihde 1998, 47), but always change these situations and are selective in the sense that technology 'both amplifies some feature of what and how something is experienced and reduces some other aspect' (47). (Consider also, again, McLuhan's view.) Thus, epistemologically, one could say that the metaphor of the mirror is not appropriate; technologies, as mediators, do not really represent reality for me, but rather *shape* reality-for-me. Technologies shape meaning and experience. Moreover, the meaning of a technology is 'multistable,' since it can be used in many contexts, indeed, 'end up' in many contexts—we cannot always predict what will happen with it and to it (48). (This point concords with the Wittgensteinian view I am developing, that use needs to be related to its contexts: to activities, games, and ultimately a form of life. I will say more about use, multistability, and the comparison with Wittgenstein below.)

Note that, according to Ihde, not all our perception is mediated by technology. We still have non-mediated perceptions, even in a 'technological textured world' (45). In addition to the mediations he describes, there are still 'direct or non-mediated perceptions,' which Ihde formalizes as '*I-world* relations' (45). Again, this shows that, epistemologically, Ihde does not go "all the way," as I put it; he maintains a distinction between mediated and unmediated relations.

Whatever the specific merits of these distinctions and theory about human-technology relations, and, indeed, of the underlying epistemology (which deserves more discussion in postphenomenology), it is clear that Ihde does something new and original here in phenomenology and hermeneutics. Whereas normally technology remains in the background in these theories focused on language (if it appears at all), he brings it to the foreground and gives it a key place and role in scientific practice and, more generally, in the way we experience, shape, and give meaning to the world. In terms of focus, with regard to language and technology, Ihde turns things upside-down. Technology, rather than text or sign, takes center stage as a medium and contributor to meaning. For Ihde, technologies and, in particular, instruments 'should be regarded as means by which our perceptions and our wider experience are modified and transformed' (Ihde 1998, 1). And, in contrast to (mainstream readings of) Heidegger, Habermas, and others, technology here is not part of a "system" or an "enframing," as opposed to the lifeworld, but is reinserted 'in all the dimensions of the lifeworld' (Ihde 1990, 41). Technologies are not neutral instruments; they transform our experiences (49) and, one could add, they transform our world.

To some extent, this phenomenological-hermeneutical significance of technology was already acknowledged by Heidegger. When I presented the embodiment relation, I already mentioned that Ihde is inspired by Heidegger's *zuhanden/vorhanden* (ready-at-hand/present-at-hand) distinction in *Being and Time*. He is also interested in Heidegger's suggestion that artefacts reveal a world; for example, a temple opens up a world (55). But Ihde rejects Heidegger's focus on technology 'with the capital T' (54), which is only seen as a danger, and rejects what he calls his 'selective romanticization;' he argues there are more involvements that can be revealed by the artwork, for example, the ancient temple may also reveal deforestation (56). Moreover, I indicated Merleau-Ponty also already wrote about perception and artefacts, for instance, when he discussed cases such as the feather on the hat (see below). Thus, there are still strong links with Heidegger and Merleau-Ponty. However, Ihde's framework was meant to enable a more refined analysis of technologies (in the plural) and make the phenomenology of human-technology relations more explicit and articulate. Ihde has succeeded in this and, indeed, has turned it into a worthwhile philosophical project on its own.

It is also a highly interesting project, given the purpose of my book, to bring thinking about language and thinking about technology into dialogue. Ihde could be read as having done precisely that, since he has transferred a particular approach in philosophy of language (hermeneutics) to thinking about *technology*. His writing also shows an occupation with culture and technology. However, in the process, it seems that he has dropped an interest in language itself, which is mainly used as a metaphor (especially the metaphor of text) and does not play any significant role whatsoever in his thinking about human-technology-world relations. Language and its relation to technology remains undertheorized.

To remedy this, one could read *Wittgenstein into Ihde*. With regard to the Wittgensteinian focus on use and forms of life I proposed earlier, it is interesting to note that (1) Ihde's account of human-technology-world relations does start from *use* of technologies, since relations such as embodiment relations and hermeneutic relations only arise in, and make sense in, the context of use, and (2) Ihde also pays attention to what he calls 'cultural hermeneutics' (124), which is about 'the way cultures embed technologies' (124). With regard to the latter, he explicitly argues that the artefact ' "is" what it is also in relation to this cultural field' (128). Or, put in a way that brings out even more similarity with my Wittgensteinian argument, influenced by a "cultural" reading of the *Investigations*: 'The technology is only what it is in some use-context' (128). He gives the example of a clock used in an astrological context (at some point in the history of China) versus the clock as time machine (in the West). There were different "readings," related to different cultures. (Note again Ihde's use of text as a metaphor.) Thus, one could say that here are different technology-cultures.

Moreover, Ihde argues that the question regarding the control of technology is therefore analogous to the question of whether cultures can be controlled (140). Since technologies are 'embedded in cultural complexes,' he argues, the question of control becomes 'senseless' (140). Technology and culture are always related. Ihde suggests here that the questions of control, determinism, etc. are only posed if we presuppose that they are two very different things. Hence, he can claim that 'technologies-in-use do not, as such, determine' (141). Technologies shape use-patterns (and vice versa, I would add). Ihde then goes back to the individual level: He gives the example of word processing, which does not determine the style or the writing, but "inclines" toward some possibilities—e.g., making many footnotes (142). But he also points, again, at what he calls 'multistability' (144) and the 'varieties of technological experience' (151). Multistability applies not only to perception—to, for instance, the perception of a Necker cube, which can be perceived in alternate ways (also known as a gestalt switch; 145)—but to what objects "are," which also varies according to their use. Technologies, considered in *use*, have different meanings depending on that use in different contexts: 'a multiplicity of users can pick up and use technologies in such different ways' (164). This means that there is diversity in use and meaning both within a culture and between cultures. Technologies also get embedded in different cultural hermeneutics. Our contemporary lifeworld is pluricultural (158) and, hence, there is not 'one world of calculative thought,' as Heidegger and others thought, but instead there is diversity (159). This diversity is 'resulting from the spread of technology across the globe' (160). In particular, he claims that pluriculturality 'is a lifeform arising out of the use of image-technologies catching up to cultures' (164). We think that these technologies are neutral, but, instead, there are different transformations and ways of relating to technology. Ihde ends with a praise of the postmodern spirit and mentions dance and technical apprenticeship. (The latter is interesting and relates to my own interests in skill and craftsmanship, and in dance, but I will not go further into those topics here.)

Thus, while generally in Ihde there is more emphasis on individual human-technology relations and less on the social and the cultural (especially in Verbeek's reading of Ihde), his focus on use and his arguments concerning cultural embeddedness and variety are in line with the Wittgensteinian argument I have been developing: The meaning of technology depends on use and on use context—on what I called "technology games" and, ultimately, on a form of life. There is variety in cultures and forms of life, and, hence, also in technology use. With Ihde, we can recognize the multistability of *use* (and not only multistability in perception and interpretation) and, more generally, point to the use dimension of technology. Against the instrumentalist view of technology and against Searle's emphasis on linguistic declaration, but in line with my use of Wittgenstein, we can endorse Ihde's claims that 'a technological object, whatever else it is, becomes what it "is" through its uses,' that 'even the technical properties [of an artefact] take on significance in the use context,' as 'in use they become part of the human-technology relativity,' that, as part of 'praxis,' objects are always ambiguous, since they can become different things and technologies in different use contexts (70), and, in addition, that 'all technologies in use are non-neutral,' since as they mediate our relation to the world, they change the situation and transform our capacities (75).

Hence, in order to understand technologies, we can employ not only a phenomenology and hermeneutics of (embodied) perception but also a phenomenology and hermeneutics of use. If we emphasize this "use" and cultural dimension of Ihde's book, therefore, there is scope for compatibility with the Wittgensteinian view I have been developing in the previous chapters. Moreover, as I will argue below, we can add a transcendental argument to the analysis, or give the analysis a transcendental twist; this will further integrate Ihde's view with my own. Nevertheless, at this point in Ihde's work, a remaining substantial problem I see with regard to my project in this book is that *language* remains out of view. In general in his work and also in his account of human-technology-world relations, Ihde focuses on the use of artefacts; the use of words remains largely out of view. Later in this chapter, I will therefore propose revisions of Ihde's framework that account for the role of language.

But, first, I propose to read a second work by Ihde that is relevant to discuss the relation between language and technology, especially in post-phenomenology, and that engages not so much with phenomenology, but this time with *hermeneutics* in a more extensive and more explicit way. (In addition, it links to Latour's anthropological approach, as I will explain.)

Ihde's Hermeneutics of Science

Second, taking distance from a humanities tradition focused on language, Ihde argues that there is also a hermeneutics in *science*: a hermeneutics not only in the sense of interpretations of texts but also involving visual and other senses. Turning away from texts and linguistic phenomena, Ihde's 'expanded hermeneutics' (Ihde 1998) shows how science, in practice, involves a 'visual hermeneutics' (137) and that instruments have interpretation built into them. Already in *Technology and the Lifeworld*, he uses the example of Galileo to argue that new instrumentation gives new perceptions (Ihde 1990, 56). In *Expanding Hermeneutics* (1998), an explicit link with hermeneutics is made. Hermeneutics is broadly construed as an *'interpretative activity'* (Ihde 1998, 2): It is not only about the interpretation of texts, but is done by the literary critic, the scientist, and, indeed, the philosopher (of technology) alike. Against the positivist interpretation of science and recognizing 'field-clearers,' such as sociology of science (144–145), the feminist critique of science, and Latour (see below), Ihde sees science as a hermeneutic practice. What does this mean?

Ihde discusses the technoscience praxis of measuring the greenhouse effect, drawing attention to *imagining* processes that show the state of the planet as the 'whole Earth' (58–59). He asks us to consider a satellite photo, thermal imagining, etc. These images, but also the earth itself, must be "read" and interpreted. Ihde proposes to understand this technoscientific praxis as 'a "hermeneutics of things" not merely of language and texts' (59). Commenting on the greenhouse effect, he also sees all kinds of manipulations that are used to 'bring out the phenomenon' (59); hence, there is first active and technological construction of measurements and the phenomenon; only *then* can the greenhouse effect be detected (60). (Later on in the book, Ihde further discusses whether images are, or can be read as, a "text." Since he claims that a hermeneutic relation is more text-like than body-like, and since images fit into his hermeneutic relations category (95–96), the answer seems to be positive.)

Ihde also tells a semi-fictional story of how Galileo had been 'tinkering with his optical instruments' (154), in particular the telescope, and then describes phenomenological shifts such as getting closer to the moon (155). He emphasizes the material-technological and bodily aspects of this science, and the selectivity involved in the discoveries (156–157). Moving away from a focus on language and text, Ihde suggests an account of scientific praxis that emphasizes (bodily) perception, 'perceptual-bodily activity,' and 'an interpretative activity with the thingly' (159). Moreover, he stresses the visual. He mentions the visualizations by da Vinci and by Vesalius, and later the use of imaging technologies such as X-rays and MRI scans. Next to language, then, scientific practice *'visualizes* its phenomena.' It is another way of embodying understanding (161). Technoscience creates depictions by means of various technologies, which each have specific phenomenological effects. In Ihde's so-called 'strong program,' perception is seen as 'whole body perception' (170) and instruments are seen as producing 'more phenomena, more "things" within the universe' (173) or as enabling interpretations of the same object (e.g., X-rays and the body). In this hermeneutics, things and visualization play a key role, next to texts. For instance, through medical-imaging techniques, the body is "sliced" and manipulated: 'Things have been prepared to be seen, to be "read" within the complex set of instrumentally delivered visibilities of scientific imaging' (183). Ihde thus proposes a hermeneutics of science that 'is not primarily linguistic or propositional [but] rather, perceptually oriented' or even 'primarily *visually* oriented' (184). There are various 'world-constituting' (185) activities, involving humans and things, and not only texts but also visual, acoustic (188), and other instrumentation. Indeed, at the end of the book, Ihde explores the idea of a 'multisensory' hermeneutic (189) and discusses VR (virtual reality) technologies.

Epistemologically speaking, this means that, according to Ihde's 'strong program,' we live in a different world today than in the past. Some phenomena, such as auras around stars or canals of Mars, no longer exist in science. New, different technologies create new, different worlds. This brings Ihde's hermeneutics close to social constructivism, although, in his view, the material receives more emphasis than the social—not to say that the social is often neglected. Ihde's hermeneutics is interested in things, perception, and embodiment at the individual level, and he makes remarks about culture, but the social remains undertheorized. Moreover, as Ihde writes, it also means that 'there are better and worse interpretations . . . but no final or perfect ones' (197). Against hermeneutic fundamentalism, he suggests another option: a creative and critical hermeneutic, which is never finished and which does not search for an origin, which accepts that it is always political (197), which points to the limits of modernist science and epistemology, and which keeps a sense of awe and is as 'fallible, contingent, and subject to change' as the human condition (198). Furthermore, hermeneutics of science means there is now room for the role of what Ihde calls a 'science critic:' someone who loves the subject matter, but is not a total insider, who looks at science practice 'with its forms of tacit and operational know-how,' who takes a critical stance, which requires distance, and who nevertheless enters into a 'multiple perspective process' by means of collaboration (134–135). (Here Ihde is close to Latour—see later in this chapter.) Finally, as is suggested by Ihde, but not very well developed: This hermeneutic character of science also implies that there are and could be different interpretative cultures. Perhaps there are not only cultural differences between parts of the world, countries, and so on, but also within science. It would be interesting to further explore this route.

Again, technology—here, technology in science—is shown to play a phenomenological and hermeneutic role. One could say that, here, technology is no longer "passively" hermeneutic, but "actively" hermeneutic: It can no longer be reduced to a mere sign or symbol or décor for symbolic exchanges, assumed to be a mute and impotent recipient of meaning (as in Searle), or banned to the shadows of linguistic order or the mysterious gothic cellar where "the real" is supposed to dwell (as in some versions of poststructuralism and postmodernism); instead, technology actively shapes meaning and experience. In this sense, the subject and the object start to draw nearer to one another, and technology "speaks."

Yet, it remains unclear what the role of *language* is in these world-making technological mediations. The focus on the artefact seems to imply the silencing of language. Ihde may not deny that language plays a role and he may be willing to recognize that there is also a hermeneutics of words and text in science, but in this effort to take distance from hermeneutics "humanities style," he overemphasizes material artefacts and their visual hermeneutics, and does not say much about the hermeneutics of language.

Given his background, Ihde is, of course, very well aware that the Greek origin of the word hermeneutics has to do with language, with 'bringing into word of what was previously not yet word,' with the 'coming to birth of word, of language' (9). He explains that, in a Jewish-Christian biblical tradition, hermeneutics became part of a culture of the word as expressed in text, which needed interpretation (10). Hermeneutics had to retrieve sacred, original meaning. In modern times, hermeneutics became a more general activity of interpretation. Ihde's own project can thus be understood as being part of that project: He broadens hermeneutics to the material. Ihde is also very much aware of the language-oriented nature of Heidegger's philosophical project, which tried to develop 'a radically new language which skirted or circumvented the terminology of "subject"-"object" and the constitution of knowledge' (14), whereas Ricoeur proposed a 'dialectic' between phenomenology and more 'objectivist' strategies (19), which does not cut off the debate with classical issues (15). Both Heidegger and Ricoeur were part of a linguistic, language-oriented phenomenology and hermeneutics. Language and text are central in their hermeneutic projects. In their work, hermeneutics is a philosophy of language. Heidegger's and Ricoeur's *poiesis* is about the word. But, as said, Ihde's turn to artefacts was a turn away from language.

Ihde's turn away from language is understandable; it is a core part of his philosophical project precisely to criticize the exclusive focus on word and text in much of the 20th century hermeneutics tradition. Ihde rightly criticizes the disembodied semiotics of the linguistic sciences and the model of language as information (24). When he argues for giving more attention to embodiment, he writes critically about language and the phenomenological tradition, about 'hermeneutic phenomenology as linguistic philosophy' (35), and about Heidegger's view that 'I am used by language as much as I am able to use language' (36). He also comments on language and perception, about the philosophical problem of language, in particular, language as 'singing the world' (63). He argues that language is always 'embodied expression,' drawing on the Merleau-Ponty view that meaning is embodied (69). For Ihde and Merleau-Ponty, embodied expression is the 'performance of the living subject;' speech is 'the *performance* of thought,' and the speech act of naming is a performance (70). Linking back to Wittgenstein, we could say that there is no first thought that then becomes speech, but, rather, that thinking happens in and through speech, as embodied and as performative. But, with Ihde and Merleau-Ponty, we must add that this speech should not be interpreted only in terms of signs or text; it is always embodied performance, performed by the living subject. In a way, one could say that "the body speaks." This aspect seems to be missing in Wittgenstein. However, as Heidegger and Wittgenstein knew, speech performances always take place "inside" language; therefore, there is still a sense in which "language speaks."

More generally, emphasizing the performative aspects of language and linking back to Wittgenstein, we can revise and develop some elements in Ihde's work into a view of language, and then perform the operation I suggested earlier: turn that view of language into a view of technology. When it comes to embodiment, there is a similarity between linguistic use and performance, and *technological* use and performance: Both can be seen as the performance of the living subject, and both take place within a larger linguistic, technological, and cultural whole—a form of life. Thus, inspired by Ihde and (more directly) by Merleau-Ponty, we can add embodiment to our account: use of technology and use of language is always embodied. (See also my remarks on embodiment at the end of the next chapter.) Again, this dimension is largely missing in Wittgenstein and must be added to the view I am developing. But, at the same time, this insight into the embodied natures of use of language and use of technology should not come at the expense of the insight that these uses are always taking place within a language and within a technological culture. This dimension needs to be added to Ihde's postphenomenology and (to use a textual metaphor to stay in tune with Ihde's own approach) needs to be *read into his work*. For instance, we could explore links between words and things in visual hermeneutics. To take up one of Ihde's examples: Seeing "canals" on Mars is an interpretation that was not only mediated by (the then available) technology, as Ihde argues; it was also mediated and, indeed, *made possible* by use of the very word "canal." I can only *see* a canal if I have a technology that reveals it, but also if I have a *language* in which this concept/word exists, a language which co-shapes my world and the world of others who share that language and that technology. (Below, I will say more about this *transcendental* argument.)

Ihde's approach also reminds us of the time dimension of hermeneutics and the need for a historical angle. Here, the link to Ricoeur's work comes to mind. In anticipation of my chapter on Ricoeur, it is interesting to note that Ihde knows Ricoeur's work very well and engages with his hermeneutics. He discusses his places within the hermeneutic tradition, his opposition to a single-perspective approach (82), and (again) links hermeneutics to religious history: the interpretation of sacred texts, in particular, biblical hermeneutics (78). In response to structuralism, Ricoeur stresses the diachronic (84). But, whereas in Ricoeur's work the notion of 'narrative' is important, Ihde does not make good use of this concept in this postphenomenology. Ihde reads narrative as a linguistic and hermeneutic strategy, which links Ricoeur to 'postmodern sympathies' (85). Later in the book, Ihde also links Ricoeur to Parfit and analytic philosophy (101–102). (Indeed, Ricoeur engages with both "continental" and "analytic" philosophy.) Thus, Ihde is very well aware of views that put language central. Ihde also recognizes and praises the openness, richness, and complexity of Ricoeur's hermeneutics. But, unfortunately, he does not connect narrativity to technology, and does not really use Ricoeur's insights about time and narrative in his material hermeneutics; in the next chapter, I will show that and how it can be done.

Similarly, Ihde misses an opportunity to bring in insights from philosophy of language when he reads Rorty as joining the ranks of 'poststructuralists and deconstructionists' (126) and says that 'language remains the guiding thread' and is 'a twentieth-century obsession for philosophers, both Anglo-American and Euro-American' (116). Ihde is right about that; there was such an obsession, and this hindered sufficient attention to materiality and technology. But this should not be a reason to *entirely* turn away from that kind of work, since this misses out on insights about language (and its relation to technology). For example, in Ihde's discussion, Rorty's sources of inspiration appear (Heidegger, Wittgenstein, and Dewey), but are not really discussed. Instead, Ihde presents his own view of what he thinks could and should be an edifying hermeneutic in Rorty's sense. This is regrettable since, at this point in the development of Ihde's work, more discussion of, for instance, Wittgenstein, Dewey, Ricoeur, and, indeed, some postmodern authors, such as Rorty, could have contributed to Ihde's own project. His obsession with getting away from philosophy of language has prevented him from developing a richer and more comprehensive hermeneutics in which both language and technology—words and things—play a role.

Thus, while Ihde recognizes the embodied and performative dimension of language (via Merleau-Ponty), praises Ricoeur's hermeneutics, and refers Rorty's sources of inspiration, his stress on *materiality* unfortunately and unnecessarily has led him to move away from thinking about language altogether. Ihde's original combination of philosophy of technology with hermeneutics (and phenomenology) has successfully bridged the gap between, on the one hand, a humanities philosophy of technology and, on the other hand, a philosophy of science focused on language (both in "continental" and "analytic" traditions) or, indeed, the engineering sciences, design, etc. But, by emphasizing the *material* side of technology, by seeing technologies as mainly instruments and artefacts that have a hermeneutic function, he has taken too much distance from (humanities) thinking about language. Ihde is explicit about this turn. He recognizes that when he did hermeneutics of science first, linguistic metaphors such as 'giving voice' and a postmodernist emphasis on 'textuality' played a role. Interestingly, he also notes that, within the science, linguistic metaphors, such as codes, play an important role (in biology and in computer science). But then he says that he has been arguing for a 'more *bodily* and *perceptualistic* mode of interpretative activity,' which moves toward the '*thingly*,' which is not just reducible to the linguistic (158). Thus, his turn to things was a turn away from words.

To be fair, it must be noted again that attention to language is not *entirely* absent in his postphenomenological work. An interesting exception, for instance, is Ihde's qualifying remark in *Expanding Hermeneutics* (which follows his claim that he moved to the more bodily and perceptualistic mode of interpretative activity and must be placed in the context of his argument about humans and animals) that 'our perceivability is intertwined with linguistic and cultural interconnections' (158). Unfortunately, this remark is not further developed. Moreover, Ihde acknowledges the role of mathematization in science (168) and writes that some visualizations are 'textlike' (166). There are journals, books, etc., but, according to Ihde, they are 'secondary or tertiary with respect to science, as we have seen from Latour;' he points to the textlike phenomena of 'charts, graphs, models,' etc. (167). However, he does not discuss the relation between textlike and other phenomena; in general, he does not sufficiently discuss the metaphor he uses throughout his work (text). Ihde makes comparisons; for instance, he compares the visual hermeneutics of science to the invention of written language. Writing is a technology that enabled an explosion in communication. In science, there is a visual hermeneutics (187). But he largely keeps language and technology separate: The textual and the visual, perception and "culture," etc. do not really touch.

Moreover, even if there are some points about language in Ihde's work, unfortunately those who have picked up Ihde's postphenomenological approach and further developed it (Verbeek, Rosenberger, Selinger, and others belonging to the postphenomenological school) have neglected Ihde's remarks on language and have only reproduced the lacuna and the problem identified here. While I fully agree with Ihde that 'we need something more than "textuality"'' (159), that things cannot be *reduced* to text, and that a universal hermeneutics is also 'a hermeneutics of things, not of language alone' (187), more needs to be said about the connections between language and technology, and any Ihde-inspired postphenomenology or any other Ihde-inspired phenomenology and hermeneutics of technology for that matter, is in danger of neglecting the linguistic dimension of hermeneutics.

According to my view, *there is no hermeneutics of things and no hermeneutics of use without a hermeneutics of language.* Even the hermeneutics of things takes place "within" language, in the sense that our relation to material things, to earth, etc. is always already mediated by language. Therefore, anything that 'things do' (to use Verbeek's phrase) also goes via language, directly or indirectly. To pick up one of Verbeek's (2008) favorite examples: It may well be, for instance, that in echography of the unborn child, visual images and the material technologies that produce them have an active hermeneutic role; but, as I argued, when we interpret the images, words play a role, too, as soon as we perceive and think about the images. When things make us see the world in a different way, when, indeed, they create a specific "world" in the phenomenological-hermeneutic sense, then they can only do this via language. In this sense, perception is not only embodied and mediated by material artefacts; it is also at the same time linguistic. Our perception and performances are not only mediated by things but also by language. Language already frames things in specific ways, which may amplify certain meanings rather than others when we deal with things. For instance, in the case of echography, which term we use—"unborn child," "fetus," etc.—matters for our perception of the unborn child and its meaning. Some parents may give a personal name. The use of words, together with the use of technologies, construct the meaning. And as I suggested before (and as Heidegger has shown so convincingly), there may be also effects on our thinking about technology in general. For instance, language itself may make us blind to the very noninstrumental aspects of technology Heidegger, Ihde, and others attend us to, and, hence, play a kind of meta-role in preventing a better understanding of technology and its hermeneutic and phenomenological role.

My criticism, then, is *not* that Ihde entirely neglects philosophy of language, but that he does not centrally use it in developing his hermeneutics of use and his hermeneutics of things, including his account of human-technology relations. With regard to Ricoeur, for instance, my criticism is not that Ihde neglects his work altogether. My criticism is that, in spite of Ihde's praise of Ricoeur, he does not make full (and the best) use of him *in his hermeneutics of things and his hermeneutics of use*. Admittedly, Ihde makes a connection between Ricoeur and discourse about technology when he distinguishes between literary fictions and technological fictions—with the first taking seriously our corporeal, embodied condition (104). He also links Ricoeur's opposition to the technological dream (which Ihde calls a technomyth) (105) and his own view, which accepts our terrestrial condition. But these distinctions—with which I sympathize and which, for instance, can be used to criticize, for example, transhumanist fantasies—do not seem to contribute to Ihde's project of a hermeneutics of things. Insofar as they concern discourse *about* technology, are about "culture" seen as a distinct issue, and do not touch Ihde's phenomenology and hermeneutics of human-technology-world relations, these remarks reproduce the gap between "culture" and "science" Ihde wants to bridge. I will propose, instead, that we include language in an account of human-technology-world relations and—with regard to Ricoeur—that we develop a concept of the "narrative" dimension of our relation to technology and the world, and, in this way, use Ricoeur in order to revise the postphenomenological project and to support the further development of the integration and synthesis I aim for here and in the next chapters.

Finally, note that Ihde's turn away from language is remarkably linguistic and textual. He mainly uses the technology of speech, writing, and text to turn to the artefact. The artefact mainly appears through the medium of writing. An artistic and/or empirical-scientific (e.g., material-archeological) practice, by contrast, would engage with the artefact in more direct way. Ihde, of course, writes about such practices, and perhaps does some of this in his free time. But my point here is that, for instance, an artistic practice does not figure centrally in his material and empirical turn, it is not used as a *medium*, and, hence, is also not so important in the *message*. His material hermeneutics aims to give things a voice, but this is usually through the medium of writing, in particular, writing books. One could retort that this is "what philosophers do." But this is not obvious. First, at the very least, it seems to me that philosophical and academic practice is not only about writing: One could highlight all kinds of material artefacts and conditions that make philosophy possible. Second, even if writing is "what philosophers do," then one may question if writing technologies are sufficient for realizing a material turn that takes itself seriously. Perhaps we also need artistic and scientific work: To know more about "what things do" (to use Verbeek's phrase), perhaps we need to do more things with things. Like Latour, Ihde went to the laboratory. But did he sufficiently think *through* the laboratory? Moreover, even if one does not want to question the exclusive use of philosophy or writing itself, there is another issue here, which I already mentioned in the previous chapter: As a philosopher, it is also important to be critical of one's own instrument—language. In this case, if one aims to give a voice to things, then one should also consider how *that* project is *itself* mediated: how our speaking about things is mediated, including the philosopher's and postphenomenologist's speaking about technology. (And, of course, the same is applicable to Latour and, indeed, to my own writings.)

To remedy these shortcomings, one could immediately turn to an alternative approach in philosophy of technology or, indeed, to art (see my last chapter). However, given the value of Ihde's insights about use, materiality, and embodiment, and the influence and impact of Ihde's postphenomenology on contemporary philosophy of technology, mainly in the form of mediation theory (especially in the work of Verbeek), it is worth trying to revise *his own framework* in order to better account for the role of language and to find a better synthesis, a better view that integrates thinking about use of things with thinking about use of language. This is what I will do in the next section. This will contribute to the further development of my view that any hermeneutics of technoscience and other technological practices must be at the same time material and linguistic. It is true that hermeneutics is and should be also 'thingly,' and this was an important contribution Ihde made to hermeneutics and to thinking about technology. But, given the lacuna concerning *language* and its relations to technology, a more complete and integrative view is now needed, one that gives a place to both things and *words*, and that—in line with Wittgenstein and Ihde—gives a central place to *use*. Let me also emphasize that *even if one were to keep the emphasis on things*, it would be worthwhile to undertake this project, since also a postphenomenology (understood as material hermeneutics and as mediation theory) that insists on its focus on things needs to clarify the role of language in its hermeneutics and theory of mediation. So far, explicit attention to language and to what *language* does, what *words* do, is mostly absent in contemporary postphenomenology. An instrumental view seems to be assumed. But this is no longer tenable, let alone that it is tenable not even to *ask the question* regarding language.

Revising and Expanding Ihde's Framework of Human-Technology Relations in Order to Account for the Role of Language, also in the Light of a Wittgensteinian Approach

One way (1) to account for the role of language in technological mediation (a minimal aim that could easily be shared by postphenomenologists and mediation theorists) and (2) to get closer to a more integrated and complete view of the relationship between language and technology (my more ambitious aim in this book), is to *expand Ihde's postphenomenological framework*, and, in particular, his theory of different I-world relations and its technological mediations, in a way that includes language and clarifies its role in mediation. I have started to do this in a preliminary reflection on the matter (Coeckelbergh 2015), but that scheme needs to be revised in the light of my Wittgensteinian focus on use and other developments in the previous chapters, streamlined for consistency, and further elaborated.

I propose the following steps to build my expanded framework, each time involving distinctions between various human-world relations in which language plays a mediating role, with the term "mediating" used both in the sense of an "in between" and in the sense of a "milieu" or "environment." In order to maintain a link to Ihde's framework, I model them loosely on Ihde's distinctions. But I will also revise the distinctions and the structure, propose new ones, and, taking advantage of the Wittgensteinian work done earlier in this book, add a focus on language and *use*—thus, going beyond Ihde. I will read Wittgenstein and Heidegger into Ihde's framework in order to transform Ihde's mediation theory into a more comprehensive one that takes into account not only technology but also language as mediator and (as I will argue below) as a transcendental condition. The first step starts from the mediating role of *language*; then I work my way up to language and *technology*.

Let us first consider the following view and scheme of human-world relations in which **language** is a medium between **humans and world**. In this view, technology is assumed not to play any role. This is the usual approach in the humanities. One can then distinguish at least the following relations:

1. (**humans—language**)—**world**: *embodiment* relation—as language is used in an everyday context, it usually remains invisible. It is, to use Heidegger's phrase, ready-to-hand. Usually, we do not notice language and the way it conditions how we think and perceive. It is a tool, but one we no longer notice, since we use it so much; it has become part of our living being in the world. It is also a milieu in the same way as air is a milieu: we use it, but usually we do not think about it. (The latter milieu interpretation is somewhat similar to Ihde's *background* relation.)

2. humans—(language—world): *hermeneutic* relation—with "hermeneutic" understood not as interpretation in general, but in a more strict sense used by Ihde (something like "similar to the interpretation of a text"), but then applied to language instead of technology: Sometimes language is what Heidegger would call present-at-hand. We may experience it as an object-to-be-read, and/or it appears as part of the world: it appears as word-objects, as text. Language is here experienced as an object or a collection of objects that is part of the world; we do not experience it as part of ourselves or as human, and we can (or have to) pay explicit attention to it. Thus, here, we do not really experience language as something we use, but more as something that "is." There is a language-world.

3. humans—language—world: *mediation* relation—here, language is experienced as an "in between" or a milieu; we become aware of it as an "in between" or milieu. In terms of use, we become aware of our use of language, and we become aware that this use shapes how we see the world. This experience is rather exceptional. We usually do not experience this in an everyday context, unless language becomes present-at-hand and we have a discussion about (the meaning of) words and about grammar, *and* become aware that our use of language matters to how we see the world. Of course, philosophers, for instance, philosophers of language in the Heideggerian and Wittgensteinian traditions, know this, but we should not forget that this is a rather exceptional experience and use of language. As philosophers, we can then have a mediation theory: We can say that language as an "in between" mediates our relation to the world, as it shapes how we perceive reality, relate to others, and act. Or it is seen as a milieu; we are "in" language. It comes "before" thinking, or, as one can say with a transcendental argument: Language makes thinking possible. (Below, I will say more about using a transcendental argument.)

Let us now add technology to the scheme: In this view, *language* mediates **relations between humans and technology.** We can distinguish between at least the following relations:

1. (humans—language)—(technology—world): *embodiment*—in everyday contexts, we sometimes speak about technology. We use language for this purpose. But language is more than a tool: Language is hermeneutically active in this speaking—it shapes how we think and speak about technology. However, we do not notice this mediation by language. We even hardly notice that we use language. Language is embodied. If I voice things, if I speak, if I perform with language, language is part of me. It is our medium, but it remains invisible. If philosophy of technology neglects language, it is because in its use of language to speak about technology, language is ready-to-hand, it is embodied. Language is like glasses or like the air we breathe. We use it and we depend on it, but

we usually are unaware of that. (The same can be said about our use of voice, which is a form of embodiment of language: Usually, that "tool" is embodied and we are unaware that we use it.)

2. humans—(language—technology—world): hermeneutic and *alterity* relation—sometimes we perceive language as an object that is part of the world. In that perception and in that world, language may be coupled with technology, for instance, in the form of text or as a speaking device. We do not notice the mediation of language, since language-technology appears as an object that stands before us (*Gegenstand*) or as an other. We may see or write a text, or we may interact with a computer or a robot that talks to us. We experience language and we see technology. But we are unaware of how both language and technology mediate and shape our perception and our thinking about the object. We think of the object as something that is independent from language, and technology is seen as a mere instrument. We do not even have the experience that we use words or that we use tools, let alone that subject and object emerge in this relation and in this use and performance.

3. humans—language—technology—world: *mediation*—here, we become aware of the mediating role of *language* in speaking about technology: We become aware that we use language, and we recognize that it mediates and shapes our speaking of technology. Once we attain such an awareness, we can try to attain a more explicit, hermeneutic (in the broad sense of interpretation, applied not only to text but to language in various forms), and critical relation to it. We can ask, for instance, if and how our use of the terms "machines" and "robots" shapes our relation to robots. We may also attend to the *grammar* of our speaking about and to the robot, and how this shapes what we think the robot "is." For instance, I have argued that the way we address the robot matters as to how we perceive and think about it (Coeckelbergh 2011). This is about grammar as syntax, but also refers to larger structures that shape our use of language and our thinking about technology. Interpreted in a way congruent with the Wittgensteinian view I have been developing, one could ask, for example: When we talk to a machine, do we use words and linguistic structures that belong to the language game of talking to a friend, or do we use words and grammar that belong to the language game of doing a scientific investigation? (This example and use of the term grammar needs more refinement, but this should suffice to illustrate the point.)

In the third view (and final step), **both language and technology** mediate our relation to the world. Here, there are also different experiential possibilities and claims, which we can model again with Ihde's scheme, and distinctions between embodiment, hermeneutic, alterity, and mediation relations, which give us a better and more integrated view:

1. (humans—language—technology)—world: *embodiment*—technologies and languages shape how we perceive, think, and act in the world, but

usually we do not notice this embodied kind of mediation. Technologies and languages are in-betweens and milieus, but phenomenologically they are ready-to-hand and embodied in everyday use, or, indeed, invisible as a background or as a kind of air we breathe, a milieu. In our use, we "are" language and we "are" technology (embodiment) and, at the same time, we also live "in" language and we live "in" technology. In Heidegger's language: Language and technology are the house of being. Our existence is linguistic and technological at the same time, and language and technology shape our world, but usually we are not aware of this because they are so much embodied and so much our milieu, so much our glasses and the water in which we swim and the air that we breathe, that they are invisible as medium. (Put in a transcendental way, one could say that, usually, the world is flat, in the sense that we do not notice language and technology as transcendental structures and conditions that make possible and shape how we see the world and how we think. Of course this already goes far beyond Ihde. I will say more about it below; for now, I limit myself to a mediation framework.)

2. humans—(language—technology—world): *hermeneutic* and *alterity* relation, with "hermeneutic" used in the broadest sense, as applied to language and technology: sometimes we not only perceive language as an object that is part of the world but also technology. For instance, language appears as a text and technology appears as an artefact, and both are seen as part of the world. Sometimes technology even appears as a quasi-other, for instance, when we "meet" a humanoid robot that is perceived as a conversation partner or even "friend." Sometimes language *and* technology become present-at-hand and appear as part of the world or as an other, an alterity. This may happen when a machine talks to us (literally), or when we attend to the 'language' of 'the system.' Then we perceive technology and language, but not as mediators and not so much as tools we use, but as objects or quasi-others. In addition, sometimes we become aware of a relation between language and technology within the world, because they are openly and practically (i.e., in use) linked to one another. We may have this experience when, for instance, a technology literally speaks (say a robot or a computer program), when we read or write code for a computer program, when we recognize "text," "script," or "narrativity" in technology (when we apply concepts from Latour or Ricoeur, for instance), or when language itself is seen as a technology (this was my interpretation of McLuhan and Wittgenstein). Thus, here, there is a hermeneutic or alterity relation to language/technology couplings, which become visible not as mediating our relation to the world, but as a "something" (e.g., code) or as an "other" (e.g., robot as other) that is part of the world. The robot or the text becomes a linguistic-technological object or even agent, which "speaks" and "does" things. Here, we no longer have the experience that we *use* language or *use* technology. Language or technology, or language-technology, appears as an object (or objects) that stands apart

from me or that con-fronts me (German: *Gegenstand*; hermeneutic rela-tion), or as a quasi-other (alterity relation). Since we see language and technology as objects or others, we are not aware of their mediating and shaping role with regard to our perception, thinking, and action. We think they are neutral external things (texts, tools, computers, etc.) or that they are others that do not influence what we think.

3. **humans—(language—technology)—world**: mediation—here, we expe-rience and recognize that both language and technology play a mediat-ing role in our relation to the world. We become aware that we relate to the world *through* and *in* language and technology. Again, usually, we do not notice this, because technology or language are either embodied (and hence are invisible) or appear as objects or quasi-others, which seem to have nothing to do with our perception or thinking. For exam-ple, I use a word processor without thinking about it, or I think about it as a text, an external object. Language itself then appears as a piece of writing, that is, an object. And technologies such as the word proces-sor and the computer appear as neutral tools. But this is also an effect of the mediation of language and technology. Our instrumental view of technology is itself given to us by our language, which constructs tech-nology as objects, things, artefacts, tools, instruments, and means. Our very language about technology hides its mediating role. And our object view of language is itself made possible by technology, such as writ-ing, paper, word processor, etc., which shapes our view of language as text, words, etc. Even the very word "word," for instance, suggests that language consists of "these things called words," that is, of language-objects. This insight, which below I will put in the form of the tran-scendental argument that language and technology are conditions of possibility for speaking about technology, effectively enables a more critical philosophy of technology, one that scrutinizes, and is critical of, the *language* we use when speaking about technology. This goes beyond "mediation" as an "in-between" or "milieu," or, rather, it gives more depth to the mediation framework.

The more integrated mediation view can thus be summarized as follows:

humans—language—technology—world

or, if we understand language as a kind of technology (as McLuhan and Wittgenstein did, and as I think is a very helpful and adequate way of under-standing the relation between both):

humans—language/technology—world

or, with an emphasis on technology and media:

humans—technology and media (incl. language)—world

If we want to open up different ways of relating to the world and to technology, therefore, we need different technologies/media and different languages. But, first, we need to better understand how language and technology mediate our relation to the world. We need to study the languages of technology and the technologies of language, and describe how they shape human experience and action, indeed, how they "speak."

Thus, in the senses developed here, both technology and language "speak." However, if we put it like this, there is something missing: We can only move towards a more integrated view if we recognize that this speaking is also done by humans, since language and technology can only have their mediating and hermeneutic functions in *use*. If technologies and languages are not used, they are dead objects and dead signs; they only live and function in use. Since often we experience an embodiment relation with regard to technology and language, we normally do not notice their mediation. And if we do, we tend to have a hermeneutic or an alterity relation to them. But neither language nor technology can be reduced to external objects or quasi-others. They are and become very human themselves, not as quasi-others, but as participants in a performative-hermeneutic process, as parts of a speaking, which involves the *use* of language and technology. Paradoxically, a focus on use has enabled us to move away from an instrumental understanding of technology. And equally paradoxically, perhaps, it enables us to *humanize* technology. This can be done by putting more emphasis on Ihde's own point about *use* and linking this use to performance, and by relating this not only to things but also to *language*—thus benefiting from the Wittgensteinian (and Heideggerian) approach I proposed. In use, humans, artefacts, and language are bound together, gathered. Meaning emerges from their entanglement and from this process of performance and meaning-making. Giving a Wittgensteinian twist to this expanded postphenomenological framework, one can say that the various meanings, human-world relations, and kinds of mediations by language and technology emerge in a process of *use*, performance, and experience, which are themselves part of a larger whole (games, forms of life), which makes possible these particular uses. (Below, I will say more about this "making possible.")

Taking advantage of Ihde's focus on embodiment (influenced by Merleau-Ponty), which is present, but less pronounced in Wittgenstein, we can then add that this performative-hermeneutic process, understood as a use of words and (other) tools and as performance, is always *embodied* use: embodied use of words, and embodied use of tools. Speech acts, but also what I have called "technology acts" are performances by living and embodied human beings. And perhaps one could go further, and, rather than stressing "the body," move beyond body-mind dualism altogether. *In* performance as lived experience and activity, always but especially when there is a high degree of skilled engagement, there is no dualism between mind and body; instead there is an embodied-mindful movement, interaction, and use. For instance, in craft work or in martial arts, such non-dualist use and experience can happen to a higher degree. But, in a sense, it always happens: There is no such thing as a "body" separate from "mind." There is an embodied mind and a mindful body. This point is also in line with embodied cognition and enactivism in cognitive science, and of course with Eastern thinking about body and mind, for instance, in Buddhism. However, since here I aim to stay close to Ihde, I mainly refer to Merleau-Ponty (and I will say more about Merleau-Ponty in Chapter 7).

However, note that already in modern Western thinking, this focus on embodiment is not only to be found in Merleau-Ponty. It is already present in Husserl, for instance, who pointed to the idealization and 'mathematization of nature' by the sciences and argued that prescientific,[4] everyday experience is always embodied (Husserl 1970, 23). But mathematics and science also involve embodiment, and, indeed, technology: Husserl already wrote about the role of 'sensible embodiment' (26) and the art of measuring (27). In general, already before Merleau-Ponty, there was a tradition of thinking in Western philosophy that recognized that we access the world via our body (German: *Leib*) and use tools, and that recognized that science is not an exception in this regard: Scientists produce knowledge by means of embodied perception and technology, including, as Ihde right says, visualizations. Already in the 19th century, for instance, Marx and, even earlier, Feuerbach already paid attention to the embodied aspects of cognition and action (Deranty 2014).

Furthermore, we can support the point about embodiment by using Wittgenstein, who suggests in the *Investigations*, against Descartes, that thinking, language, and embodiment are intimately connected. He writes: 'Thinking is not an incorporeal process which lends life and sense to speaking, and which it would be possible to detach from speaking' (Wittgenstein 2009, §339, 116e). It is done by 'a living creature,' it is '*not* a hocus-pocus that can be performed only by the mind' (§454, 140e). Thus, for Wittgenstein, thinking is a corporeal, embodied process that is *performed* by a *living* being. Use of language is, hence, always linked to that embodied performance; it is not something separate. Similarly, when we use technology, this technology is connected with that embodied performative process; it is part of it. There is skilled and embodied performance, in which both language and technology play a mediating role and—as I will argue below—*make possible* this performance by functioning as transcendental conditions.

In order to better understand what it means that, in their mediating role, language and technology "speak," let me give some examples. An electric guitar is, of course, a tool, an instrument—a musical instrument. But it is more than that. It shapes the kind of music that is played. For example, rock music could only emerge because of the development of electric guitars, which made possible a different kind of music. But, in play, musicians are not aware of that. They are not even aware of the instrument they use, at least not when *in* use, when the instrument is being used. The guitar is either embodied, when players do not even notice it, that they are in the flow of use, or it is experienced as an object, as a thing that is part of the world.

For example, a specific guitar may be experienced as a thing one wants to have, own. Or as a thing that needs tuning or repair. Moreover, in skillful and engaged use and performance, the player is not only unaware of the instrument as an object; one is also unaware of one's "body" as opposed to one's "mind;" instead, there is embodied and mindful performance. And the kind of knowledge we have of guitar playing is a know-how that is embodied. While, first, we need instruction, as "expert" users we have a more intuitive, non-explicit understanding of using and playing the guitar. Consider here Dreyfus and Dreyfus's model of skill acquisition, which runs from novices being instructed about the rules, to competent routines, to expert intuitive grasp, to tacit understanding of situations and the ability to improvise (Dreyfus and Dreyfus 1980).

But not only technology speaks; language also speaks. Playing the guitar is a technology act and a speech act. Not only literally (in cases when the player sings, for instance, as a singer-songwriter or a rock singer who also plays the guitar at the same time); language is and does more than that. The player uses musical language, for instance, in written form. And often, metaphors from use of language are used by musicians to describe what they do. For instance, one could say that the playing involves gestures, but also various forms of musical "utterances." There are "phrases" (think about blues and jazz) and patterns, "grammars" that are used and responded to, for instance in improvisation. Moreover, words are used to speak *about* the guitar and, in lyrics, words are used to speak about other things. These words are linked to activities and games to which they belong (e.g., declaring love, expressing despair, etc.) and to a form of life. All these uses of language shape the message. Music, understood as language and technology, shapes the music-as-message. Particular music games, music grammars, and forms of life shape concrete musical uses of language and technology. However, musicians, as users of language and users of technology, are also shaped by (their use of) languages and technologies, and by the games, grammars, and forms of life to which these (uses of) languages and technologies belong. One could say that when the musician plays music, music (as medium) also plays the musician.

Thus, music-language and music-technology shape the musician and the music. When there is play, there are already musical patterns, there is a specific musical language and style, there is already musical form before the player starts playing. There are already rock styles, for instance, but there are also older musical and cultural "grammars" in which the playing is embedded. When the musician lets the guitar speak, and, indeed, when the guitar "speaks" with the help of the musician, then this speaking is embedded in larger wholes. This is also true for the language of rock. Language is not only used; language also uses the speaker. Rock musicians are not the first humans on earth to declare their love, for instance. Our language and culture already contain forms—linguistic and technological—to do this. There is already a certain know-how, which has the form of linguistic structures and technologies (e.g., use of writing or use of voice as a tool, and, indeed, use of the body). Language in music and about music is not hermeneutically and normatively neutral or isolated, but belongs to a specific style and thinking about music, and belongs to activities, games, and a form of life. The moment one uses one's guitar, all this comes into play, all this plays. This is also true for improvisation: There are already patterns, there are games, there is a culture. Improvisation *responds* to that.

Another example is social media. Once we consider their *use*, we see that their content is also shaped by the medium: the technology and the language. The specific program or app shapes the kind of content. For instance, Twitter has its own kind of format and style for messages and Facebook has its own kind format and style of messages, and these formats and styles are not neutral towards the content of the message. For instance, if a message is limited to 140 characters (Twitter), this influences the content. Knowing social media means having the know-how to use these media, knowing the formats, styles, and patterns, and having the ability to adapt the message—including its content to the medium. Or, rather, there is not first a neutral or "authentic" message, which then gets adapted to the medium; rather, the message takes shape as one uses the medium. Furthermore, there are different uses of language in different media and different communities, which may or may not be different from "offline" use of language—but may shape that "offline" use of language, and are certainly also shaped by other ("online" and "offline") uses of language. [I prefer not to use the terms online and offline, since, as many authors have noted, both are increasingly mixed up today; see, for instance, Floridi's use of the term "onlife" (e.g., Floridi 2015).] In principle, the influence of the medium can be analyzed by using "mediation" terminology. One could use the expanded human-technology/language-world scheme I proposed, for instance, to analyze and show how uses of social media are mediated by language and technology.

However, here, it already becomes clear that mediation theory/language does not seem to enable an analysis that does sufficient justice to the more holistic environment in which specific uses and mediations take place, and to the role of language as "grammar" and form of life. A richer, expanded analysis, one that certainly goes far beyond Ihde's scheme of human-technology relations and that takes into account Wittgenstein's insights, would need to analyze the different language games and technology games that make possible and shape these uses. For instance, the activity of bullying and the language game of insulting take specific forms through the various technological platforms. To understand what goes on in specific cases, it is not sufficient to understand the "material" artefacts and its hermeneutics, including its mediating role, such as the smartphone and the code (if that is material at all). One also needs to understand the language games and (older) technology games. One needs to understand, ultimately, an entire (social media) culture. (I will further address this issue by means of a transcendental argument in the next section.)

Furthermore, as already became clear in the examples, in such a revised approach, an historical perspective and acknowledgment of the aspect of "givenness" is mandatory. The present uses and forms are not entirely new. For instance, before contemporary rock music, there was older rock music, blues, etc. and, before Facebook and Twitter, there were already language games and (older) technology games, such as writing e-mails, reading the newspaper, etc.; earlier, there was already letter writing. There was already a form of life, a changing one, but nevertheless a form of life in which a set of games were available. There were already linguistic and technological structures that shaped uses of language and technology (e.g., writing or use of voice). Uses of language and technology in rock music, for instance, are made possible by these preexisting forms and patterns. The games remain largely the same; what differs are the media and the uses. Perhaps one could say that mediation theory has emphasized the new so much, that it forgot about the old: It has shown how new technologies shape how we perceive (and what we do), but, lacking a transcendental and structural approach (see below), it has neglected the grammar of language and the grammar of technology. It has neglected the form of life that was already there before the new technologies kicked in. It has neglected the given. It has neglected what makes possible particular uses of technology and, hence, also what makes possible particular "mediations" by technology—and I added: by language. Of course, rock music and new social media also alter the larger whole. Think about the counterculture of the 1960s and 1970s, which then altered the mainstream. Consider also the influence of social media on politics today. Forms are adapted, they change. Cultures change, including music cultures and social-media cultures. But the magnitude of these alterations should not be exaggerated. New uses and cultural changes always tune in to what is already there.

To conclude, the proposed modifications of Ihde's framework help us to take into account the mediating role of language. However, I have started to argue that this expansion is not enough. I have proposed to revise the post-phenomenological approach by giving it a Wittgensteinian use-oriented and holistic twist, adding a structural argument concerning games and forms of life. This promises to give us a richer and more comprehensive analysis than can be achieved by only talking about mediation in terms of relations of embodiment, alterity, etc., or, indeed, by only talking about mediation as such. In the approach I propose, these relations and specific mediations are not rendered obsolete, but are now incorporated in a larger web of relations in which, next to Ihde's scheme and terms, also concepts such as language (words, grammar), use, activities, games, and form of life play an important role, and in which language and technology liaise and combine in various ways. This opens up a new kind of phenomenological hermeneutics of technology: one that is not only material but, at the same time, *also* linguistic and cultural.

Moreover, an additional advantage of the proposed Wittgensteinian revision is that it renders postphenomenology not only more linguistic but also more social. So far in Ihde and Verbeek, the stress is on the technological mediation of how individual subjects relate to the world. This is a limitation if we want to better understand the social and intersubjective dimensions of how language and technology shape our world. As Van Den Eede has suggested, mediation theory insufficiently takes into account the social, in the sense of an "in between" (Van Den Eede 2011). Use and the social seem at least disconnected. By contrast, the view I propose here acknowledges that our use of technology is also always social, as it is related to language games and technology games, and hence to a form of life.

Let me now further develop the proposed approach by taking a next step: a transcendental argument. I already used "make possible" and "transcendental" at various points in my analysis and discussion, but these suggestions need to be developed. Let me explain why I think **we need to take *a* transcendental turn to the question regarding language and technology.**

A Transcendental Turn

The expanded postphenomenological framework I proposed, which still focuses on mediation but takes into account various mediating roles and phenomenologies of language, did not yet take a full-blown transcendental perspective. At the end of the previous section, I made some suggestions in this direction by using Wittgensteinian terms such as "grammar" and "forms of life," but I had to hold back in order to stay close to Ihde. Now it is time to do more work on this and move beyond Ihde, who rejected a transcendental turn. If we add a transcendental approach to the framework, or, rather, transform it by giving it a transcendental twist, we can see language and technology not only and not so much as mediators, but, rather, as *transcendental conditions*: as conditions that make possible a particular human-world relation. I already proposed a transcendental turn in the previous chapter. Let me now further explain the nature and benefits of such a turn.

A transcendental approach, as I understand it, focuses not only on phenomena but also on the conditions of possibility of these phenomena, on what makes possible and structures the phenomena. It is a, by now, well-established traditional philosophical approach that can be found in Heidegger, Husserl, and Kant—among others. Without discussing this approach extensively (let alone discussing all variations of this approach), let me explain my working conception of "transcendental" as I will use it here, by using Heidegger and Wittgenstein, and apply this to language and technology.

Let me start with Heidegger. A lot could be said about Heidegger's use of transcendental arguments, but, for the present purpose, I will take a short-cut via Heidegger's philosophy of technology, which shows Heidegger's transcendental reasoning. Heidegger argued in 'The Question Concerning Technology' (1977) that modern technology makes us see the world in a specific way: as a standing reserve and an enframing. Now, this is a very specific view of what (a specific kind of) technology does, which has been criticized by Ihde and Verbeek as too pessimistic. But, apart from that specific view, and whether or not this is an adequate and fair understanding of Heidegger, for my purposes here, his *approach* is interesting: Apparently, technology *makes possible*, and provides constraints for, a way of thinking. And it does so not as a kind of agent that is "in between" me and the world, but, rather, as an ontological structure and a condition of possibility, as a whole culture, a technological culture in which I speak and think, and which shapes that speaking and thinking. The term "mediation" does not do sufficient justice to this "broader" or "deeper" role of technology, and it is at least worth exploring what it would mean to employ a transcendental approach to technology in contemporary thinking about technology. It seems that postphenomenology's out–of–hand and hasty rejection of the transcendental approach is unwarranted.

This transcendental move and the related criticism of postphenomenology is in line with Smith's arguments (see the previous chapter), but also with Zwier et al.'s (2016) Heideggerian objections to Ihde's postphenomenology. The authors argue that, in taking the empirical turn, postphenomenology has forfeited a phenomenological way of questioning. In particular, drawing on Heidegger's distinction between ontic and ontological, the authors argue that postphenomenology, with its mediation theory, has neglected the ontological and has taken a theoretical attitude, which is itself part of a technological way of revealing. In other words, according to them, postphenomenology as mediation theory is no longer proper phenomenology. The transcendental argument I propose here can contribute to a rehabilitation of the ontological, as proposed by the authors, since it attends to what they call 'an ontological structuring of reality'—with "ontological" understood in Heidegger's sense of transcendental conditions as opposed to the "ontic" (rather than "ontology" as a subfield of philosophy concerned with what "is"). Using this understanding of "ontological," we can say that technologies are conditions of possibility or "ontological" conditions that shape our thinking and speaking. It is questionable if this still can be put in terms of "mediation" at all, given its "ontic" connotations.

Similarly, influenced by Wittgenstein, I argued previously that our use of technology is always embedded in games—language games and technology games—and a form of life. We can reformulate this thesis in a transcendental way: These games and form of life *make possible* and structure particular uses of language and technology. Again, "mediation" does not really capture this, at least if it is interpreted as an "in between." Perhaps the "milieu" interpretation is better. But, then, without using a transcendental argument, it is not clear how use is related to this "milieu," and how this "milieu" is made up. A transcendental approach attends us to the *structure* of milieu. The Wittgensteinian terms I introduced seem to enable a more refined analysis of the relations between, on the one hand, use (of language and of technology) and the way we relate to the world in practical contexts, and, on the other hand, all kinds of larger wholes and structures that make possible that use and that relation, including the phenomenological relations Ihde talks about.

But is such a transcendental turn Wittgensteinian? There seem to be diverging opinions among Wittgensteinians, but if we read the *Investigations*, Wittgenstein is remarkably clear about his approach:

> our investigation is directed not towards phenomena, but rather, as one might say, towards the 'possibilities' of phenomena. . . . Our inquiry is therefore a grammatical one.
>
> (Wittgenstein 1953/2009, §90, 47e)

Wittgenstein, thus, clearly uses a transcendental approach. He uses the term "grammatical" rather than "transcendental," but it is clear that he is concerned with the possibilities of phenomena, in other words, with conditions of possibility. He is interested in the grammar that makes possible phenomena. Moreover, his concept of 'form of life' plays a key role in this transcendental method. As Gier (1980) and others have argued, forms of life 'perform a transcendental function' (257); they are not phenomena, but 'the grounds of phenomena, much like Kant's *Bedingungen der Möglichkeit der Erfahrung*' (242). Forms of life are 'the patterns in the weave of our lives that make a meaningful world possible (257). However, contrary to Kant and many other uses of the concept of transcendental conditions, Wittgenstein sees *language* as a condition of possibility. I propose to expand this and understand these conditions to include both language and *technology*; in particular, forms of language and forms of technology, linguistic-technological "grammar," so to speak.

Yet, transcendental conditions are not "grammar," in the sense of syntax (as in linguistics or as in Searle, for instance, who shows the grammar of declaration); they concern a "deeper" ground. Wittgenstein (1953/2009) distinguishes between 'surface grammar' and 'depth grammar' (§664, 176e-177e). We can see language games, technology games, and the form of life of which they are part as transcendental conditions, as forms, patterns, or (depth) "grammars" that shape, structure, and make possible our concrete uses of language and uses of technology, and, hence, our concrete thinking, experiences, uses, and performances with language and technology.

These linguistic and technological grammars are "underlying" in the sense of "making possible," but that does not mean that they are totally hidden or unconscious, as in Lévi-Strauss's structuralism. In a Wittgensteinian manner, we should stress that these grammars are not "hidden," but are open for everyone to see; we usually just do not think about them, since *in* use, we do not think about the linguistic and technological structures and infrastructures that make possible our use. Furthermore, as I already stressed in the previous chapter, a transcendental turn does not necessarily divert away attention from the materiality of technology. On the contrary, it is the materiality of technology (next to other aspects) that enables technology to play its transcendental role, next to its role as thing-to-be-used (tool). Technologies in their materiality, for instance in the form of infrastructure, but also in the form of the artefacts we daily use and in which we live, make possible our uses and performances. But it is always a materiality and functionality that manifests itself in and through use, since it is that *use* that establishes the link with the transcendental conditions: In and through our use of language and technology, we are also, at the same, time conditioned by language and technology. Materiality, as such, cannot "make possible," unless it is connected to us in our uses, interactions, and performances. (Similarly, material artefacts cannot mediate our perceptions and our actions, unless we use them.)

Using a transcendental argument, we can, thus, see language and technology not only as mediating within the subject-object relation, or human-world relation, but also as constituting the conditions under which we speak of that relation—and, indeed, of anything whatsoever. This linguistic and transcendental turn—but one which does not discount materiality—has implications for doing philosophy of technology. Any analysis of how technology mediates between humans and their world, and, indeed, any account of what things 'do' (Verbeek 2005), remains incomplete and not *critical* enough, unless one attends to, and questions, the influence of language and technology on the philosophical language one uses as a philosopher of technology. In this case, Ihde's and Verbeek's mediation language and, hence, also the mediation language I have used in my expanded version, should be analyzed critically. The term mediation suggests a kind of in-between: The metaphor is a screen, a line, or an other object that stands in between humans and their world. But (1) there are other ways of interpreting the concept, in particular the more ecological way I discussed earlier in this chapter, in which "medium" means, rather, *milieu*, although, as I said, this has its limitations, and (2) when we use technology and talk about technology as if it was, for instance, an instrument or a quasi-other (to take two very different views), then this language is itself embedded, shaped, and *made possible* by deeper structures that are already there in our language, in the social, and in our form of life as already shaped by material technologies. In the case of "social" robots, for instance, our language about robots relies on the ways we talk about, and deal with, human-human relations, and it is only because of particular material technologies, such as humanoid robots, that this kind of language use is triggered and, indeed, made possible. And if people respond to the discourse about social robots by saying that robots are "mere machines," this is, in turn, made possible and shaped by a language in which there are clear distinctions between subject and object, humans and machines, etc.

Moreover, instead of using both the language of mediation and transcendental language, "adding" this transcendental argument to the analysis, as in "put object A next to object B," one could also go further and give a transcendental "twist" or "turn" to the postphenomenological framework itself. One could reinterpret the role of technology—and language—as a "making possible," or "underlying" human-world relations, *rather* than as "mediating" and being "in-between." In other words, one could (to continue with place metaphors) *verticalize* the framework, bringing verticality to it, and drop the "horizontal" analysis, in terms of mediation.

Consider again a favorite example used by Verbeek (2008) when explaining his mediation theory: the ultrasound (echography) of the unborn child. His postphenomenological analysis is meant to show that the way parents see and interpret the child, think about having the child, indeed, their entire experience of the pregnancy, is shaped by the technology. This is then understood in terms of mediation: The technology (the screen, the instrument used to scan the womb) is literally in between the parents and the child, and this shapes how one thinks about the child and, indeed, which moral *decisions* and actions one takes. Verbeek (2008) argues that ultrasound technologies play a mediating role in moral decisions about abortion. In the subject-object relation, both subject and object are constituted by the technology as mediator: the (moral) subjectivity of the parents and the (moral) status of the object.

But another, transcendental way to put this, is to say that the technology *makes possible*, and therefore shapes, the parents' experience of the pregnancy, that it is not so much a mediator "between" parent and child, and between the human and their world, but, rather, a transcendental condition under which subject and object appear, emerge, and are constituted in the first place. This implies that both the subject (here, the parent) and the object (here, this could be the unborn child and/or the artefact) are constituted by technology as a transcendental condition, which makes possible that parents think and speak of the unborn child in a particular way.

This leads us to consider the role of language. Language plays the same transcendental role. The way the parents think and talk about the unborn child is made possible and shaped by both technology and language. "Before" they experience, before they put their experience into words (which presupposes that there is first something "mental" which then gets expressed, which is problematic from a Wittgensteinian point of view) or, better, *before they experience-think-speak*, there is already a language that has certain structures and meanings ready to use, and there is already the "grammar" of the technology to structure use and experience of the technology. Without the words and expressions "one" uses when talking about unborn children and without the technology, which lets the child appear in the first place, the parents *could not speak* about the unborn child in these ways at all. Both language and technology let the unborn child appear as a "child." From a Wittgensteinian point of view, they do not so much express inner thoughts, which are then translated into language and speech; instead, their speech is already structured by language and (I add) by technology. The expressions are already given, so to speak, with the language and the technology. While what they will say, discuss, and decide is not *determined* by language or technology, there are transcendental structures available, which they have to draw on in order to speak and think about the child—including making moral decisions—and which shape that speech and thinking, and those decisions. Furthermore, linguistic and material structures are connected: Because of the new technology, a new language game has emerged (e.g., talking about the option of abortion when going to get an ultrasound), which was not there before the technology. And because of a certain discourse (e.g., ethical discourse), designers may change the technology. Furthermore, these structures are part of a form of life, which is co-shaped and co-constituted by language and technology, and which, in turn, makes possible and shapes linguistic-material practices. Here, decisions about the unborn child are taken within a culture that emphasizes individual decision autonomy, for instance (here, the autonomy of parents).

This approach gives us a different kind of scheme, a more "vertical" one, which expresses that the ways we relate to the world, including our uses of technology and our use of language, are made possible, "carried," so to speak, by language and technology as conditions of possibility: They are transcendental conditions that make possible, structure, and constrain our relation to the world and to others.

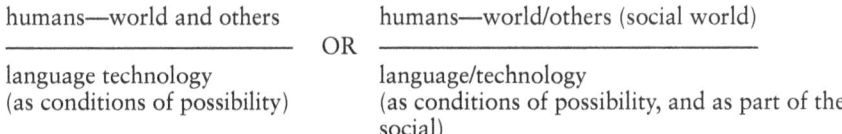

Figure 6.3.1 Language and technology as conditions of possibility.

The second version recognizes that the world cannot be related to independently from the social, that the social, like the body, is the way we access the world, or, rather, *is* our world. This social aspect is missing in Ihde's scheme; it only appears in the alterity relation, suggesting that the social is only one compartment of human experience and use, rather than that the social *permeates* human experience and use and is *made possible* by it. The second scheme also recognizes that language itself is a technology, understood as a tool (both are used), or is very much linked up with technology in use.

However, the horizontal lines still suggest too much a "carrying," without shaping and mutual influence, even without contact. Let me therefore render the scheme in a more Wittgensteinian and holistic form, with "|" expressing relations in two directions, since conditions of possibility are given, but not entirely unchangeable; they are also shaped by what happens on the ontic level, shaped by what we do, including our use of language and our use of technology. We can express this and put more emphasis on use in the following way:

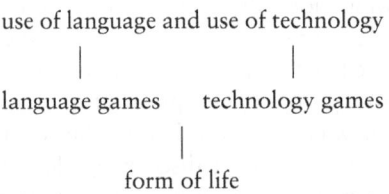

Figure 6.3.2 Use, games, and forms of life.

This scheme is not meant to imply one can no longer speak about and analyze specific human-technology relations. But these relations can now be studied in the context of an expanded framework, which connects the level of individual use of technology (and individual use of language) to activities, games, and a form of life. Moreover, I do not wish to deny that there is a sense in which we also "make" the social; but this making is always made possible, structured, and shaped by the social and the cultural as given and their existing games and forms of life. Perhaps one can summarize this as follows:

ontic level

language and technology used as tools (my reading of Wittgenstein)

human-technology relations in use (Ihde)

making the social by using words (Searle) and by using things

|

ontological level (in Heidegger's sense), transcendental structure

language and technology as transcendental conditions, conditions of possibility

language and technology games (Wittgenstein)

form of life (Wittgenstein)

(the social is already there, form of life is given)

Figure 6.3.3 The ontic level and the ontological level.

Thus, there are two levels, which are very much related to one another, but are conceptually distinct: one is ontic and relational in a "flat" sense, whereas the other is ontological and is about the conditions of possibility of experience and appearance; it is about structure. Moreover, we should add that use is the connecting term. Another way of presenting the different kind of thinking, which stresses the holism and puts use and performance in a central position, is therefore:

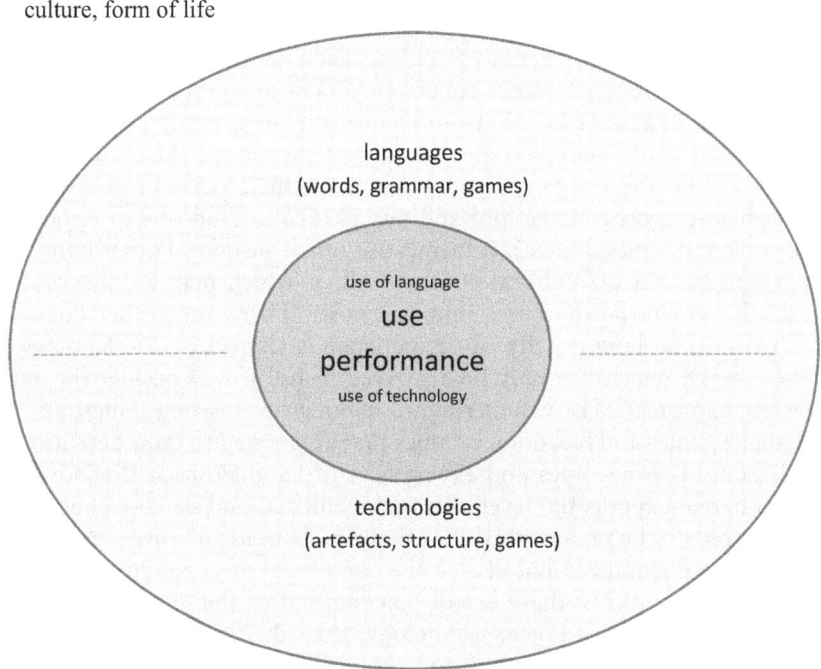

Figure 6.3.4 A holistic view of use and performance.

The scheme as a whole is relational in the sense that all uses of language and technology are related to larger structures, in the sense that they are embedded in them: Use and performance are embedded in structures of language and structures of technology, and these grammars are, in turn, embedded in a form of life. But both the graphic representation with two levels and the one with circles still have an important limitation: They may suggest a "state," whereas use of language and use of technology are not static matters; rather, they are processes, from which subject and object emerge. To address this issue, I suggest attending to temporality and adopting a process approach. Use of language and use of technology understood as performance take place in time, and the structures must also be placed in a temporal setting, or, to avoid using a spatial metaphor, they must be set in motion—they must be moved and they must move. In the next chapter, I will discuss narrativity and, in the last chapter, I will return to this point about process. Note also that I do not make a sharp distinction between "culture" and "nature," or "language" and "technology" in this scheme. In the next section, I will discuss Latour's nonmodern approach, to develop this.

Although this is a claim *about* use of technology and performance with technology, which is meant as a conceptual tool to guide analysis, and, hence, operates at a meta level and creates distance, the analysis of use and performance with technology itself need not be more abstract or less empirical. On the contrary, it even makes the postphenomenological analysis *less* abstract, since the holistic analysis reflects back on the individual experience and the way that experience is concretely linked to a practice and a culture. To take the hammer example from Heidegger: We can talk about embodiment in use of the hammer, that we do not experience the hammer when we use it. But this is a very limited analysis, which focuses entirely on the hermeneutic agency of the tool and one particular kind of experience and phenomenon (embodiment). It blends out other meanings, other individual experiences, and the cultural context—all of which may be phenomenologically present more or less simultaneously. There are further questions with regard to how exactly our experience is shaped by the hammer (in other ways), which can only be answered in full if we consider the entire activity and practice of hammering, of making or repairing things, and the language games and technology games that are related to these activities. To understand the meanings and experiences of hammering at the individual and concrete experiential level, this social-cultural and social-technological whole needs to be understood. This requires "empirical" investigation, not only of the phenomena but also of the conditions of possibility of the phenomena. Moreover, if there is still "mediation" of the human-world relation at all, by, for instance, technology, then this mediation is not only, and not so much, to be understood as an "in between" or as involving a kind of agent doing something to our perception and thinking (and, indeed, our morality, as Verbeek argued) but, rather, as a kind of carrying or making possible, which is at the same time also structuring and constraining. Interpretations of Ihde in terms of mediation and agency obscure this more holistic picture, and rejections of transcendental approaches close off this promising path of inquiry.

Moreover, as I suggested previously, the transcendental argument can also be applied to philosophy of technology itself. Understood as a speaking about technology, philosophical discourse about technology needs to be critically examined, since its language is not neutral, but constructs technologies in specific way. The transcendental role of language explains why Heidegger could only move beyond the instrumental understanding of technology by inventing a new way of speaking about technology. Even if one does not embrace his particular view of, and language about, technology, there is a need to talk about technology in a noninstrumental way, that is, to use a different language and to use language differently. Further development in the field will have to respond to this need, which starts with asking what kind of language we should use to talk about technology. Moreover, in hindsight, one can indeed interpret Ihde's and Verbeek's postphenomenology as introducing new ways of speaking about technology. In Ihde's writings, hermeneutic vocabulary is applied to things; this is an innovative use of language, since it makes a metaphorical bridge from language to technology. In Verbeek, speaking about "what things do" (Verbeek 2005), under influence of Latour, also introduces a new way of speaking about technology, accomplishing a metaphorical transfer from the language of agency (usually applied to humans) to language about technology. However, these innovations seem to have happened in an uncritical way, that is, they have been achieved in a way that was insufficiently aware and critical of its very toolbox: language. Ironically, in spite of their emphasis on materiality and things, their innovations and contributions to philosophy of technology were *made possible by language.*

Thus, in contrast to what Ihde and Verbeek think, a transcendental argument does not entail hovering high above material or "empirical" reality. Instead, this kind of argument helps us to further deepen (to continue my use of the place metaphors) the postphenomenological analysis, which remains focused on concrete technologies and experience and the lifeworld. In particular, expanding the framework in a Wittgensteinian and transcendental way enables us to connect particular human-technology relations and experiences to a larger technological-cultural whole and encourages us to think about the many concrete, empirical ways language and technology are entangled in concrete technological-linguistic practice and use/performance. Moreover, philosophers of technology should not be excluded from this picture of use and performance: They are also language users and technology users. To exclude them, indeed, to assume that they can stand entirely outside the picture, as a theoretical approach does, is both unempirical and uncritical.

Consider also my excursion to Dewey's view of language in Chapter 2: Dewey's claim that language makes possible the social (and "mind") can be interpreted as a transcendental claim: language is a condition of possibility for social interaction and institutions, and, indeed, for the "inner" life,[5] since it make possible and shapes that social life and that "inner" life. But using this argument does not go at the expense of anything "empirical," in the sense that, with Dewey, we can focus on concrete experience and interaction. With Wittgenstein and Dewey, we can go beyond the dualist assumption that use of language (e.g., philosophical language) would somehow be entirely disconnected from materiality—an assumption Ihde and other philosophers of the empirical turn did insufficiently take distance from when they shifted from language to technology at the expense of language. For Dewey, the cultural and the material are bound up in concrete use and experience, in lived use and social cooperation, in the practical social life that is made possible by all kinds of tools (including language) and from which all kinds of tools emerge (including language tools).

Note also that both Ihde's postphenomenological framework *and* my expanded, more integrated and complete framework, which accounts for the role of language, cannot only be used to theorize the phenomenology and hermeneutics of use and experience in an everyday context or contexts in science (which are also contexts of use and performance). They can also be used to better understand and guide design and innovation processes, which is another context of use and performance. Since it is focused on use and experience of technology more generally, the framework I propose can be applied across contexts to describe relations between humans, language, and technology, to describe various kinds of hybrid human/nonhuman "speaking," and to make explicit the normative implications of these forms of speaking, understood as uses of language and technology. This renders the expanded and integrated framework and approach I propose at least as "empirical" and widely applicable as the design-oriented postphenomenology of, say, Verbeek.

To conclude, analyzing the role of language, using Wittgensteinian concepts, and taking a transcendental turn, need not lead to the language-obsessed approaches in the humanities Ihde rightly criticizes. Instead, postphenomenology's project (and the empirical turn's project), to take seriously material artefacts and to build bridges between the humanities and the technological-scientific world, can actually be *supported* by this approach, at least if it succeeds in moving more towards a synthesis or integration between thinking about language and thinking about technology. I have shown that such an integration is possible, and that it is even possible to transform the (post)phenomenological approach to philosophy of technology in a way that opens up new insights into what it means to use technology, understandings that take into account language in its mediating and transcendental roles.

In this section, I used Wittgenstein and Heidegger. Let me now return to another thinker who did say more about the role of language, but, at the same time, also took part in the larger project of what one may call "rehabilitating the object" and in thinking about the material aspects of science and technology: Latour.

Ihde and Latour

Another way to elaborate a more integrated view or synthesis with regard to language and technology is to connect postphenomenology to posthumanist thinking. Consider again the posthumanist interpretation of McLuhan (Stephens) or, indeed, the nonmodern view of Latour. One could connect the hermeneutics of things to the hermeneutics of language by seeing both things and words as belonging to the same relational and social whole— environment, ground, world, or ecology—which shapes and makes possible uses, performances, and, ultimately, the (post)human being-in-practice, part of a whole in which use, practice, and what we call "the human" grow and flourish. Let me briefly discuss some links between Ihde and Latour to start developing this thought; then, in the next section, I will further read and discuss Latour's view on its own merits.

Ihde himself claims that his own view is compatible with that of Latour. He endorses Latour's claim that we have never been modern, says that Latour's conclusions are commensurable with his own, and suggests that, like his own work, Latour's work intends to bridge the gap between the human sciences and the natural sciences (Ihde 1998, 49). He even sees Latour as one of the 'field-clearers' for the expanded hermeneutics he proposes. He discusses what he calls 'Latour's hermeneuticization of the laboratory' (149), stressing that texts themselves are insufficient for understanding truth in science. Instead, Ihde argues, one has to go to the laboratory where instruments produce inscriptions or visual display (149). Reading his material hermeneutics into Latour, Ihde helpfully reads Latour as showing that 'the instrument is already a hermeneutic device' and that 'hermeneutic practice lies at the very heart of the laboratory' (149). He then argues that, in the laboratory, the 'scriptorium' of the scientists, objects are made readable, and adds a focus on 'scientific visualism.' Like Latour's approach, Ihde's expanded hermeneutics also enters into the realm of scientific things. Although Ihde emphasizes that his formulation draws on Husserl and Merleau-Ponty, that is, on his own research history in phenomenology and hermeneutics, he writes in a very Latourian way:

> 'The postmodern hermeneutics of things must find ways to give *voices* to the things, to let them *speak from themselves*. . . . For both Husserl and Merleau-Ponty, voiced language is a bodily and fully perceptual activity. It is that materiality of perceptual activity which I seek in this thingly hermeneutic.'
>
> (Ihde 1998, 151)

Thus, here we find a very explicit claim about things that speak, one that, indeed, seems to come close to Latour's. It seems that, here, Ihde "thinks through the laboratory," as I call it. However, it soon becomes clear in the text that Ihde understands the claim that things speak in a literal way. When he briefly discusses under which conditions things can have a voice (which is very interesting for thinking about voice and language), he distinguishes between two conditions: Things can be given a voice by humans when they are struck, for instance, by musicians (151), or things can have a voice, but we cannot hear it; it may be already "sounding," but we cannot perceive it, unless the "sound" is mediated by an instrument (152). Latour, however, has a very different, more metaphorical understanding of things that speak, and is less interested in embodiment. He thinks through things. He has let things enter his language and has given language to things. This has then influenced the "things" language of Verbeek and others.

I already discussed Akrich and Latour's semiotic approach and its implications for understanding technology; let me first revisit that view, but then discuss Latour's later (or "middle") work on non-modernity, and explore how we can use and interpret that work in order to arrive at a synthesis, according to which not only language but also humans and *technology* speak—with "speaking" understood in ways that go beyond Ihde's literal interpretation, ways that connect language and technology in a more intimate way.

6.4. What Technology Tells Us (To Do): Latour's Nonmodern View

When Things Speak (1): Akrich and Latour

In Akrich and Latour, there is already a sense that artefacts "speak," since artefacts have inscriptions and can even prescribe to humans. Here, there is a strong link between technology and language. Things tell us things, and things tell us what to do. In contrast to Ihde, who mentions these ideas but has little interest in further discussing links between (his) hermeneutics and semiotics, here, artefacts can be seen as speaking in the sense that they take up a linguistic-performative role. Retrieving the linguistic-performative dimension of this part of Latour's work, in a more explicit way than I have done so far, can help us to establish a synthesis in which not only artefacts and humans but also *language* "speak." Let me explain.

Employing the expanded mediation theory proposed above (but not the transcendental argument), one could use the second scheme and interpret this part of Latour's work as implying that "things speak," in the sense that *language mediates between humans and artefacts*. Consider, first, what kind of speech and agency artefacts have. It makes sense to say that, in Verbeek's words, things "do" things, but the way they "do" things is linked to, and highly dependent on, both humans and language. The performativity and (written) speech of things is a kind of derived performativity and speech, since *humans* remain the main users of language and, in this sense, the main performers, or at least indispensable co-performers. In Akrich and Latour, things only "use" language and "speak" in a derivative sense; humans are the "first" users/performers. Second, language as grammar always mediates the relation between humans and things. Through linguistic forms and language games, such as prescription and translation, artefacts are empowered to speak, to tell things, and tell humans and things what to do. (In Akrich and Latour, this is mainly understood metaphorically, but it can also be taken literally: Consider today's ideas concerning "Internet of things"— communication is always via code, that is, it is linguistically mediated.) Thus, in this interpretation of Akrich and Latour, there is still more or less a balance between the speech power of humans, language, and things.

Taking into account the roots of Akrich and Latour's view in narrative semiotics (and, according to my interpretation in Chapter 4, its link with Wittgenstein), one could also use this interpretation to helpfully add a *narrative* and *performative* dimension to Ihde's postphenomenological account of mediation: Mediation by technological artefacts has a narrative structure as it is part of an action program involving human intentions and nonhuman actants as helpers. There is a sense in which the artefact speaks as it *performs* with words, in the process of which it acquires competence. Postphenomenology and mediation theory either neglect or do not further develop this aspect of performance with words, since they downplay the linguistic dimension in Latour (and elsewhere).

However, the "semiotic" part of Latour is very much focused on the technologies of writing (script) and text—consider terms such as "inscription," "prescription," and so on. The materiality of things is present, but language is dominant. Postmodernism and poststructuralism, here in the form of semiotics, reign. (This also means, for instance, that embodiment is largely absent. For instance, the voice is absent in Akrich and Latour. As we will see, in later work, Latour writes more on voice, but even then this voice is only understood metaphorically, that is, disembodied and dematerialized. It is a political "voice"—speech is disembodied language.) Moreover, whereas here, human language users are still in the foreground, in the later, more explicitly nonmodern part of his work, humans are no longer in the driving seat and language also retreats to the background. *Things* take center stage. Let us now turn to this part of Latour's work.

When Things Speak (2): Latour's Nonmodern View and Pickering's Emphasis on Performativity

In *We Have Never Been Modern* (1993) and *Politics of Nature* (2004), we see that there is more emphasis on things, and, in particular, on what things say, on the speech and voice of things. There is much less emphasis on semiotics and written language (text, script, etc.). Moreover, things speak, but neither their speech (understood as language) nor the speech of humans receives much attention; language is no longer of central concern to Latour. Moreover, humans and things are on the same horizontal ontic plane. It is unclear if, in this view, there are still concrete, living and performing language users at all—human or nonhuman—and if there are such conditions, like games and a culture, which make possible that use. Let us look more in detail at this part of his work.

In *We Have Never Been Modern* (1993), Latour criticizes our modern ontological distinctions between humans and nonhumans, and between nature and culture. In modernity, including modern science, we make these distinctions when we describe what we do. But, Latour argues, scientists have always created hybrids, in spite of modern practices of 'purification' that try to maintain distinctive 'ontological zones' (Latour 1993, 10). (Latour uses "ontology" not in a Heideggerian way, but in the sense of thinking about what "is," common in philosophy.) Inspired by Shapin and Schaffer's history of Boyle and Hobbes, Latour argues that science is 'based on forms of life, practices, laboratories and networks' (note his use of the term 'forms of life') and is neither situated on the side of the object (things-in-themselves) nor on the side of the subject (society, language game, culture, etc.; 25). Instead, epistemology and social order, science and society, objects and subjects mix and merge once we see things from a nonmodern perspective—and, according to Latour, we have always been nonmodern, anyway. There are networks of humans and nonhumans. Science is political, and politics—as we will see soon—also involves things.

Latour explicitly rejects postmodernism, which he sees as a symptom of the problems of modernity. We cannot be postmodern because we have never been modern in the first place: 'we have never really left the old anthropological matrix behind' (47). He now also turns away from semiotics, semiology, 'linguistic turns,' and so on (62), since they aim at creating 'a mediator independent of nature and society alike' (62). Latour rejects postmodern approaches that 'limited their enterprise to discourse alone.' which 'has taken over the entire space' (63). Like Ihde, Latour understandably leaves this 'Empire of Sings' (63) behind. He seems to agree with the postmoderns that 'texts and language make meaning' (63), but, when 'everything becomes sign and sign system' (63), he rightly asks, where is nature and the social? He writes:

'Are you not fed up with language games, and with the eternal scepticism of the deconstruction of meaning? Discourse is not a world unto itself but a population of actants that mix with things as well as with societies, uphold the former and the latter alike, and hold on to them both. Interest in texts does not distance us from reality'

(Latour 1993, 90)

Instead of creating an autonomous sphere of language to avoid the 'traps of naturalization and sociologization' (64), Latour sees nonmodern hybridity, in the laboratory and elsewhere. He sees scientific objects 'circulating simultaneously as subjects objects and discourse' and machines that are 'laden with subjects and collectives' (66). He claims that there never has been a modern world (67).

Latour's non-modernism offers an interesting view to respond to with regard to the integrative, synthetic aim of this book: One could conclude from it that, here, neither language (discourse) nor humans (subjects) nor technological artefacts (objects) are reduced to one of the other poles. In a nonmodern world/view, it is not problematic to say that things speak, that discourse speaks, or that humans speak. They all seem to speak (and do things) simultaneously, without, however, language and speech taking over everything. This view is achieved not by means of purification, not by keeping nature, society, and discourse separate, but by acknowledging hybridity. Latour's view enables us to understand not only the politics of the vacuum pump but also today's politics of genes and cyborgs. It is a view inspired by anthropology, which comes home from the tropics, as Latour says (100), and is then applied to the world of science and technology. Like the so-called "savages" studied by earlier anthropology, humans in the West and other so-called "modern" cultures also live in a nonmodern way, in spite of the modern language they use. There is no Nature or Culture (at the expense of bracketing off one or the other); instead there are natures-cultures (96). And I propose to add: natures-cultures-*discourses*. One should not bracket off discourse, either.

Latour also emphasizes that there are always mediations and relations. He proposes a 'relationism' (114) and—in line with his earlier actor-network theory (ANT)—he thinks in terms of networks. The human becomes 'a weaver of morphisms' (137), a human that can no longer be threatened by machines, since 'it has put itself into them' (137). There are no longer objects, since they have become quasi-subjects. Once there is no longer a Nature or a Society (or a God), and, once the human is seen as 'delegated, mediated, distributed, mandated, uttered' and so on, there is no longer a threat (138). (Note his use of the term 'uttered;' here, theory about language is still in the background.)

At the end of the book, Latour proposes a nonmodern constitution and a 'Parliament of Things' (144). Instead of a parliament of mutes, the laboratory, where scientists speak 'in the name of things' (142), separate from what is considered 'politics,' a parliament of humans that speak, he proposes to also accept political hybridization, a parliament in which representatives and mediators speak in the name of citizens, societies, and natures—again, a hybridization, which is not new but has always been going on. The representatives speak about 'the object-discourse-nature-society' (144). Thus, here, Latour includes discourse, but discourse no longer takes all the space, as in postmodernism; discourse is mixed with objects, nature, and society. It is more material, natural, and social than postmodernism's discourse has ever been. The signs of the postmoderns get wet and dirty.

What does this approach mean for the relation between human speech acts and the "speech" of artefacts? In contrast to the earlier Akrich and Latour work, here humans do not delegate speech to artefacts; instead, human speech represents artefacts. Perhaps one can also interpret Latour as saying that artefacts speak *through humans*, that, if there is delegation at all, it is a delegation from things to humans, rather than (only) the other way around. According to this interpretation, artefacts do not mediate human speech, script, etc., but humans mediate the "speech" of artefacts. Then, humans are not only mediated (by technology and by language); they are themselves mediators—mediators in the service of things. And I propose to add: in the service of language.

This takes us to *Politics of Nature* (2004), in which Latour further develops his view. Let me pay specific attention to language and speech, and their relation to technology.

In the book, Latour further criticizes the modern dichotomies, which leads him to the claim that *political ecology has nothing to do with nature* (5). Instead of a 'division of labor between human politics and the science of things' (5), and holding on to a notion of nature, he imagines a new collective, which no longer separates nature from cultures, or science from politics and the social world. In this new political epistemology and political ecology, there is room for beings, such as 'rules, apparatuses, consumers, institutions, mores, calves, cows, pigs, broods,' and so on (21), and for a mix of facts and values. Beings such as snails or earthworms (25) can have political agency. There is no longer a split between the assembly of things (nature) and the assembly of humans (society); instead, there is a 'pluriverse' (40), a collective, which, like non-Western cultures, is not interested in 'nature' (43), but, instead, is inspired by anthropology of non-Western cultures, is conceived as consisting of both things and persons. Latour writes:

> 'There are only nature-cultures, or rather collectives. . . . We now see the reversal of perspective: the savages are not the ones who appear strange because they mix what should in no case be mixed, "things" and "persons"; we Westerners are the odd ones, we who have been living up to now in the strange belief that we had to separate "things" on the one hand and "persons" on the other into two distinct collectives'
> (Latour 2004, 45)

Latour imagines a social world in which nonhumans are no longer objectified and humans are no longer subjects that resist objectification by science (51). Instead, in that world, there are 'assemblages of humans and nonhumans' (52). In a single collective, humans and nonhumans can exchange properties; thus, there is pairing of humans and nonhumans, which was forbidden by the subject-object opposition. This means, also, that all kinds of beings can have 'speech' (61). Indeed, it is interesting for my present purpose that Latour extends speech to nonhumans, to science, and so on. This implies that citizenship and politics, which was always defined in those terms, can also be extended to nonhumans. Arendt (1958), for instance, connects politics with action and speech in her articulation of the ancient Greek conception of politics. Latour can thus be interpreted as retaining the old idea of politics, but arguing for a wider application. The collective is still defined as 'an assembly of beings capable of speaking' (62). What changes is the border of who or what can speak. Previously, things were not supposed to speak, they belonged to 'the nature of mute things' (62). But Latour liberates them, or, rather, observes and conceptualizes their liberation. The crossing of the nature/politics and science/politics barriers happens in the form of controversies between scientific experts; suddenly, science is also a world of 'endless discussion,' that is, a political world (63). And politicians speak about things and get mixed up in scientific controversies, such as climate change.

But Latour goes further: There is also a sense in which things *speak*. For Latour, 'speech is no longer a specifically human property, or at least humans are no longer its sole masters' (65). What does he mean by this? First, in the lab, it is not scientists that speak; instead, they 'have invented speech prostheses that allow nonhumans to participate in the discussions of humans, when humans become perplexed about the participation of new entities in collective life' (67). Technology is used to let things speak and spokespersons speak for them. Thus, things speak, but not on their own; they need spokespersons and technology. For Latour, not even humans speak on their own; they always speak 'through something or someone else' (68). Nonhumans also have 'voices' (69). Facts speak for themselves through the scientist and through technology. And things act. There are a series of social actors and *actants* (76–77; my emphasis). There are surprising associations (79). Latour suggests that it may even be pleasant for humans to experience the presence of new nonhumans (80).

Thus, actors and actants speak, human and nonhuman. Each has their competences. Scientists use their instruments and laboratories 'to make the world speak, write, hold forth' (137). The sciences thus socialize nonhumans (235). Politicians produce voices 'that stammer, that protest, that express opinions' (145). And politicians deal with things and people (145). Economic calculation makes diverse things, such as 'black holes, rivers, transgenic soy beans, farmers, the climate, human embryos, and humanized pigs' commensurable (151–152). Moralists look out for 'invisible entities and appellants' (162). Indeed, this politics has, of course, also a moral and normative side: Propositions and voices should be taken into account. An entity can knock on the door of the collective to 'demand that it be taken into account' (125). Moreover, being part of the collective means speaking its language. Latour uses a narrative form that mixes human and nonhuman speakers and the metaphor of political and legal proceedings to show the appeals and demands that knock on the door of the collective:

> Thus the propositions are now already almost involved in the collective; in any case, *they are beginning to speak its language*: "I cause a deadly and unforeseen illness," says this virus and its virologists; "I demand the means to modify cosmology profoundly," say that pulsar and its accompanying radioastronomers; "I pay and yet what I want is not taken into account," says this consumer and his means of calculation; "I propose to modify cosmology even more profoundly," say that flying saucer and its ufologists; "I cast spells," say this fetish and its fetishist.
> (Latour 2004, 168)

In response to these demands and appeals, there are proceedings. There is inquiry, consultation, and so on. Propositions have to be translated (172). Then the following problem needs to be addressed: 'How can these contradictory beings be made to live together? How can a world be produced that is common to them?' (173). This construction of the Leviathan is a difficult task, since modernism, which eliminated most beings or excluded them as fictions, can no longer be used (173). There are negotiations and attempts at stabilization. There are 'pains of the progressive composition of the *cosmos*' (187). There is triage and experimentation, 'collective experimentation' (197). But, on the way, we have to keep track of 'what we have been able to absorb and what remains outside' (200). Administration is needed (204) with the power to follow up, as are diplomats. At the end of the book, Latour says he wants to rediscover common sense (222) and redefines equality as a term that 'asks us to take responsibility for nonhumans without knowing in advance what belongs to the category of simple means and what belongs to the kingdom of ends' (227). Political ecology thus becomes 'experimental metaphysics' (235).

This nonmodern view questions, for instance, Searle's very modern assumption that there is a clear distinction between a physical and a nonphysical, social reality. If things "speak," then the work of social construction, of meaning-giving, is no longer necessary; or, rather, humans are not the only ones who perform speech acts and they no longer are in full control as linguistic agents. There are no longer dead things that need to be animated by human subjects. They are already meaningful, or already have or acquire what we could call "speech subjectivity" through humans and through technologies. With regard to political speech, subjects are a bit less subjects, perhaps, and objects are a bit less objects. Furthermore, while compatible with the claim that things also have agency, Latour's nonmodern view also questions Ihde's and Verbeek's mediation theory, since it is not only the case that humans are mediated by technologies (and by language) but also that humans themselves become mediators.

Yet, one may even wonder if Latour goes *far enough* when it comes to offering an alternative view of the social. According to Latour (and also according to Searle and the social construction of technology approach), the social still needs to be produced and there seems to be a kind of ontological "state" (with a "constitution"). But, if we take the nonmodern approach seriously, perhaps the very business of ontology, including social ontology and social construction, is problematic. In a nonmodern view, things are already meaningful, there is already a social bond. Then construction or agreement, as in Searle, but also still in Latour, is no longer needed, *at least not for establishing the social bond*. The production of the social is not a matter of collective intentionality, as in Searle. And there is no longer a Hobbesian problem of social order, as in Latour. Influenced by Wittgenstein and Heidegger, I already pointed to the given: There is already a form of life, a grammar, which includes linguistic and technological conditions of possibility that make possible, shape, and structure our use of words and things—and, more generally, the "ontic." Therefore, I propose to interpret and revise Latour, not so much as proposing an ontology, but, rather, as holding a *process* view of social reality: the social is constantly being (re)produced by means of humans and nonhumans, including technologies. And in this process, we have to recognize that there is a lot already given.

This *process* interpretation of Latour's view of the social and the political is compatible with Harman's, who interpreted Latour as presenting politics as a continuous process of formation of the collective, which always extends the number of those who can "speak" (Harman 2014; see also Heaney 2015). Harman (2014) writes: 'Latour recommends a constructivist path . . . in the sense of a theory that treats nature and culture as both constructing through due process' (73). This process includes administrators and bureaucrats. However, in response to Latour and Harman, I wish to emphasize that a lot also remains the same and stress the structural and transcendental role of this process. When we speak and act, this is *made possible and shaped* by this underlying political and social process. Moreover, later I will propose a *performative* reading, which highlights the role humans play as users and performers.

But, even if we limit ourselves to Latour's own view as expressed in the book, we should ask: What exactly is the role of technologies, and what is their relation to *language*? What does Latour's view imply for the relation between language and technology? In his book, not much attention is paid to presuppositions about politics and language. But, for my project, it is essential to make these explicit. On the basis of the text, we can conclude that, according to Latour, technological artefacts and technologies have at least two functions with regard to speech: (first) they can be instruments to help things acquire a "voice," to speak, and (second) they can be those "wanting-to-speak" things themselves, as they are part of the collective of humans and nonhumans. In Latour's political imagination, technological artefacts are liberated from the mute kingdom of things; they are given a voice or they help to give other things a voice.

More generally, it turns out that language is of crucial importance in Latour's view of politics, since politics depends on it. Politics is all about who gets a voice, and about giving voice. The "only" modification to this old model of politics Latour undertakes is that he includes nonhumans in the political-linguistic collective and dynamics. As I indicated, this assumes an old, even ancient, model of politics as speech. And, indeed, Latour admits that he uses 'outmoded' terms such as 'speech' and 'discussion' (221). In this old and perhaps outmoded view of politics, language, thus, appears as "voice," mainly, but also in the form of propositions and other nonhumans (or: other technologies). This gives us another (third), perhaps surprising way in which language and technology are connected in Latour's text: Language itself can be understood in terms of "nonhumans," in terms of objects. Latour sees *propositions as nonhumans*. This makes a strong connection between language and technology: It suggests that language is a technology, that *language consists of artefacts*. This enables us to connect Latour's view to the use-oriented Wittgensteinian view developed earlier and to McLuhan's equation of technology and language. With Wittgenstein, I have argued that we use words in the same way as we use tools, and, therefore, that both can be analyzed in the same way. And with McLuhan, I have argued that language is itself a technology. But, in contrast to Wittgenstein, Searle, Austin, and so on, Latour gives not only speech to humans but also to these nonhuman linguistic and technological objects. If there is use and performance, the agency is now not only on the part of the human but also on the part of linguistic and technological objects. This is revolutionary indeed. *Things* also perform and speak, and these "things" include not only technological artefacts but also linguistic tools. Thus, in this view, we use tools and words, sentences, etc. as much as we *are used* by them. In non-modern social performance, there is a dance of humans and nonhumans, which are all performing, using, and *speaking*.

This "performative" reading of Latour is in line with Pickering's interpretation of Latour. Let me briefly discuss his view to further elaborate my own view, and also to connect back to the discussion about social constructivism in Chapter 4, since Pickering writes as a sociologist of science. In *The Mangle of Practice* (1995), Pickering argues for an approach to studying science, in particular, an approach to the sociology of scientific knowledge, which is focused on performativity and emergence. Like Latour, Pickering argues that scientific practice has a material and social dimension. Artefacts and, indeed, machines are used to produce knowledge. But they are not passive things; they also perform. Pickering proposes a 'performative image of science, in which science is regarded as a field of powers, capacities, and performances, situated in machinic captures of material agency' (7). Like Latour, Pickering's performative understanding of science adds the agency of things to human agency. But not only science but also everyday life and society are 'a field of human and nonhuman (material) agency' (11). His view takes seriously the fact that much of everyday life 'has this character of coping with material agency' (6). Performance, then, is not only about what humans do but also about what things do (to use again a phrase from Verbeek 2005). For Pickering (1995), the world is about 'doing things' (6). Like Latour's, his metaphysics is 'symmetrical with respect to human and nonhuman agency . . . each is constitutive of science' (11) and—we may add, with Latour—constitutive of the social. Furthermore, Pickering points out that this Latourian view, which takes material agency in science as seriously as human agency (14), is based on semiotics:

> Semiotics, the science of signs, teaches us how to think symmetrically about human and nonhuman agents. In texts, agents (actors, actants) are continually coming into being, fading away, moving around, changing places with one another, and so on.
>
> (Pickering 1995, 12)

This raises the question regarding the difference between human and nonhuman agents. In contrast to Latour, Pickering thinks saying that there are no differences is problematic. He acknowledges that 'as agents, we humans seem to be importantly different from nonhuman agents like the weather, television sets, or particle accelerators' (15). He prefers to say, not that both are the same, but, rather, that there are parallels and intertwinings (15).

In any case, by pointing to semiotics in this context, Pickering helpfully connects the earlier, "semiotic" phase of Latour with the later "nonmodern" work: He points out that the latter is based on the former, since semiotics gave the model for a more posthuman way of thinking in which humans and nonhumans are part of the same process and space. Indeed, Pickering sees 'a posthumanist space, a space in which the human actors are still there but now inextricably entangled with the nonhuman, no longer at the center of the action and calling the shots. The world makes us in one and the same process as we make the world' (26). He argues that this is how things are *in practice*. This is interesting for my purposes, since the idea that "the world makes us" can be linked to my transcendental point—which, again, turns out to be very *practical* and empirical. To take up Pickering's example: Machines also "make" us, in science and (I add) everyday life. Moreover, it is also, again, a view of performance that includes both human and nonhuman performers. This comparison between language and the social enables Pickering to emphasize performativity, process, and emergence in his view of science as one 'in which the performances—the doings—of human and material agency come to the fore' (21). Furthermore, in this view, the human and, for instance, human goals do not disappear. According to Pickering, humans and machines 'collaborate in performances' (16) and he pays attention to humans skills and also to time: He says that 'we humans live in time in a particular way' (18). As humans, we have plans and goals, for instance—even if these goals are part of 'the plane of practice' and must be understood as themselves emergent (20). He stresses the temporal dimension of practice: Goals are part of, and are transformed in, 'real-time practice, which includes sensitive encounters with material agency' (20). He offers an image of science as 'a dance of agency' (21):

> As active, intentional beings, scientists tentatively construct some new machine. They then adopt a passive role, monitoring the performance of the machine to see whatever capture of material agency it might effect. Symmetrically, this period of human passivity is the period in which material agency actively manifests itself.
>
> (Pickering 1995, 21–22)

This manifestation includes resistance of the machine, to which humans then respond with accommodation. Pickering sees a 'dialectic of resistance and accommodation,' which can include revisions of goals and intentions (22). In this 'mangle' (23), then, various reconfigurations of humans and nonhumans emerge. We cannot fully predict beforehand what will emerge—human goals, but also material agency. Pickering gives the example of Glaser's bubble chamber and stresses that the scientist did not know in advance how the material configuration of the chapter would evolve—things 'just happened' (52). In this process, emergence of material agency is entangled with human agency, and neither is reducible to the other (54). One could say that it is not only the case that what humans do, plan, intend, etc. shapes what technology does but also vice versa: What technology does also shapes what humans do, plan, intend, etc. He writes about the scientific practice of Glaser:

> Though Glaser formulated the goals of his practice as a classically human agent, the field of existing detectors in which he formulated those goals was a field of material agency. There is, then, a temporal and posthumanist interplay here between the emergence of material agency and the construction of human goals.
>
> (Pickering 1995, 56)

Human goals are thus formulated by humans, but they are revised in the process and are also to be seen as emergent. Human goals and technologies transform one another. Our human goals are not independent from technologies; machines shape our aims and our plans. Using a transcendental argument, one could say that human goals are also made possible by technology-in-practice, as a condition of possibility, including by what Pickering calls 'machines and their material performativities' (20). Thus, Pickering sees the social, as a whole, as emergent and as 'in the plane of practice' (61), with its posthumanist mangles and struggles, with its 'human and material performances' (58), with its processes in which knowledge, and material and human agency, are 'all configured at once' (67).

In recent work, Pickering says more about his (social) ontology of emergence and becoming: He speaks of 'an ontology of decentred becoming' (Pickering 2017, abstract) and further elaborates the nonmodern dimension of his view by comparing science with shamanism. He argues that, if one moves from a representational idiom to a performative idiom, there is not one world (as Latour's Actor Network Theory suggests) but different worlds, and that we cannot predict where performance will take us: 'Who knows where dances of agency can take us?,' he asks. Responding to the ontological turn in STS and anthropology, and in line with Latour's symmetrical view, he says that his ontology 'is a symmetric one of a multiplicity of reciprocally coupled emergent agents, human and nonhuman.' Technology is, then, one of these 'performative agents;" in the world of shamanism, spirits are also performative agents. His view is also relationalist, but with an emphasis on temporality: His view 'foregrounds the temporal evolution of relational entities' (Pickering 2017).

This seems a helpful elaboration and revision of Latour's view; the latter does not emphasize practice and, at first sight, is less dynamic, seems to suggest a more static and less multiple ontology. Latour may well be sympathetic to process and emergence approaches, but this is not always clear in this work, which lacks sufficient attention to the temporal dimension. His ontology of networks of actors and actants also seems to suggest one world rather than many. Moreover, in his efforts to let things speak, he also tends to disregard or deemphasize human intentions, goals, etc. Pickering's view *explicitly* accounts for this human side.

This performative, emergent, process, and time dimension of technological and scientific practice is also not sufficiently theorized in Ihde, whose view of technological hermeneutics is perhaps too static, since it does not sufficiently take into account the process dimension and is not sufficiently pragmatic and focused on use, performance, and practice, understood as a process that takes place in time and also depends on human users having goals, intentions, etc. And while Verbeek, influenced by Latour, has stressed the agency of things (Verbeek 2005), his mediation theory is not linked to the concept of performativity and is divorced from its source in semiotics. Pickering and Latour can be read to retrieve this performative dimension and to add the concepts of process and emergence to our conceptual toolbox. Pickering also helpfully pays attention to the time dimension of scientific and technological practice; in the next chapter, I will say more about this. Finally, Pickering at least takes into account what humans intend and plan; it is, thus, more appreciative of the fact that humans also speak—even if this speaking is always also emergent and (as I added) made possible by technology.

However, in Pickering's view, it is not clear what the role of *language* is and what the relation is between "what things do" and "what words do." Pickering unnecessarily restricts nonhuman agency to *material* agency. In Pickering and in Latour, language is mainly used as a metaphor. In both Pickering and Latour, semiotics is a metaphorical toolbox that is applied to technological and scientific artefacts. In Latour, it requires work to reveal and articulate the speaking and agency of language. Pickering's view is sufficiently performative and helpfully adds the concept of emergence (see also my last chapter), but does not really say much about language. Like in Latour, the stress is on the agency and performativity of things; we can only guess about the agency and performativity of *words* and language, and about the relation between doing things with things and doing things with words. To remedy this, I propose to *read more language into Latour*. Both words and artefacts perform. We use material and linguistic things. And this not only happens in scientific practice but in any kind of social practice. If we must have a posthumanist view, in which the world is a matter of both human and nonhuman agency (of humans and nonhumans doing things) at all, then we should not only acknowledge that artefacts do things, as Pickering and Verbeek say, but we should also include *what language does*. If nonhumans perform and dance with us, then those nonhumans should not be restricted to material things. Without making the mistake of setting up, again, a universe of signs and forgetting about things, we should take the insight that Latour's view is routed in semiotics as an encouragement to read language back into Latour-style posthumanism; in other words, we should include language in the performative process, reinvite words to the dance and performance of humans and things.

To conclude, the linguistic and performative dimension is currently receiving too little attention from contemporary readers of Latour, such as Ihde and Verbeek. And although as Pickering also shows, there is a performative dimension in Latour's work, the link with, and attention to, *language* has slowly disappeared. Both in Ihde and Latour, there is already some recognition of the importance of language and discourse, even if not always very explicit or lengthy, and, as I have shown, we can interpret and revise these aspects in order to gain a more comprehensive understanding of the relations between humans, language, and technology, and, hence, also between subject and object, culture and nature, and so on. However, due to empirical turn philosophers' reaction against the postmodern and humanities emphasis on humans and language, Ihde, Latour, and their contemporary readers (e.g., Verbeek) have employed the rhetorical strategy of inversion: In their urge to draw attention to artefacts, *as opposed to the postmodern stress on language*, they have overemphasized artefacts (things, nonhumans) and their materiality (and politics and "speech"), and have thereby given the impression that they muted humans and language. In their turn to artefacts, they have nearly bracketed off language. Even in Pickering, *language is in the background*. His reading of Latour stresses the performative, emergent, and process aspects of science (a move that, to my knowledge, was not picked up by Ihde; note also that Pickering is not even mentioned by Verbeek (2005)), but it is usually the performance of things and machines, not of language and linguistic elements. Although at least Pickering pays some attention to human agency, goals, intentions, etc. (and, hence, "speech"), he shares with Ihde, Latour, and Verbeek a one-sided focus on material agency and material hermeneutics.

But we can and should do better. In my reading of Ihde and Latour, I highlight the role of language and I propose interpretations, expansions, and revisions that take us to a better synthesis, a more integrative view of the relations between humans, language, and technology. Let me summarize what I have done in this chapter and indicate remaining tasks.

6.5. Towards a (Better) Synthesis

Let us return to the questions of this part of my book and this chapter: Can technology speak? Can artefacts speak? Can things speak? What sense can we even give to these phrases? And can we support postphenomenology's project of reestablishing or liberating the materiality and politics of technology without diminishing or neglecting the role of *language* in mediating our relation to the world and to technology?

I have shown that McLuhan can be interpreted in a way that makes it sound as if technology determines us, and in a more "linguistic," "textual," and, indeed, dualist way, but also in a more posthumanist, non-dualist, and (media-)ecological way. Seeing language as a technology, and seeing both humans and nonhumans as part of a larger ecology, give us an opening for saying that not only the medium/technology speaks but also language and humans. It makes room for acknowledging both the symbolic-linguistic dimension and the biological-material dimension of (post)human existence. Human beings—especially considered as users of language and technology/media—are understood as emergent from, and made possible by, a whole, field, or network of relations, an ecology of linguistic and nonlinguistic elements, human and nonhuman beings. In this sense, humans and non-humans (words, animals, technologies, etc.) speak. Moreover, I have shown that, in *Laws of Media*, the McLuhans take a "linguistic turn" and connect language and technology in interesting ways by connecting language, metaphor, and technology, stressing performance, arguing how technology reshapes both users and culture, and offering a "grammatical" approach, which is in tune with the Wittgensteinian and transcendental approach I develop in this chapter.

Ihde wants to bridge "engineering" and "humanities" approaches with his postphenomenology and expanded hermeneutics; this project has succeeded. His work also enables us to add an important dimension to my proposed approach: embodiment. I will take this up again in the next chapter. But his turn away from language also created a lacuna: Language and its relation to technology, for instance its role in technological mediations, remains undertheorized. I suggest that Ihde could make better use of relevant work in humanities—in particular, philosophy of language—to avoid or at least address this problem. For instance, he could use Ricoeur or Rorty, whose works he knows very well. In response to this gap, I offer a way to expand his framework in a way that includes language as mediating our relation to the world and to technology. I argue that we should not only give things a voice (what things do) but also say more about what language does and how this shapes our thinking about technology. Moreover, benefiting from Heidegger's concept of ontology and using a transcendental interpretation of Wittgenstein, I have added a transcendental argument to Ihde's mediation theory, and even proposed to give a transcendental twist to the entire postphenomenological framework. This has enabled me to arrive at a richer, more comprehensive, and arguably *more* phenomenological-hermeneutic conceptualization of the roles of language and technology in how we relate to the world.

When I suggest that Ihde could do more with Ricoeur, I have in mind the aspect of narrativity, but also the social (in general), which is largely missing in Ihde. But this can also be remedied. In the next chapter, I will use Ricoeur to connect the hermeneutics of things to the hermeneutics of language, which brings out the *narrative* and *social* dimensions of language, and not just language use in the sense of words or propositions, or even grammar in the sense of a kind of fixed spatial structure. There is also narrative structure, narrative grammar, which helps us to "set in motion," "move" the framework.

Latour is also a good candidate for achieving a synthesis of use of technology and use language, since, as we have seen, his work with Akrich is influenced by semiotics and, at the same time, gives materiality its due (with regard to the latter: consider also Verbeek (2005), who reads Latour in this "material" way). However, already then, perhaps, but especially later in Latour's work, the stress is more on *things* than on language. Like Ihde, Latour turns away from (postmodern) thinking about language. He re-establishes things, or better: nonhumans, as political (speech) agents: things, which are never mere "things" or "objects," but are political and knock on the door of the collective, having been given a voice by scientists and others. Yet, the precise role of language and discourse in his view remains unclear. In my readings of Latour, I highlight the role of language in his text and propose interpretations and revisions that further develop this language dimension and the idea that language (also) "speaks." I do so in relation to the Wittgensteinian framework I am developing. If there is performativity, process, and emergence in technological practice, as we can infer from Pickering's sociology of science and reading of Latour, then it is a performativity, process, and emergence in which not only material nonhumans but also linguistic nonhumans play a role as actants. In use and practice, both humans and nonhumans "speak."

Perhaps we can conclude that, after this operation and interpretation, the concepts of language and language games acquire new meanings. The very concepts of "agency" and "speech" seem expanded. To say that language and technology "speak" is a metaphor; but, as always, it is also more than a metaphor, or, rather: Metaphors do an important job by connecting and shifting, hence, helping new meaning to emerge. Here, the metaphor serves to connect language and technology in a new way. With Latour and Pickering, we can say that to sharply distinguish between the agency and "speech" of humans, on the one hand, and the mute passivity of material and linguistic things, does little justice to the posthumanist dance of performativity and emergence in which subject and object are produced and emerge. And even if one does not accept this particular conclusion, at the very least I show that a (post)phenomenology and (post)hermeneutics that enables us to better understand technology cannot suffice with rethinking how material artefacts (for example, machines) shape our practices and our thinking; it should also include an account of what *language* does in this respect.

This chapter contributes to this task. I show that we can use Wittgenstein, Heidegger, and McLuhan (and posthumanist media ecology) in combination with Latour and Ihde (and Pickering) to reveal and rethink the relations between language, technology, and humans. In this and the previous chapters, I make proposals for how to fuse and assemble the linguistic *and* material in a way that takes philosophy of technology into new directions. Soon, I will be in a position to weave together the various threads explored in this book. Heeding Latour, this may well take the form of a relational, ecological, holistic, and posthumanist view in which words, things, and humans are all connected and in which humans and nonhumans, language and technology get a voice. But it is still an open question in what sense they "get a voice." For instance, on the basis of my reading of Ihde it is not yet clear if we should understand this as an "ontology" or a process. Adding Pickering helps me to emphasize the process dimension, but more needs to be said. We also need to further discuss the implications of these interpretations for the social, a dimension of technology use that is also not very prominent in Ihde, who is mainly interested in embodiment. For understanding the social, we can add a Latourian twist, which seems to imply that the social itself can be redefined in more-than-human terms, encompassing not only humans but also nonhumans: We use linguistic things but also material elements. But, for a more adequate and comprehensive view of the social, it is not enough to give things, understood as technological artefacts, a voice. Words and text are also part of the social. Moreover, use of language also has a structure, which is, in turn, related to larger social-cultural wholes and forms of life, which, as I argue, function as conditions of possibility for our use of language (and technology). What exactly is the relation between language and the social? And how does this connect with an understanding of technological use and practice, and, indeed, of the social, as happening *in time*?

In order to further think about language and the social, and investigate the temporal dimension of technology use and culture, let us turn to Ricoeur in the next chapter. With his focus on narrativity, he offers a perspective on language that helps us to redefine the social and also better understand how technology is related to language. I will argue that the concept of narrativity can helpfully complement the Wittgensteinian approach, based on concepts such as language use, language games, forms of life, and performance, since it directs us to attend to the temporal dimension of use and attends us to the narrative structure of the social. Using the concept of performance itself already adds the time dimension, but narrative says more about the structuring of time, and perhaps it should also be added to the transcendental architecture I propose: Narratives could also be considered as transcendental condition. Reading Ricoeur may thus help us to further develop the performative and emergent aspects we find in Pickering and Latour: The dance of humans and nonhumans has a structure, a *narrative* structure, which makes possible the constitution and emergence-in-practice of subjects and objects.

Furthermore, in my discussion of Pickering, I suggest that attention to the "speaking" of language and technology does not necessarily come at the expense of the human. But, in this respect, there is further work that still needs to be done at the end of this book. There is a danger in McLuhan, Ihde, Latour, etc. that they are *interpreted* in such a way that the human is muted. It sometimes seems that only the medium speaks (McLuhan) and that only things speak (Ihde and Latour; perhaps Pickering, to a lesser extent, since he recognizes human goals and intentions). This muting of the human needs to be avoided, even if "human" is understood in posthuman terms. Another, related danger is that if language is presented as a large ecological or structural whole, the individual language user gets out of view. More generally, when the stress is too much on structures and wholes, such as narratives, games, collectives, or networks (as, for instance, in Latour), the social risks becoming disconnected from language users and personal *experience*. As a social theorist, Pickering also looks at scientific practices and the social from a helicopter or god's-eye point of view. To avoid this objection, I will need to interpret Wittgenstein in a way that connects Wittgenstein to Ricoeur by adding the dimension of narrativity, but, at the same time, does not entail a neglect of individual use, performance, and experience—even if "individual" should always be understood as made possible by the social. Furthermore, in the process of elaborating my framework, I should not forget the aspect of embodiment, as emphasized by Ihde (following Merleau-Ponty). If there is such a thing as "individual" experience and performance at all, the subject of that experience and the actor of that performance is not only and not so much a disembodied Winch-Wittgensteinian rule-follower, a Latourian node in a network, or an abstract actor interacting with abstract "nonhumans" but a living, *embodied* human being interacting with a living and dynamic environment, in which there are *other* embodied beings— it is both a social and a living environment. Ihde recognizes embodiment, but the *social* dimension of use and performance is largely missing. Latour helps us to bring back thinking about the social. Bringing together Ihde and Latour in order to give a *social* interpretation of performance fills the lacunas. When I introduce the concept of narrativity in the next chapter, this is also meant as a contribution to addressing this problem.

Notes

1 I use a phrase here from Verbeek (2011), who argues in *Moralizing Technology* that technologies have moral significance, as they mediate our experience and actions. He rightly criticizes an 'externalist' approach, which places technologies outside the realm of society and fails to recognize that 'humans are technological beings' (4).

2 See, for instance, McLuhan's remark that, 'Prior to writing and print, words and utterances were still endowed with the magical power to form and transform existence' (McLuhan and McLuhan 1988, 69). The authors then connect this difference with their interpretation of differences between different hemispheres in the brain.

3 Another interesting philosophical context to discuss this question of technological otherness or quasi-otherness would be Hegel (see, for instance, recent work by Nolen Gertz or my own remarks on automation and the master-slave dialectic), but Ihde does not follow this route.

4 Husserl writes: 'Prescientifically, in everyday sense-experience, the world is given in a subjectively relative way. Each of us has his own appearances' (Husserl 1970, 23). I would add that there is also the intersubjective: We also share appearances.

5 I use quotation marks here, since, although one should not deny that there is the phenomenon of "inner" life, I share Wittgenstein's and Dewey's views that this use of language should be problematized, since, in use and performance, "inner" is always related to "outer."

References

Achterhuis, Hans, ed. 2001. *American Philosophy of Technology: The Empirical Turn*. Bloomington/Indianapolis: Indiana University Press.

Akrich, Madeleine and Bruno Latour. 1992. "A Summary of a Convenient Vocabulary for the Semiotics of Human and Nonhuman Assemblies." In *Shaping Technology/Building Society: Studies in Sociotechnical Change*, edited by Wiebe E. Bijker, and John Law, 259–64. Cambridge, MA: MIT Press.

Arendt, Hannah. 1958. *The Human Condition*. Chicago, IL: University of Chicago Press.

Coeckelbergh, Mark. 2011. "You, Robot: On the Linguistic Construction of Artificial Others." *AI & Society* 26(1): 61–9. doi:10.1007/s00146-010-0289-z.

Coeckelbergh, Mark. 2015. "Language and Technology: Maps, Bridges, and Pathways." *AI & Society* (online first). doi:10.1007/s00146-015-0604-9.

De Preester, Helena. 2010. "Postphenomenology, Embodiment and Technics". *Human Studies* 33(2–3): 339–45. doi:10.1007/s10746-010-9144-y.

Deranty, Jean-Philippe. 2014. "Feuerbach and the Philosophy of Critical Theory." *British Journal for the History of Philosophy* 22(6): 1208–33. doi:10.1080/09608788.2014.974139.

Dreyfus, Stuart E. and Hubert L. Dreyfus. 1980. *A Five-Stage Model of the Mental Activities Involved in Direct Skill Acquisition*, research report. Berkeley, CA: Operations Research Center, University of California.

Feuerbach, Ludwig. 1980. *Thoughts on Death and Immortality*. Berkeley/Los Angeles, CA: University of California Press.

Floridi, Luciano, ed. 2015. *The Onlife Manifesto*. Cham/Heidelberg/New York/Dordrecht/London: Springer.

Gier, Nicholas F. 1980. "Wittgenstein and Forms of Life." *Philosophy of the Social Science* 10(3): 241–58.

Gordon, W. Terrence. 1997. *Marshall McLuhan: Escape into Understanding*. New York: Basic Books.

Harman, Graham. 2009. "The McLuhans and Metaphysics." In *New Waves in Philosophy of Technology*, edited by Jan Kyrre Berg Olsen, Evan Selinger, and Soren Riis, 100–22. Basingstoke: Palgrave Macmillan.

Harman, Graham. 2014. *Bruno Latour: Reassembling the Political*. London: Pluto Press.

Heaney, Conor. 2015. Review of *Bruno Latour: Reassembling the Political*, by Graham Harman. *Theory, Culture & Society*, February 2. www.theoryculturesociety.org/review-of-graham-harmans-bruno-latour-reassembling-the-political

Heidegger, Martin. 1977. "The Question Concerning Technology." In *The Question Concerning Technology and Other Essays*, edited by Martin Heidegger, translated by William Lovitt, 3–35. New York: Harper & Row.

Husserl, Edmund. 1970 (1954). *The Crisis of European Sciences and Transcendental Phenomenology*. Translated by David Carr. Evanston, IL: Northwestern University Press.

Ihde, Don. 1990. *Technology and the Lifeworld: From Garden to Earth*. Bloomington/Indianapolis: Indiana University Press.

Ihde, Don. 1998. *Expanding Hermeneutics: Visualism in Science*. Evanston, IL: Northwestern University Press.

Ihde, Don. 2007. *Listening and Voice: Phenomenologies of Sound*, 2nd edition. Albany, NY: State University of New York Press.

Ihde, Don. 2009. *Postphenomenology and Technoscience: The Peking University Lectures*. Albany, NY: State University of New York Press.

Ihde, Don. 2016. *Acoustic Technics*. Lanham, MD: Lexington Books/ Rowman & Littlefield.

Latour, Bruno. 1993. *We Have Never Been Modern*. Translated by Catherine Porter. Cambridge, MA: Harvard University Press.

Latour, Bruno. 2004. *Politics of Nature: How to Bring the Sciences into Democracy*. Translated by Catherine Porter. Cambridge, MA/London: Harvard University Press.

McLuhan, Marshall. 2001 (1964). *Understanding Media: The Extensions of Man*. London and New York: Routledge.

McLuhan, Marshall and Eric McLuhan. 1988. *Laws of Media: The New Science*. Toronto: University of Toronto Press.

Ong, Walter J. 2002 (1982). *Orality and Literacy: The Technologizing of the Word*. London/New York: Routledge.

Pickering, Andrew. 1995. *The Mangle of Practice: Time, Agency, and Science*. Chicago, IL/London: University of Chicago Press.

Pickering, Andrew. 2017. "The Ontological Turn: Taking Different Worlds Seriously." *Social Analysis* (forthcoming).

Postman, Neil. 1993 (1992). *Technopoloy: The Surrender of Culture to Technology*. New York: Vintage Books and Random House.

Postman, Neil. 2000. "The Humanism of Media Ecology." *Proceedings of the Media Ecology Association* 1: 10–16.

Stephens, Niall P. 2014. "Toward a More Substantive Media Ecology: Postman's Metaphor Versus Posthuman Futures." *International Journal of Communication* 8: 2027–45. doi:1932-8036/20140005.

Van Den Eede, Yoni. 2011. "In Between Us: On the Transparency and Opacity of Technological Mediation." *Foundations of Science* 16(2): 139–59. doi:10.1007/s10699-010-9190-y.

Van Den Eede, Yoni. 2012. *Amor Technologiae: Marshall McLuhan as Philosopher of Technology—Toward a Philosophy of Human-Media Relationships*. Brussels: VUBPRESS.

Van Den Eede, Yoni. 2015. "Exceeding Our Grasp: McLuhan's All-Metaphorical Outlook." In *Finding McLuhan: The Mind, the Man, the Message*, edited by Jacqueline McLeod Rogers, Tracy Whalen, and Catherine G. Taylor, 43–61. Regina: University of Regina Press.

Van Den Eede, Yoni. 2017. "Formal Cause: McLuhan's 'Objective Turn'?" In *Taking Up McLuhan's Cause: Perspectives on Formal Causality*, edited by Corey Anton, Robert K. Logan, and Lance Strate, Bristol, UK and Wilmington, North Carolina: Intellect Ltd.

Verbeek, Peter-Paul. 2005. *What Things Do*. University Park, PA: Pennsylvania State University Press.

Verbeek, Peter-Paul. 2008. "Obstetric Ultrasound and the Technological Mediation of Morality: A Postphenomenological Analysis." *Human Studies* 31(1): 11–26.

Verbeek, Peter-Paul. 2011. *Moralizing Technology*. Chicago, IL: Chicago University Press.

Zwier, Jochem, Vincent Blok and Pieter Lemmens. 2016. "Phenomenology and the Empirical Turn: A Phenomenological Analysis of Postphenomenology." *Philosophy & Technology* 29(4): 313–33. doi:10.1007/s13347-016-0221-7.

7 What Technology Tells Us (to Do) (Part 2)
Narrative Technologies, or Interpreting and Materializing Ricoeur

7.1. Introduction

Language use is not only structured in terms of words and sentences; it also has a time structure and, often, a *narrative* structure. One can approach meaning and experience synchronically, at a given moment in time, or diachronically, over a period of time. Humans understand and structure their lifeworld and experience along narrative lines, which are also time-lines. And embodied use of language, understood as speech, including use of words and use of voice, happens in time. What does this temporal dimension of language use and semiotics, and the narrative aspect of the lifeworld, imply for thinking about technology? How can a narrative approach to language (use) help us to better understand technology (use)?

Theory discussed so far in this book lacks sufficient attention to time and narrativity. In Wittgenstein and Austin, thinking and language use is performative, as we use words to do things and to tell others to do things. But they pay less attention to the narrative dimension of language (and, hence, thinking). In Akrich and Latour, the notion of script, and, in the later Latour, the notions of actors and actants, already refer to the performative and the theatre. I also mention in Chapter 4 that Latour is influenced by Greimas's narrative semiotics. Yet, in most of Latour's work, for example, actor network theory or his work on nonmodern thinking, not much attention is paid to the narrative and time aspects, and the application of the theatre or performance metaphor is not really developed—although, connecting to Pickering's interpretation, which emphasizes temporality and real-time practice, emergency, and the interaction between human and material agency, helps to partly remedy this. Finally, in Ihde's postphenomenology, attention to performativity, time, and narrativity seems to be lacking altogether—in spite of Ihde's knowledge of the work of Paul Ricoeur.

Ihde's playing down of language and narrativity in his postphenomenology and expanded hermeneutics is curious, since, earlier in his career, Ihde even wrote a book about Ricoeur's 'hermeneutic phenomenology' (Ihde 1971). This book includes a preface by Ricoeur, who suggests an analysis that is both linguistic and phenomenological (xv). And in contrast to his later work, here, Ihde *does* engage with philosophy of language and relates to an academic context in which linguistic analysis is popular: He discusses Ricoeur's 'turn to the problems of language' (xiv), his 'moving phenomenology toward a linguistic focus' (4), his philosophical anthropology, which claims that the human 'is language' (23), and even his philosophy of language (24). Later, in the 1980s, Ricoeur writes his article on narrative time (1980) and his volumes on *Time and Narrative* (Ricoeur 1983, 1984, 1985). But, in the 1990s, Ihde's postphenomenology turns away from his own earlier focus on Ricoeur's language-oriented hermeneutics, and neglects *Time and Narrative*.

Indeed, in the 1960s, Ricoeur already turns to (philosophy of) language. For instance, in his lectures at Johns Hopkins University (Ricoeur 2014), he compares the later Wittgenstein with the later Husserl, points to Wittgenstein's emphasis on use, language games, and forms of life as a way to bring us to the lifeworld, and turns to a semiological paradigm inspired by de Saussure, in particular, the distinction between "language" and "speaking:" Ricoeur claims that Wittgenstein's theory (if he has one or even wanted to have one at all) is not about language as a system of signs, but about *speaking* (37). This interpretation is interesting for my project, as it supports my use of the terms speech and *speaking* by humans, language, and things as a way of saying something about the lifeworld, about the concrete ways of using and relating to language and technology, rather than about abstract sign systems. A focus on language need not make the mistake of talking only about signs, as poststructuralists and postmoderns did in the 1970s and 1980s. With Wittgenstein, we can see language in its concrete use. We can also see it as embodied and as connected to power (see later in this chapter). Moreover, like language, technology can be regarded not only as a system of objects external to human experience and use but also as tools we use, as social and embodied beings, and as something(s) that are part of the lifeworld. In the lectures, Ricoeur distinguishes between being engaged in practical activity—being in the game, so to speak—and theoretical enquiry, between life and reflection (41). For Ricoeur, "transcendental" means no longer being part of the game, but thinking about it. This raises the following question about a "transcendental" philosophy of language, but also a "transcendental" philosophy of technology: Is such a philosophy necessarily alienated from the concrete lifeworld? Do we need to adopt Ricoeur's and Ihde's conception of the transcendental, or can we understand "transcendental" in a different way?

In the previous chapters, I have propose, against Ihde (and, as it turns out, also against Ricoeur), a notion of the transcendental in which conditions of possibility are not to be situated in a different realm "above" concrete experience. They make possible experience, but they should not be understood in an abstract way, as signs, logic, or categories. Against Kant and Husserl, we must see conditions of possibility as words and material artefacts that are part of the lifeworld, but then not only as phenomena but as conditions that make possible the phenomena, conditions that make possible experience, thinking, and talking in that lifeworld. Once we consider language and technology *in their use*, we can link that use to transcendental structures that shape and make possible that use, and interpret these structures also in a material and empirical way. If we really want to take a material or empirical turn, we have to partly *materialize the transcendental*. Language and technology have their narrative and other hermeneutic functions *as they are used* and experienced by humans, and these functions are dependent on conditions, which are also linguistic and technological. (With regard to language as a condition of possibility, see also my book *Growing Moral Relations*; Coeckelbergh 2012).

Moreover, at the end of the previous chapter, I already suggest that we could give a more dynamic twist to our framework, bringing in the time dimension, by attending to process, for instance. However, the term "process" sounds rather removed from the concrete use of technologies, from human experience, from the lifeworld. Like Latour's view sometimes suggests a static metaphysics (an ontology of humans and things), focusing on process might suggest a process metaphysics, say à la Whitehead or, more recently, Deleuze, if interpreted as a metaphysics (but not, perhaps, à la Dewey, who connected meaning to what humans do together in interaction and in situations). A process metaphysics may be an improvement, as compared to more "static" ones, but it remains a metaphysics nevertheless. As a metaphysics, it is too alienating, too far removed from the lifeworld and the concrete use of words and things.

Perhaps this can be avoided by focusing on *narrative* time, which enables us to bring in a time and process dimension, but which is connected to the lifeworld. This chapter uses Ricoeur's theory of narrativity to say more about what technologies do and tell us (also: tell us *to do*), phenomenologically and hermeneutically. I will argue that technology cannot only be read as a "text," as Ihde suggests, and that there are not only stories *about* technology, as Kaplan (2006) writes; technologies also have a much more active *narrative* hermeneutic role, to the extent that they actively configure actors (and actants) and events into a plot. As they co-constitute narratives and our understanding of these narratives, technologies become what Wessel Reijers and I call 'narrative technologies' (Coeckelbergh and Reijers 2016). This account of the active hermeneutic role of technologies, which brings back language, but *not* at the expense of (the hermeneutics of) material objects, does not mean that humans are muted, that there is technological determinism, or that human experience is neglected. On the contrary, humans, *in* their use and experience of technology, interpret technologies and cowrite and codirect the narrative (which includes events with human and nonhuman characters). I will show what this means by discussing the case of social media and by discussing questions regarding power (making an excursion to Foucault) and embodiment (using Merleau-Ponty).

Nevertheless, one may question what the precise role of language is in these hermeneutic uses, experiences, and configurations. To what extent is language really part of this? Is it used as a metaphor, or is this narrative process really (also) linguistic in nature? And is human experience and the lifeworld really taken seriously, given that this may seem like a third-person, detached view of what is happening? Here and in the next chapter, a discussion of these criticisms will further contribute to a synthesis, which tries to give a voice to humans, language, and technology.

7.2. Narrative Technologies

Let me start with recapitulating some gaps in contemporary philosophy of technology, especially in postphenomenology. In Ihde's postphenomenology, technologies have a hermeneutic role, as they mediate human-technologies relations in a way that frames our world. Yet, in postphenomenology and related mediation theory, no attention is paid to the time dimension of human existence, let alone to the *narrative* structure of human experience and meaning. They neglect MacIntyre's and Ricoeur's view that human life and existence have a narrative character (MacIntyre 2007; Ricoeur 1983). Moreover, as argued previously, the role of language in relation to technological mediations remains undertheorized. Divorced from Heidegger and its other own sources (Ricoeur), postphenomenology becomes a theory about humans (subject) and things (object), with language being largely absent. More generally, so far, philosophy of language has seldom been used by contemporary philosophers of technology, let alone theory about time and narrative. The same can be said about the social and social theory; the level of analysis seems to be limited to individual subjects and individual worlds. While it makes links to STS, when it comes to conceptual work, postphenomenology mainly focuses on the individual level of human experience; a theory of the social is absent. For instance, in Verbeek's work (2005 and subsequent work), the focus is mainly on how individual subjects are mediated and constituted by technologies.

Moreover, in philosophy of language—Wittgenstein and Searle, for instance, but also poststructuralist and postmodern theory—there has, of course, been attention to narrative, for instance, in the work of Lyotard and Ricoeur, but, here, the problem is the opposite: The use and hermeneutics of language is emphasized at the expense of the use and hermeneutics of technology. A material perspective and attention to technology seems to be lacking altogether. Can we avoid these one-sided views and integrate insights from philosophy of technology and philosophy of language?

In order to address these gaps and work towards an integrative view, Wessel Reijers and I propose to use Ricoeur's narrative theory to account for (1) the role of language in technological mediation, (2) the narrative aspects of the hermeneutic role of technologies, and (3) the social dimension of that hermeneutics (Coeckelbergh and Reijers 2016). Let me use this recent work and connect it to my previous arguments in this book, in particular, my development of a Wittgensteinian-Heideggerian approach to technology use and performance, in order to further develop yet another understanding of "what technology tells us" and "what technology tells us to do"—in other words, to further elaborate the notion of technology "speaking," but, this time, elaborating the aspects of time and narrative.

Like Wittgenstein and many other so-called "philosophers of language," Ricoeur is not concerned with technology, but with human language. Influenced by Heidegger and Wittgenstein, he has his own version of the claim that language mediates human experience, has a 'grasp' on it (Ricoeur 2014, 29). According to Ricoeur, humans interpret their actions within a temporal setting, and, more specifically, in a narrative way. Interestingly, not only in his lectures in the 1960s but also in his famous article 'Narrative Time' (Ricoeur 1980), he draws on Wittgenstein to make his point. He compares the relation between narrativity and temporality with the relation between a language game and a form of life in Wittgenstein:

> My first working hypothesis is that narrativity and temporality are closely related—as closely as, in Wittgenstein's terms, a language game and a form of life. Indeed, I take temporality to be that structure of existence that reaches language in narrativity and narrativity to be the language structure that has temporality as its ultimate referent. Their relationship is therefore reciprocal.
>
> (Ricoeur 1980, 169)

Like our use of language is part of language games, which, in turn, are set within a form of life, our use of language and our interpretations of our actions are set within a narrative structure, which, in turn, is set in time. In other words, it turns out that *the structure and grammar of language and our form of life are temporal and narrative.* I propose to say the same about technology: **the structure and grammar of technology, and the form of life in which it is embedded, are temporal and narrative.** In order to develop this idea, we need to say more about time and narrative—here, in particular, Ricoeur's understanding of time and narrative.

Let me start with *time*. Using Heidegger, Ricoeur analyzes different notions of time. I will not further discuss this use of Heidegger here, but emphasize Ricoeur's insight that narration is always done within a social and public context: 'the time of narrative is public time' (Ricoeur 1980, 175). Ricoeur then makes a distinction between scientific and narrative time: In the time of science, there is a distance to human experience, whereas, in narrative time, the interactions between people are present; it is time of 'being-with-others' (Ricoeur 1980, 188). In text, Ricoeur finds that public, social time. It is also human time: when time is articulated through a narrative mode, it becomes human (Ricoeur 1983, 52). Ricoeur, thus, strictly distinguishes between speaking humans (narrative) and mute, nonspeaking objects described by science, between subjectivity and objectivity, between humans and nonhumans. I have already questioned these modern dichotomies by using Latour;

in this chapter, I will question them by focusing on the narrative functions of technologies. But let me first further unpack Ricoeur's concept of *narrative*.

What happens when we narrate? Ricoeur argues that the narrative mode is about what he calls 'emplotment.' According to Ricoeur, the plot is 'the intelligible whole that governs a succession of events in any story' (Ricoeur 1980, 171). In the first volume of *Time and Narrative* (1983), Ricoeur says more about plot by using Aristotle. Based on Aristotle's notion of *mimesis* in the *Poetics*, he claims that the plot is the *mimesis* of action (Ricoeur 1983, xi; see also Aristotle 50a1 as quoted in Ricoeur 1980, 34). It is an imitation of action, and, more particularly, it entails the organization of events in a plot. Ricoeur distinguishes between three senses of *mimesis*: 'a reference back to the familiar pre-understanding we have of the order of action; an entry into the realm of poetic composition; and finally a new configuration by means of this poetic refiguring of the pre-understood order of action' (xi). Thus, one of Ricoeur's main points concerns configuration, order, and composition, in other words, concerns *grammar*. It makes us think of the relation between Wittgenstein's forms of life and concrete language use: "Before" we use language, there is already a pre-understanding given within a language, language games, and a form of life. There is already a given grammar. But, at the same time, our use of language also changes the form of life—even if only marginally. There are configurations and reconfigurations.

This reconfigurative and transformative role of narrative is interesting: Narrative turns out to much more hermeneutically "active" than one would think, since it is not just *about* what we do and experience, but transforms lived experience and practice. Like in tragedy, for instance, the plot gives us a sequence of events (chronological dimension of narrative time), but also configures characters and events into a meaningful whole (a-chronological dimension of narrative time). This gives us a new experience of time and a new meaning, transforming our understanding, which nevertheless refers to our pre-understanding of the human world (prefigured time). One could say that, in Ricoeur, the whole mimetic, narrative-hermeneutic process starts with practical experience and ends with practical experience. An emplotment mediates 'between a stage of practical experience that precedes it and a stage that succeeds it' (53). In terms of time, prefigured time becomes 'refigured time through the mediation of a configured time' (54). Moreover, Ricoeur stresses the social character of the narrative process. When he writes about action, Ricoeur also refers to Greimas's semiotics and its actants (56), which I mention in my readings of Latour, and says that to act is always to act with others (55). Moreover, in narrative, actual characters and events (so-called first-order entities) can be abstracted into quasi-characters and quasi-events (second-order entities), for instance, when, in historical discourse, nations are said to attack other nations, rather than real persons; nations are then quasi-characters. Similarly, society appears in historical discourse as an entity, a quasi-character (197). What remains hidden in these histories are real characters and events, characters and events that are part of the lifeworld. Thus, narratives can be more or less alienated from our concrete lives.

For Ricoeur, this conception of narrative (and time) has little to do with technology. As said, he makes a strict distinction between the speaking of humans and what happens in science and technology. Being critical of modern science, he sees a gap between the time of science and the time of narrative, or between technological rationality and the social world as expressed in fiction and history (see also Kaplan 2002, 2). Thus, narrative is placed on the side of "culture" and the social, whereas technology is placed on the side of "nature," science, and rationality. When we go beyond this dichotomy, however, new possibilities to theorize technology open up. Kaplan argues that narrative theory can be used to study how humans interpret—"read"—technology (Kaplan 2006, 49). Indeed, as I say earlier in this book, humans have narratives *about* technology. However, there is also a sense in which technology "reads" and "writes" *humans*. What does this mean?

Reijers and I use Ricoeur's terms 'configuration' and 'emplotment' to flesh out this more active-hermeneutical role of technology. We argue that *technologies themselves* configure time into a plot, organizing characters and events, and that this is an interesting way to understand and conceptualize the key insight in contemporary philosophy of technology since Heidegger, that technologies are no mere instruments, but co-shape human meaning and human action (Coeckelbergh and Reijers 2016). Technologies are not merely like a text in the sense that they can be "read," but are much more active *narrative technologies*, as they co-author narratives, which (so we may add) contain human and nonhuman elements. **Technologies actively reconfigure and shape our narratives** and, indeed, shape the narrative structure of our actions and experience. Moreover, through their mediations, technologies can give us narratives that either stay close to the world of human action, or can be more abstracted from it. Ricoeur already gives the example of modern clocks that give us an abstract representation of time, divorced from human, existential time (Ricoeur 1983, 63). But this idea can be applied much more widely. Let me explain these configurative and abstracting functions by means of two examples.

The first example is an old wooden bridge, an example we borrow from Heidegger (1977), but reinterpret by using Ricoeur. This bridge can be "read" by embedding it into stories about historical events and characters. It is a "narrative technology" in the sense that it is a passive element within a human narrative. In this conceptualization of the narrative dimension of the bridge, there is still a dichotomy between technology and lifeworld. One could say that the bridge "speaks" of the historical events and characters, but the speech agency is mainly on the side of the humans, who construct a narrative *about* the bridge. Humans speak of the past, and the bridge is an element within their narratives. The bridge is already part of the human lifeworld for ages. It is an element in our prefigured time; it is part of life as we know it. However, when we consider the bridge *at the time when it was built*, we get, literally, a different story. When the bridge was built, it was not yet part of the lifeworld—it was still to be integrated in the lifeworld by means of narratives and actions—but, next to this passive role, it also had a much more active role: It *changed* the way people did and experienced things. In particular, it (re)configured human narrative time by organizing characters and events (e.g., merchants, armies, trades, and wars) into a new narrative whole. The technology configured interactions and social relations, and events. It was not only, or not yet, a passive text that could be "read," but, rather, an active script—to use, again, a term from Akrich and Latour, which is appropriate, since it is also a theatre metaphor. And at the time when the bridge was built, there were, not so much abstract events and characters (for example, "Germany attacked France"), but concrete people (characters) and events. Scientific and historical narratives (as in, "the history of France") tend to abstract from concrete social relations, tend to create distance, and the technology—the bridge—can play a role in either the concrete histories or the abstract ones. Thus, in these two senses, the bridge "speaks:" At this moment in time, it speaks of the past, it tells us a narrative about the past, whereas, at the time it was built, it "spoke" more actively, in the sense that it created a story, created a narrative whole. It was not only spoken about, but was itself speaking, since it restructured human narrative (time).

The second example is computer games. When a game is well known, it is a "text" in the sense that it is spoken *about* and interpreted by humans (e.g., gamers, journalists, philosophers). And after a while, it may become part of the lifeworld of specific game communities, in which narratives *about* the game circulate, narratives that are part of that particular *culture*. But, when and as it is played, the game is a "narrative technology" in a much more active sense. Of course humans play the game. But their experiences of time and of social relations are actively mediated by the game, which configures a specific narrative structure in and with the game play. There is a narrative plot, which organizes virtual characters (avatars) and virtual events into a meaningful narrative whole. Of course, the gamer influences the unfolding of the narrative with their actions. But their actions are also configured and structured by the technologically mediated organization of events and characters "in" the game. Thus, here, the technology also "speaks," as both humans and technology coauthor the game narrative. For example, the game Pokémon Go is an augmented reality game played on smartphones that involves, among other things, a hunt for virtual creatures that appear on the screen as if they are in the real world; gamers thus also search for Pokémon characters outdoors. On the one hand, there are narratives *about* the game. For example, journalists write about gamers trying to locate Pokémon. There are also stories about accidents, when people are so busy with searching for Pokémon that they do not pay attention to, for instance, a car. On the other hand, the game actually structures the time (or action-in-time) of gamers and organizes virtual creatures (Pokémon), gamers, and other characters and events (appearance of a Pokémon) into a plot, a narrative whole. This changes the way players do things in their lives, their lived time is reconfigured. The gamer becomes an actor in a plot that is about hunting the Pokémon. The game is thus not only "spoken about" but also "speaks," in the sense that it actively shapes the narrative time of its players. (Note that, next to organizing gamer's time, the game also structures spatiality: It "tells" gamers *where* to go.)

Instead of conceptualizing this active narrative-configurative role of technologies in terms of "mediation," as Ricoeur does, one could also conceptualize this role in a transcendental way: The game as technology makes possible and shapes the actual game play, understood as a narrative with a plot and characters. We then have a notion of "narrative technologies" that, in contrast to Ricoeur and going beyond the mentioned work on narrative technologies (Coeckelbergh and Reijers 2016), not only considers technology as a configurator, narrator, or author but puts more emphasis on pre-configuration and, hence, on technology as (providing) an underlying structure and grammar, which is not so much "in between," but "carries" and shapes the concrete, real-time and lived time and narrative of the use of the technology (here: the game play). The lived time is prestructured by technological (and linguistic) conditions of possibility. Moreover, like all transcendental conditions, these are more or less "given;" they are already part of a (game) culture, of existing social games, and of a form of life. Recognizing this gives us a more holistic understanding of the configuring role of technology.

One may remark that the narrative structure of the game is made by humans. This is right, since human developers make the narrative setting of the game and human players decide if they use it, go outdoors, etc. Moreover, *its narrative structure already relies on a pre-configured understanding of the world shared by players and game developers.* Players are only able to play the game because they have experience of the real world and their real lives, including its social relations. In that sense, "imitation" is at work. This is Ricoeur's first sense of mimesis: There is already prefigured time, there is already experience, and, we may add: There are already (other) games and there is already a form of life. Indeed, with Wittgenstein, one could say that computer and smartphone games, as language games, but also as what I called "technology games," are embedded within a form of life and are linked to other activities and games. For example, the game of hunting, made possible by language and technology (to give, again, a transcendental twist to Ricoeur and Wittgenstein), has a long history, and, once, there were concrete experiences that shaped an entire culture. Hence, when a gamer hunts for a virtual creature, they rely on structures and practices that already there: The activity of hunting is related to games that are already available in their society and culture. But these more concrete human actors, technological structures, previous experiences, social relations, game rules, and larger cultural wholes are not immediately visible or experienced by the gamer. When one is "in" the game, when one is "in" the phenomenology of playing a computer or smartphone game, there is active configuration: The game, *as technology*, pre-structures the narrative and co-configures and refigures the plot. This "player time" and "player narrative" is usually phenomenologically divorced from other times and other narratives. The game has its own time and narrative, and that time and narrative is shaped by both human authors and nonhuman authors (including virtual characters and the game technology). In augmented reality, perhaps there is less of a gap between "virtual" game time/narrative and "real" time/narrative. But, through the playing, both kinds of times and the entire narrative are reconfigured. People start living Pokémon narratives and Pokémon lives. Their world becomes a Pokémon world and their life is narratively structured and prestructured, made possible by the game-technologies and the narrative structure and lifeform it co-constitutes and belongs to. In that sense, the narrativity-at-play is both human and nonhuman.

Thus, using Ricoeur's narrative theory, one can make sense of the idea that technologies have a narrative character, not only and not so much because they can be "read" by humans and "speak of" "exclusively human" narratives (here is the dichotomy technology-lifeworld, again) but also because they actively configure human time and narrative, because they structure and prestructure, make possible, the lived time. Narrative technologies, then, have the capacity to *change* the human lifeworld (including human experience, action, and practices), as this lifeworld, with its social relations and interactions, is prestructured and always again restructured, recounted, re-narrated, and reenacted by humans and nonhumans. (Note that repetition is important in Ricoeur's work on narrative.) So, in this sense, it is also true that technologies "speak:" they speak insofar as they are not only the object of narratives but *are narrative themselves*. Technologies not only make us see the world in a different way (as a novel does, which is meant to make us see the world in a *new, novel* way)—this is still the model Ricoeur sometimes seems to use when he talks about refiguration and transformation—but also change the way we *do* things by configuring and pre-configuring (making possible) our lived real-time narratives. Technologies coauthor and coact our stories, the stories we tell about ourselves, but also the stories we continuously experience, (en)act, repeat, and *live*.

For thinking about the relations between language and technologies, this means that now we have given more flesh to the performance and theatre metaphor, which now not only suggests a Latour-style ontology of humans and nonhumans but also opens up the aspects of time and narrativity. We have also seen that there are parallels between, on the one hand, Wittgenstein's holistic view of meaning and language use, and, on the other hand, Ricoeur's holistic view of narrative hermeneutics: Ricoeur's thinking enables us to add a time and narrative dimension to the Wittgensteinian framework. Moreover, this focus on hermeneutics and narrative, including a transcendental interpretation, has not come at the expense of giving the hermeneutics of (material) technology its due. We identify another *intrinsic* or *internal* relation between language and technologies: Any particular technology can play an active hermeneutic role, rather than being merely the object of interpretation or, rather, than being a mere instrument for hermeneutic work, as the technology of writing and technology of text are usually interpreted—in which case, there is an extrinsic or external relation. Like in Ihde and Latour, technology itself becomes hermeneutic, and not only by shaping experience (Ihde) and action (Latour) at a given moment in time by mediating that experience and action (synchronically, so to speak) but also by co-shaping the experience of time and the action *in* lived time itself, in the *course* of time (diachronically), by configuring and pre-configuring our time and narratives. This use and interpretation of Ricoeur therefore renders our framework more comprehensive: It now includes a phenomenology and hermeneutics of technology that also has temporal, diachronic, and narrative aspects.

We can conclude that this focus on narrative and time is, thus, an important addition to, and revision of, Ihde and Latour, since it highlights key dimensions of human experience, of the human lifeworld—time and narrative—without which the postphenomenological imaginations and the Latourian scriptings remain an incomplete and impoverished picture of the social life and, indeed, of human existence with technology. We now have gained a better understanding of what it could mean that technologies "speak"—next to humans and language, or, rather, *with* humans and language—and, indeed, "tell us" things. Moreover, with the notion of narrative comes a more *social* and cultural understanding of human existence, including a more social understanding of use of technologies, which is in line with the "cultural" Wittgensteinian approach. That narrativity technology structures and makes possible has a public and social aspect, and the concept of narrativity enables us to link concrete uses of technology and words to what Wittgenstein calls games and forms of life.

Moreover, insofar as technologies have this narrative character, it is clear that, like in Latour (and in Verbeek), they also have a *normative* dimension. They not only tell us things, they also *tell us what to do*. I already point to this in the context of my discussion of Latour. But this idea can now be further developed with regard to *narrative technologies*, in a way that complements Latour's approach, and retrieves and develops the narrative dimension of Latour's metaphors. First, narratives *about* anything are never neutral; because of their hermeneutical nature, they always give a specific interpretation. Hence, narratives about technologies, or narratives about anything in which technologies play a role, also always give a specific interpretation, including normative aspects. Second, when and to the extent that technologies *actively* configure plots and narratives, together with humans and language, they also script the human role, they also (re)organize humans and events into a new plot, and this is not normatively neutral. They also shape acts, habits, and characters—good ones and bad ones. They make possible particular plots and narratives, rather than others. For instance, a computer game will also tell humans what to do—not so much or not always literally, but by configuring the characters and the events into a narrative whole, which encourages and trains specific actions and experiences rather than others, and excludes other acts, plots, narratives, and experiential possibilities. It is then likely to lead to some stories rather than others, some social relations rather than others, some characters rather than others, etc.

This point about the configuration of characters suggests an interesting link with virtue ethics, which also focuses on character, and, hence, provides a suitable ethical complement to narrative-technologies theory: As narrative technologies, games will shape certain characters rather than others, and this includes the moral dimension of character. For example, violent computer games may configure and make possible more violent narratives and characters; even if they do not lead to actual violence in the so-called "real" world, they may structure human experience, plots, and human character in a more "violent" way, shaping less virtuous characters and narratives. The way the technology structures and shapes the narrative, the way it organizes the characters and events thus also influences the characters and their moral qualities. It is not only the case that being a virtuous person leads to a virtuous narrative; it is equally the case that a virtuous narrative—as made possible by particular technologies—also "makes" virtuous characters. Similarly, vicious narratives—in the context of a game or elsewhere—will also include vicious characters. Gaming technology and other technologies may thus structure and pre-structure characters and narratives in morally relevant ways.

Furthermore, to the extent that technologies, such as books or computer games, create *distance* from real characters and events, they are also not normatively neutral. Technologies do something with our experience of the world and of others when they turn real characters and events into more abstract entities, such as avatars, wars, national citizens, etc. This abstracting away from concrete, lived narratives and lives may lead to a lack of understanding—or, at least, *different* understanding, since the understanding is of a more detached and abstract kind—on the part of those who "comment" on the narratives of others. This is problematic when these people influence or make decisions that impact the narrative-lives of others. For example, in a computer game, real-world power games may get abstracted into characters, events, and, hence, the result is plots and narratives that may be very remote from real characters and events. This is problematic when the same narrative distance is then, in turn, transferred to activities and games outside the computer-game context. For instance, a fantasy game may suggest that morality and politics can be reduced to narratives about good versus evil. It may make possible a pre-structuring of (our understanding of) real-world narratives. When evaluating new technologies, such as computer games, it is therefore important to identify, better understand,

and take into account the active hermeneutical functions of the technologies *as narrative technologies*.

Note, finally, that this view is very different from Searle's, since Searle's view of language use and the social is non-hermeneutic, entirely based on human intentionality, and more closed in the following sense: It seems that Searle assumes that there is a social agreement on one meaning, and then, through a status function, that agreement in meaning is given to an artefact. Once this meaning is given, this is, so to speak, the full story and the end of the story. But, with Ricoeur, we see that meaning-giving and, indeed, meaning-finding are social-hermeneutic processes, which are not so direct and are more open. As narratives are (re)configured and mediated—and, I added: made possible—by language and technology, they are not simply the product of individual or collective human intentionality. What the social is, is prefigured; there is already a form of life. Moreover, narratives are continuously transformed by humans *and* by technologies; narratives and plots are rewritten, reinterpreted, and so on. In this sense, there is a two-way process: Technology shapes the social, and not only the other way around. Artefacts are not mere receivers of (linguistic) meaning but also shape our language use and our narratives. To take a Searlean example: The meaning of money is not only given to material artefacts by humans; those material artefacts also shape the meaning of that money (something that is not intended), and this meaning is part of a culture. For instance, in some places, electronic money or electronic currencies are more trusted and accepted by people than in other places, and this is not only so because those people collectively intended this; it is also a result of the specific ways the technologies shape the social narratives and interact with the form of life that was already there.

If one wants to use the term "mediation" at all, one could say that technology mediates and shapes our "world," understood not only in Ihde's synchronic sense as a worldview but also as *narrative*, which highlights the diachronic dimension of hermeneutics and gives it a social dimension. Ihde and Verbeek might say that, through technology as mediator, we see, feel, hear, etc. our world. We make objects present in a specific way (Verbeek 2005, 141). But this way of putting it misses the social, time, and narrative aspects of this hermeneutics and phenomenology: Through technology, we also *narrate* our world, and thereby co-constitute it. As we use the technology of writing, as we use artefacts, computers, and phones, these technologies shape the message we see and tell, but they also shape the plots and narratives we live and repeat. They shape our world-experience and our acting in place, but also our time-experience and our acting in time with others. In a sense, too, as they configure and reconfigure our narratives, these technologies actively contribute to *making the social*, making reality-for-us. A transcendental and more performance-centered way of putting this, which, again, goes beyond Ihde and Verbeek's mediation language, is that technologies function as transcendental conditions that make possible and shape our *performances*—here considered as lived narratives that are pre-configured and configured by technology and language. Let me say more about performance and narrativity.

7.3. Further Discussion: Performance and Narrativity

Some critical questions may be asked at this point. Is this focus on narrative technologies technological determinism? Do we still have a voice in the stories technology writes for, about, and with us (as characters)? And is it a deterministic view to say that technology "speaks?" What exactly is the relation between humans, technology, and language? What exactly is the role of language? Is it just a metaphor (e.g., the text metaphor) or is there more to it? Furthermore, is the (first-person) experience of people really taken into account in these postphenomenological and hermeneutic accounts, or is theory about narrativity itself problematic insofar as it abstracts from humans and their experiences, as perhaps all kinds of structuralism does? And what is the relation between individual experience and the social? Let me start with the question regarding technological determinism and give an answer that may also shed some light on the other questions. More importantly, however, this discussion and "reprise" of the previous argument will say more about performance and take us towards a more integrated view of relations between humans, language, and technology, which I will conclude in the next and last chapter.

The question regarding determinism presupposes that there are two entirely separate entities or spheres, a human one and a technological one. This split is also assumed by Ricoeur when he distinguishes between human narrative time and scientific time (and perhaps also by Postman, when he argues about culture "surrendering" to technology). But the relation between humans, technology, and language is more complex. Using the interpretation and revision of Ricoeur I present in the previous section, which moves away from his emphasis on text by arguing that, next to language/text, there are more technologies that mediate, structure, and shape our world and play a hermeneutic role. Using the language of "voice," and keeping in mind McLuhan's phrase about media and messages, we could say that humans not only use languages and technologies as instruments but in and through their use of these technologies *acquire a voice*. Both language and technology are media, but they also have their own "voice" in the sense that they co-shape the message. Now, this shaping has a deeply *social* character, with the social being structured as performance and narrativity. Together with technology, and, indeed, by using technology, humans coauthor narratives and co-perform, repeat, and modify the social.

From Austin, Searle, and Wittgenstein, we learn that language use can be performative (Austin, Searle) and that it is governed by rules (Wittgenstein). We can add that this is the case, since the social itself is performative and governed by rules. It has a performance and gamelike character, or stronger: It is *all about* performance and games. Particular language uses and social performances constitute, and are embedded in, language games and, ultimately, in a culture, in a form of life. Moreover, as I show, we can apply this Wittgensteinian view of language use to technology: Technology, like language, which is also a technology, mediates our social performances. Particular uses of technology are part of a larger whole that renders the technology use meaningful and structures it, that renders it successful, and that must be presupposed for it to be meaningful and successful. Thus, meaning is "made" and "performed," but, at the same time—*because* it is performed—meaning is already given in language and other technologies. Meaning is already given in a particular culture, which structures the performance, understood as socially shaped use and narrative, and which renders it successful. There may well be collective desires and beliefs, as Searle thinks; but individual or collective desires and beliefs (if the latter exist at all) do not determine what the social is and becomes. "Before" there are collective desires, beliefs, and agreements, there is already a meaning whole to which we must relate in order to make meaning and to be successful in and with our performances. There are already narratives. This brings me to the next point.

The social life is not only performance-like and gamelike; if we attend to its phenomenology and hermeneutics, it also has a temporal structure and, more specifically, a narrative structure. The performance takes place in time, and individual performances have a narrative structure and are part of larger narrative structures. From Ricoeur, we can learn that narrativity means that characters and events are organized in a plot. This is done by means of language, but—I add—also by using (other) technologies. Narrative technologies also reconfigure characters and events. Ricoeur's model of the social is that of ancient Greek tragedy and is also inspired by the concept of mimesis, which is not simply "copying" or "representing" what is already there in the social but also has the capacity to reconfigure the social. Whereas Plato was suspicious of art and performance, linking it with untruth, with appearance as fake, as opposed to a detached, eternal truth (consider the cave metaphor), here—if we must speak of truth at all, rather than meaning and experience—truth is linked to what makes sense to a community, to the social whole, to the narratives of a culture. The concept of "narrative technologies" suggests that technologies are not mere instruments but, through their use, participated in this social sense-making.

Technologies, then, are not to be understood as "things" or "artefacts" divorced from the human and the social, but, instead, as embedded in them as tools for human voice and speech acquisition, human existence and social life, human performance and interpretation, and social reconfiguration and narration. But they are never mere tools; as narrative technologies and tools for performance, they also shape the human voice and the social. This is also true for language, which is, after all, a technology (see also McLuhan, again) and is combined with other technologies. Texts, for instance, refigure our understanding of the world; other technologies also have that capacity. As narrative technologies, language and other technologies configure how we understand our world and what we do by structuring our lived time and narratives—not only narratives *about* us but also our performances *in* time, "real-time" performances.

This process-centered and performance-oriented approach also gives us a more developmental and process view of personality, identity, the human, the social, and the political. Humans become persons by becoming characters in narratives and by "sounding through" (per-sona), which is always a mediated and relational matter. The human voice and speech, understood as performance with a narrative structure, sounds through and speaks through language (games), through communities and traditions (forms of life), through other narratives, and, indeed, through technologies and media, which mediate and make possible. For language, this means that our logos is always a techno-logos. It is word and it is discourse, but it is not mere word and discourse. It is, at the same time, human and technological.

This view replaces the focus on individual subjects, on static subject-object configurations, and on "things" (as found in postphenomenology and mediation theory, which focus on subject-technology-world relations) with one that is more holistic, social, and process-oriented, and takes into account language. Yet, individual humans and their experiences do not disappear; far from it. Drawing on Wittgenstein and others enables me to ensure that, in this view, individual human experience and action is there. But, just as "instrumental" use of technology is never *merely* instrumental, what we call "individual" experience and action are never *merely* individual: It is now understood as performance and as language and technology use, and this use and performance is seen as embedded within a social-technological context of language games, technology games, and narratives, which are themselves embedded in a form of life. Performances-with-language and performances-with-technology are social and narrative at the same time. A performative and narrative view of the social-with-technology, thus, enables us to repair what Van Den Eede calls a failure of mediation theory: a failure to theorize 'being-with-each-other' relationships (Van Den Eede 2010), that is, intersubjective relations. And it does so in a way that also remedies postphenomenology's lack of attention to language (Coeckelbergh 2015) and brings in the time and narrative dimensions by using Ricoeur.

Of course, this approach is far from complete and comprehensive. For instance, in addition, we may ask *political* questions: Who gives a voice to whom or what, and what is the role of technology in this? What is the relation between rhetoric and technology? Latour's work enables us to ask the political question (see his *Politics of Nature*, again), but, in the interpretation of Ricoeur's work as presented, so far this aspect seems far less developed, if not absent. What could it mean to talk about the politics of hermeneutics and narrativity?

I will turn to the issue of power later in this chapter, but let me already ask some questions and indicate some potential avenues for further discussion about this political aspect. Perhaps paying attention to the politics of hermeneutics and narrativity could mean that we have to acknowledge the possibility of disagreement. More generally, without having to accept a fully Hegelian or Marxist approach with their specific narratives of recognition and struggle, one could at least acknowledge the antagonistic dimension of human social relations and recognize the political and power dimension of society. For example, there may be competition between authors (human/nonhuman, human/technological?) and between narratives. If meaning is more open in this hermeneutic model, it may be contested by all kinds of parties. There does not seem to be any guarantee that such social-hermeneutic processes end up in closure or consensus; there may be an antagonistic and agonistic aspect to the political hermeneutics of human-technological-linguistic speaking. And on the basis of the Ricoeur-inspired view articulated in the previous section, narrative processes appear perhaps too much as apolitical, in the sense of having nothing to do with power. But we can point to power differences, which may be related to differences in knowledge and understanding. These power issues can be put in narrative terms. For instance, someone may feel disempowered when they are not aware of or do not understand a narrative, which nevertheless influences their life, or an abstract narrative may be used to exercise power over others. Maybe some people are more empowered than others to write their own narrative and, indeed, to write the narratives of others and of their community. More needs to be said about the relation between hermeneutic processes, power, and political processes, here (see later in this chapter) and elsewhere. Such political questions are also lacking in mediation theory and postphenomenology; but, if narrative theory is truly a theory about the social, then it should not be excluded from this aspect. Moreover, paying attention to power may also offer an additional answer to the question if first-person experience is taken into account. Narratives—concrete narratives, but also the meta-narrative of this book, which is, in itself, abstracting from real characters and events—may be told *about* people and may be used to structure and pre-structure their lives without taking them into account as (potential) co-narrators. Narratives may hide other narratives and they may hide their own active hermeneutic-narrative role. They are not politically neutral. Furthermore, the politics of hermeneutics and narrative is not just in "discourse" but also in technologies: If technologies make possible particular discourses and structure our narrative existence, then this has normative consequences. As I will argue later, technologies make possible exercises of power; in this sense, too, they are normative and political. Grammars of technology and grammars of power are connected. This gives us yet another reason why it is important to understand how humans *and* technologies shape narratives (what others and technologies tell us and what they tell us to do) and what these narratives do to our lives—also in terms of power and politics.

To show what this performative and narrative view of relations between humans, technology, and language means for understanding technologies, how it relates to my previous (Wittgensteinian) arguments and readings of authors such as Searle, Ihde, and Latour, how it can contribute to a more integrated view that lets humans, language, and technology speak, how it is related to thinking about power and embodiment, and how it works in practice, let me discuss the case of social media.

7.4. Case: Understanding Social-Media Performances

Social-Media Performances

Social-media technologies, for instance, in the form of Facebook, YouTube, LinkedIn, WhatsApp, Instagram, Twitter, and Skype, are Internet and World Wide Web-based technologies that enable users to communicate and share information. In the past two decades, they have become the main interface for many people to use the Internet. Typical for social media is that *users* generate and modify content and profiles—in contrast to earlier Internet technologies, like websites maintained by companies and organizations, which are monologic instead of dialogic.[1] This makes possible that users not only "reveal" their identity (misleadingly assumed to be only created "elsewhere," "offline") but also create it in and with their use of Internet technologies. The same is true for reputation, which is also created by using social media. Social media have become important technologies in many people's private lives, but have also become crucial for career opportunities, business, and (other) professional activities.

The phenomenon of social media is itself not merely "technological;" rather, it strikes me as an excellent example of how technology, language, and humans merge in a way that lets all three "speak." Human users speak in the sense that they write or talk on social media, sharing their messages. But they do that by using language and by using (other) technologies, including computers, smartphones, and specific programs and apps, which, in turn, include various kinds of code and programming languages. These tools are not neutral; the media shape the message. To understand this noninstrumental dimension of language and technology, we can use the concepts discussed in this book and other concepts from philosophy of language and philosophy of technology. This gives us a recapitulation of what I already say in the previous chapters, but, more importantly for the purposes of this book, it also helps to take the next step: Using the insights from previous discussions and running through the chapters, we can work towards a more *integrated* view and gain a better understanding of the inter-relations between humans, language, and technology.

First, based on Wittgenstein and Heidegger, language use in social media can be understood as the use of tools, in particular, the use of words and the use of code, computers, and smartphones, and this use is rule based and social. This use of words and (other) tools in the context of social media is always a social and cultural use, in the sense that the use is part of language games/technology games and, ultimately, a form of life. Perhaps different social media have different language games and different ways of using the medium; to the extent that this is the case, then these differences in use are connected to different language games and technology games that are already part of our society and culture. Thus, social media are "social" in this deeper sense, not only because they are used by *humans* to communicate with one another (not only because they are used as communication tools) but because *language*-as-connected-to-social-relations and *technologies*-as-connected-to-social-relations also "speak" *through* humans as social beings, through the social.

Second, the meaning of "virtual" artefacts is not *determined* by the technology. Their status is given on the basis of social agreement (Searle) and created in use, but this use is not merely a matter of use and human (individual or collective) intentionality. It is connected to larger wholes such as language games and forms of life (Wittgenstein) that are not intended at the point of use; the meanings of particular technologies or (other) objects are already partly given and pre-structured in and through language and the social (as Heidegger says, language "speaks"). At first sight, there is collective intentionality: Many "likes," for instance, means that there is a collective desire, emotion, etc., which gives meaning to whatever object (image, video, text) has been posted. But if there is such a thing as collective intentionality or status function at all, these "constructions" of the social are only possible because, in particular social contexts (groups, societies, cultures), people already have a meaningful relation between the object and the social whole. For instance, "virtual" money is only possible and meaningful since there is already such a thing as money we use in society, which already has a meaning (in and through use). Similarly, use of images on social media is socially and culturally embedded. For instance, images of pets or political revolutions already have meaning before we explicitly form a collective intentional state and ascribe status to them, and before we use them. These given meanings belong to particular kinds of activities and language games, which are also played "outside" the context of use of the particular technology-environment (e.g., "outside" of Facebook), or, rather, are played with different, older technologies, such as telephone, TV, and newspapers—technologies that, of course, also have similarities to the more recent media. I use quotation marks for "outside," since the Wittgensteinian approach calls into question the very distinction between inside and outside: Both the so-called "online" and the so-called "offline" worlds are actually *one* world, once we consider the *use* of and performance with technologies, and connect this use and performance to language/technology games and a form of life.

Third, this "cultural" interpretation of the later Wittgenstein (and Heidegger) *can* be interpreted in a way that shifts the focus entirely to language. In a poststructuralist and postmodern interpretation, the world of social media is recast as a world of signs and images. Perhaps, then, the world of contemporary social media becomes a world of simulacra, signs without any link to "the real"—whatever that is. The materiality of the technology disappears; a similar observation can be made about embodiment and materiality: It seems that there is only a kind of hyperreality, which is both immaterial and unconnected to the body. But I argue that this conception of social media and, more generally, of (technological) culture as media, of signs and as a culture of signs, is problematic. Although we can learn some things from these discussions (e.g., we can accept a non-representational view of social media, we can use a transcendental argument, we can acknowledge the limits of philosophical discourse), these poststructuralist and postmodern approaches are too focused on a linguistic, often textual reality that seems entirely unconnected with materiality, body, etc., which, if they are acknowledged at all, are rendered entirely mysterious. A focus on concrete use and performance, by contrast, enables us to focus on materiality and embodiment without losing track of language.

Fourth, with Ihde, we could (re)turn to the medium in its *materiality*, take into account the body and embodiment in technology use, and analyze the mediations between subject and world in social media by applying his scheme of human-technology-world relations to social media, albeit stressing that different relations are connected to different *uses*. Sometimes, in some uses, social media play a hermeneutic role to the extent that they shape our world, reality-for-us. Social media are then like Ihde's thermometer: We look at it and experience a technology-world. The world is presented through and as technology. When we use a thermometer, and perhaps even when we do not use it, we experience warmth and cold *as* temperature, because of the technology. Similarly, through social media, we may come to experience the world *as* posts, likes, followers, etc.—even if we do not use our smartphone or computer. At other times, social media are used and experienced as embodied and transparent, insofar as the media are no longer visible in our use of them. We are so busy with the message that we no longer notice the medium, since, in use, we are "in" social media, we are bewitched by it and under its spell. Yet, what is missing in this phenomenology and hermeneutics is a more social and narrative approach: How do others appear when social media mediate? What is the "social" of social media? And where are narratives in this picture? More generally, the role of language in these relations and mediations is unclear. To remedy this, in chapter six, I propose a revised and expanded scheme of human technology-world relations, which includes language as a mediator. I also add a transcendental argument, which understands language and technology not only as mediators but also as conditions of possibility that make possible, structure, and shape concrete uses of words and things. For social media, this means that language (1) always mediates our use of social media as an "in between" that plays a hermeneutic role and (2) also functions as a condition of possibility, which structures and shapes particular, live uses of social media. Considered as living experience and concrete uses, social media have their own language/technology games, which makes them more than mere media or instruments: The (particular) language and technology also shape the message, and these games, languages, and technologies belong to a form of life, which already pre-structures our use and experience through its linguistic-technological culture. For instance, the activities and games of "showing off to someone" or "praising someone" are (part of) social games that are already available in our culture, and particular language/technology games within social media are made possible by the availability of these forms in our culture; these forms and structures shape our use of social media and must be presupposed for social-media games to work.

Fifth, to give a more social and linguistic twist to our framework and approach, we may also turn to Latour's work, which gives us a posthumanist ontology of the social that includes humans and nonhumans, such as technologies, and which accounts for the role of language. As Latour and Akrich argue, technologies inscribe and give a script to humans. Applied to social media, this means, again, that when we use social media, the relevant technologies actively shape and script the messages. A social-media program/app also "speaks," in the sense that it tells us what to tell others. For example, through the way the medium works, with its likes, followers, etc., we post and share what receives positive feedback (in the form of likes) from others. Technology in conjunction with the social, thus, both shape our message. And conversations have to take place within the formal constraints and rules given by the medium and scripted in the technology, for instance, use of short messages in Twitter. This rule scripted in the technology is not neutral in terms of content, but influences our messages, which have to become a kind of mute soundbite (if we have to use a sound metaphor—perhaps "textbites" is more appropriate here) and which are structured by particular language games (e.g., congratulating someone, expressing one's contempt for an opinion or for someone, making a joke and showing wit, etc.). Assuming that users of social media seek to be liked by others and, insofar as this is scripted into the technology/medium, the social itself becomes reconfigured as a pleasing game, an entertainment game. And perhaps in this way these media also reveal something important about the social, in general: Among other things, it is a pleasing game. People try to please others, seek social approval and esteem. This has always happened, also before current social media. But it is not merely a pleasing game. Does the technology reduce it to a pleasing game? With regard to media and technology, the question is how the new technology shapes these desires and practices, repeats or changes the game. The technology has a "voice" and "speaks" in the sense that it co-scripts our messages and conversations. As philosophers of technology and researchers of new media we can reveal this voice and this speech, showing the noninstrumental dimensions of technology and media use. If we take seriously Latour's social ontology, then we should also recognize, and beware of, nonhuman "actants" in social media. At a superficial level, this means: Perhaps, when we use social media, our conversation partner is nonhuman, is a bot, whereas we might have assumed that (s)he is human. But, more profoundly, and even if there were only human users, nonhumans take part in the conversation anyway in various ways: All kinds of nonhumans "speak" and are "actants," in the sense that the messages of the postings are mediated by the particular technologies (app, smartphone, etc.), as I explain, but also in the sense that, through new postings and shares, new nonhumans are constantly entered in the conversation and compete for attention and approval—consider images of animals (e.g., cats), scientific results, videos of robots, pieces of text, such as opinions, one-liners, jokes, short poems, text, etc. There are also many combinations and pairings of humans and nonhumans, such as text, images, etc., which are humans, signs, and media at the same time. Hence, the world of social media turns out to be less modern than may be expected: It is a world that is full of hybrids, a world in which actors and actants, humans and nonhumans, nature and culture, seamlessly mix and remix, engage in new associations and networks, and, in the process, acquire a voice. This approach also enables us to ask the political question: Who or what acquires a voice? Who or what delegates speech to whom or what? Who or what is silenced in social media and other media? Which humans and nonhumans are more empowered than others?

Furthermore, we can elaborate and complement this approach and its speech, theatre, and performance metaphors by focusing on time and narrative. Drawing on Ricoeur, we must emphasize the time dimension, the process dimension, and the narrative dimension of social-media use. The language of social media, or, rather, language use and performance, and *technology* use and performance in social media, and the social life on which it depends and that it helps to constitute, have a temporal dimension. More specifically, they have a narrative and performative dimension. Social media exist as use and performance, and provide a site for performance. That performance on social media is always social and *narrative*. Through social media, we tell stories: We (re)tell the stories of others and we tell our story, and, thereby, constitute our self and identity. At the same time, our story is constituted by others. There are many coauthors of our narrative. But, not only humans speak; technology also speaks as it coauthors our narrative. Facebook, for instance, provides a very specific format for our narrative and conversations about our narratives. By recording, collecting, and ordering our posts and their "likes" chronologically (and reminding us of past posts), it tells a story about us, configuring a specific, linear time-narrative. The result is, again, a mix of humans, language, and technology. Social media and their narratives are human, since humans write the narratives; but, at the same time, the media and narratives are shaped by the technology and by language. A social medium such as Facebook is thus human, linguistic, and technological at the same time.

This can be understood as "mediation" by social media, but also as social-media technologies that make possible and pre-configure our narratives and performances. Let me explain this more (pre-)structural and transcendental role. The medium Facebook, for instance, offers a specific configuration of the social, in which social interaction and approval is linguistically and technologically translated into likes and comments, which are also quantified and "plotted" by the technology. In particular, using Ricoeur's analysis of narrativity, we can say that the technology, together with the humans, organize characters and events into plots. The characters can be humans or nonhumans. They can be first order or of a higher order, further removed from real characters and concrete events. But, in contrast to what Latour suggests, there is not just an unstructured or merely networked collective of actors and actants; instead, there is also a temporal structure, a narrative one, and this structure is not only built into language (use) but is also present in, and made possible by, the technological framework. If people can have a Facebook history or create a YouTube narrative, this is made possible by the availability of a narrative structured and provided by the technology. Moreover, it is made possible and shaped by the culture in which it is embedded. Hence, social media such as Facebook are never culturally neutral.

Of course, we can still use Ihde's scheme to say more about how social media shape us and our world. We could say, for instance, that, here, what Ihde would call the hermeneutic role of social media is at work again, but it is a hermeneutic that is not only about "world" but also about "self" and "other:" By using the technology, we come to experience our world as a Twitter-world, we come to experience our self and others as Facebook-narratives. But this has not only to do with this specific human-technology relation or with the medium role of the technology; that we experience our world in these terms is also made possible by the technologies *as connected with an entire culture that makes possible this technology use and shapes that specific experience.* With concepts from Wittgenstein, Ricoeur, and McLuhan, we can conceptualize this "grammar." Moreover, my analysis of these media focuses not so much on vision and other senses and modes of *experience* (as in Ihde); here, the emphasis is on acting (Latour's metaphor) and, influenced by Wittgenstein, on use and *performance*. Social media do not (just) "do something with us." They are not so much external to us. We also *perform* our narratives, live and are our narratives. In addition, there is more emphasis on the *social* than in Ihde. We do not perform alone. What we do on social media is also coauthored, cowritten, and co-performed by others and by the *technology* and medium, which stimulates and creates a particular narrative structure and a particular temporal configuration for our experience and our performance.

Let me stress, again, that this focus on use, performance, and narrativity enables us to cross virtual/real and online/offline dichotomies. The technology not only gives us a "virtual" or "online" narrative; it also structures the narrative of our (real) lives. It does so, for instance, when and to the extent that we come to see the world through images-for-Facebook (this is the visual, experiential, Ihde-type of effect, again) *and* when we start to adapt, rewrite our own narrative-as-lived accordingly: "I have to do this, or I have to do this in this way, I have to meet this person, go to this event, etc., so I can then share it via social media, make it *worthwhile* to share it." In other words, insofar as the medium speaks, we start performing in a way that responds to the medium and does what the medium "asks" or even "demands." We play the social-media game, online and offline. More generally, as users of social media, we may come to see our lives as a sequences of events in a plot, with ourselves as the main character. At the same time, we are also the codirector and the camera. "Camera," since we look at ourselves, others, and the world through social media. "Codirector," since others also write and direct our narratives, as does the technology as a medium: As said, a particular social medium shapes the narrative and the performance in a particular way. The same is true for narratives about others: By using particular social media, we also experience and act towards others in the way of these social media—when we use them and when we do not use them. Insofar as we use particular social media, our social performance is shaped by the technologies involved as much as by others, that is, other users of these social media. It is also a social performance in the sense that it is structured—and made possible, I like to add—by the social language and technology games available in our culture. In this sense, our use of contemporary social media is not significantly different from our social performances "outside" of these media or "after" these media. Stronger: There is no longer an inside and an outside; rather, there are social performances, mediated by various language and technology games.

Our current social media are narrative technologies, then, not only because they give us a specific form of narrativity "within" the medium and when we use it but also and especially because (1) they shape and structure our lived narrative time, understood as use and performance, everywhere and always, and (2) because they reveal the narrative and performative-theatrical character of the social and of human existence itself. Social media show the truth in Shakespeare's lines 'All the world's a stage/And all the men and women merely players.' They reveal and constitute us and the world as narratives, as plots, and as characters and events. (Compare with Heidegger's claim that modern technology reveals the world in a particular way, as enframing; whether or not his particular point about enframing is true, the useful insight is that *technology is a way of revealing*; one can then discuss what kind of revealing.) Social-media technologies, then, provide "exits" and "entrances" for people to play their parts on the technology-mediated social stage. And if it is true that there is no (longer) an exit, no world outside social media, then this is not only because of the pervasiveness of the current media and technologies but also because, once it is revealed that all human action is social performance mediated and made possible by technologies and media, we come to recognize that what we call "social media"—the specific technologies—are only part of a general social-media existence, which existed already long before the term and technology "social media" was used. Social media reveal that performance is *always* social and technologically mediated and structured; the narrative is cowritten/co-told by technology. Similarly, it is always mediated and structured by language. Note also, again, that this analysis of social media can be interpreted as implying that social media also have a normative effect: As they cowrite our performance and narrative, they not only tell others about us (and tell us something about ourselves) but also tell us what to do. As codirectors of our lives, social-media technologies coscript our activities and tell us what kind of games to play. They make us adapt what we do and how we do it. (This reminds us, again, of the political and power aspects, which I mention at the end of the previous section and which I will further elaborate later.)

This view of language and technology as coscripting our performances does not imply, however, that technology "takes over" or "determines" us. Humans have always used technologies and media to tell stories about themselves to themselves and to their community, and to tell others what to do, whether by means of material artefacts or by means of writing and printing technologies. And these technologies have always been "social media," in the sense of enabling communication between people and in the richer and narrative sense elaborated here. New are the specific ways in which current ICT-based and Internet-based social media co-shape our use of language, our language and our language games, our status declarations, our scripts, and our narratives. But humans and language are still important as co-users, co-declarers, co-narrators, and coauthors, and have always been. In order to perform its role as coauthor and co-narrator, a particular technology needs humans and language as much as the humans and language need the technology. Whether it is good and desirable to have that specific technology and medium, rather than others, is a different question.

But, while technological determinism is not true, so is the idea that we can simply choose our technology and that our performances are entirely in our hands. Surely, if humans-language-technology have moved and are moving in a particular direction, then one could try to redirect this process. But, if the framework articulated here makes sense, then this redirection cannot be done by humans alone. If "user" means "single author," then the user of technology is dead—as dead as the single author. As users, we can, of course, try to use the technology in a different way. But this project always takes place within language and within technology/media. Alternative use of technologies is also a social performance, which is made possible by, and already shaped by, language and technologies. There are already games. There is already a larger whole, a given, a horizon, which structures our different uses, performances, and narratives. The same is true for thinking about technology, and, hence, philosophy of technology. Today, our language/ thinking has already changed through the new media; our lives, as narrated and reconfigured by technologies, have changed. When we make arguments about social media, that is, when we use language and (other) technologies, there are already "givens." There is already a language (and, hence, way of thinking) and there are already specific technologies, such as the word processor I am using now and the smartphone that tries to grab my attention. These technologies and the language games and form of life they are related to make possible and shape individual use. If change is desirable, therefore, it can only happen in conjunction and promiscuity with language and technology, and, indeed, with the social contexts and cultures we inhabit.

In the next and last chapter, I will further discuss how humans, language, and technology are entangled and all "speak," and what this means for philosophy of technology. Before ending this chapter, however, let me return to issues concerning power and embodiment, in general and, specifically, in relation to social media.

Power

There are various ways to discuss the relations between power and social media, depending on the social and political theories one uses. For example, inspired by Marxian thinking, one could study the political economy of social media and reveal its power structures and relations (see, for instance, Fuchs 2014). And using Foucault (1975), one could investigate how people are disciplined by means of contemporary social media. These are valuable and fruitful approaches. What the proposed analysis, inspired by Ricoeur, can add, however, is the focus on narrative and performance. This gives us a range of new questions, which can be merged with a Marxian or Foucaultian perspective: What is the political economy of *performance* in a Facebook context, and what is the role of technology in this? How does disciplining by others and self shape our *narratives*, and how is the technology used for this purpose? Is performance mediated by social-media labor? Is it work? Who or what writes the narrative of my life? To what extent can I choose my character, my roles, and my plot? How do social media influence and shape *characters* and events? Who has the power to *direct* my life? Who has the power to play certain roles and script certain narratives, rather than others, and who has the power to revise my role (and those of others) and script new narratives in light of new circumstances?

Furthermore, connecting back to Wittgenstein, Lyotard, and other writers that enable us to bring in the role of *language*, one could ask: What is the relation between *language games and power games*, linguistic structures and power structures, in the case of social media? How is rhetoric on social media shaped by the media/technologies? What kinds of language games are played? Do social media empower us to write our own narrative, or is the power in the hands of others, and what is the role of language in this? Are all "others" equal? How does social-media status and reputation relate to one's general social status and reputation, and how is this mediated by language? How does the medium reconfigure power relations, power structures, and, indeed, power *language* in society? Do social-media use, performance, and theatre have transformative power at all, and, if so, what is the role of language and *technology* in this? And is this transformation just about the use of language or technology as instruments, or are there larger structures and grammars, which also relate to language and technology as mediators and as structures, grammars, and conditions of possibility? The latter question suggests, again, a transcendental turn: It seems that the question, "Who or what speaks?," must also be related to the question, "Who or what has and exercises power, and what is the role of language and technology in making possible and shaping the power structures that enable, constrain, or hinder that speaking?"

This transcendental question is in line with Foucault's approach to power. Let me say more about Foucault, in order to explore what his approach could imply for understanding language, technology, and power.

In his history of sexuality, Foucault writes that he is interested in power's 'condition of possibility' (Foucault 1976, 93), in understanding what makes possible its exercise. Power is often thought about as being about a person or a state coercing someone else, but Foucault showed that power is everywhere: in discourse and in concrete social relations, in families and in institutions.

> The omnipresence of power: not because it has the privilege of consolidating everything under its invincible unity, but because it is produced from one moment to the next, at every point, or rather in every relation from one point to another. Power is everywhere; not because it embraces everything, but because it comes from everywhere.
>
> (Foucault 1976, 93)

According to Foucault, power relationships have a 'strictly relational character' (Foucault 1976, 95): There is no single locus of power or of resistance (96). Everywhere there are battles for truth. In an interview, Foucault uses the term 'micro-mechanisms of power,' for instance, mechanisms of exclusion, such as 'the apparatuses of surveillance' and 'the medicalisation of sexuality' (Foucault 1980, 101). The system of 'micro-powers' is not 'installed at a stroke,' but slowly develops. He wants to know how power and power "struggles" become operative in particular cases, in the everyday. Even top-down exercises of power, according to Foucault, depend on these mechanisms:

> 'Generally speaking I think one needs to look rather at how the great strategies of power encrust themselves and depend for their conditions of exercise on the level of the micro-relations of power.'
>
> (Foucault 1980, 199)

Again, it strikes me that Foucault is interested in the *conditions* of the exercise of power, which he localizes on the micro level. To clarify what this approach means in practice for analyzing power, it is instructive to look at feminist interpretations, which have made good use of Foucault by pointing to 'micro-mechanisms of power' (McCallum 1996, 93) with regard to gender. Amy Allen, for instance, has applied Foucault's approach to focus on 'microlevel power relations,' power relations 'at the level of the everyday' (Allen 1996, 271) to the issue of sexual harassment. Sexual harassment is not imposed by the state, but takes place at micro level and depends for its workings on everyday relations between people (here: men and women). Furthermore, discourses—for instance, about gender—also produce, and are produced by, relations of power. And, of course, Foucault says a lot about power and the body, for instance, in his histories of sexuality and in his lecture on technologies of the self. (I do not have space to elaborate this here.)

One could use this approach and add not only language but also *technology* to this analysis, and argue that *both discourse and material technologies are conditions of possibility of power: They make possible the exercise of power*. Disciplining, for instance, is made possible by the use of words to get people to do things (perlocutionary use of words), and, more generally, by rhetoric and discourse. Similarly, sexual harassment is made possible by use of words and sometimes use of the body. And specific institutions and forms of social organization support this disciplining and these uses of power, for instance, the prison or the modern workplace, understood as forms of social organization. However, we must add that the exercise of power is also made possible by use of material technologies. For instance, disciplining is made possible by technologies of surveillance and punishment, which (we should add in the light of my focus on performance and narrative in this chapter) constitute specific power performances and narratives. It is also made possible by specific material technologies and architectures. Foucault's example of the panopticon (Foucault 1975), for instance, can be interpreted in this way. As a building designed by Bentham so that prisoners can be observed without the prisoners being able to know if they are being watched, and specifically aimed at enabling a new mode of exercising power, the panopticon is a *technology* and architecture that makes possible the exercise of power by individuals and the state. It constitutes a specific configuration of power relations at micro level, but does so by means of technologies, understood as material artefacts and (infra) structures. More generally, analysis of micro-relations should include not only the use of language and techniques of organization but also concrete material technologies. For Foucault, the panopticon is still mainly a metaphor, a symbol for modern disciplinary society. Contemporary readers of Foucault have, instead, pointed to concrete technologies such as CCTV cameras and surveillance on the Internet: all technologies that are used to watch people. Indeed, one could also add social media as a new kind of panopticon, albeit one that replaces "vertical," centralized surveillance with more "horizontal," decentralized surveillance, involving humans and nonhumans: One never knows who or what is watching, who or what has access to the data. (For a brief overview of some discussions about the Internet, social media, and the panopticon, see, for instance, Fuchs et al. 2012, 1–8.)

But how Foucaultian is this interpretation? It is in line with his interests in the conditions of power, and he talks about 'technologies' (Foucault 1988). However, Foucault's "technologies" are not very material. He mentions 'apparatuses' (Foucault 1980, 101), but, often, techniques and technologies are used as metaphors; his focus is on discourse and the social. The same is true for his use of the term 'technologies of the self' in a seminar he gave at the University of Vermont in 1982. When he speaks of 'the hermeneutics of technologies of the self in pagan and early Christian practice' (Foucault 1988, 17), he includes technologies of 'production . . . which permit us to produce, transform, or manipulate things,' but also signs, systems, power, and self, and he mainly attends to the last two. Of course, one can read more (material) technology into Foucault's analysis of power and self here, stressing, for instance, the technologies of writing used in the ancient practices of letter writing as practices of the self. But this technological and material dimension remains largely implicit in Foucault. It seems to require further steps in bringing out the role of technology and, indeed, in *materializing* Foucault. This could be done, for instance, by combining postphenomenology and Foucault (see also what follows), or by using other empirically oriented approaches in philosophy of technology and integrating them with a Foucaultian focus on (micro) power relations.

In general, I see a lot of scope for using Foucault for philosophy of technology, indeed, for reading him *as a philosopher of technology*. Some useful steps have been taken in this direction, for instance by Jim Gerrie (2003), who asks if Foucault was a philosopher of technology, and by Dorrestijn (2011), who links mediation theory to Foucault's writings on power and subjectivation. With regard to the project of this book, however, let me explore how we could further discuss the relations between things and words, and, more generally, technology and language, in light of Foucault's work on power. I see at least the following points of connection.

First, the approach presented in this book focuses on the use of things and the use of words and is hence also concerned with the everyday and the micro level. But more work could be done on the *power* dimension of the use of things and words, especially in practice, in everyday life. This could be done, for instance, by discussing the rhetorics of the use of words but also the "rhetorics" of the use of things: how we use not only words but also things to make others do things, for instance. I have already made some suggestions about this earlier in this book. Consider again my interpretation and expansion of Searle and Latour, or McLuhan who in his *Laws of Media* asks us to look not only at grammar but also at rhetorics. We could discuss uses and performances with technology as perlocutionary "technology-acts", as technology "utterances" (see also what I said about McLuhan), which have a particular grammar as embedded in technology games and forms of life, but which are also micro level forms of power and exercises of power, having effects on others and making others (and one self) do things. This does not only include intended effects (as a Searlean would have it) but also unintended effects. In what one could call a hermeneutics of things understood as a "rhetorics of things", one could further study how the use of things, next and together with the use of words, constitutes, and is constituted by, power relations.

However, language is not only about the use of words and about rhetoric but also about narrative, for instance. As I suggest in the beginning of this section and, taking into account the reading of Ricoeur I present in this chapter, we can add a narrative dimension to this approach to power. Narratives also make possible power, and narrative technologies constitute ways in which power is exercised. When technologies structure our time and narrative, when they configure plots and shape characters, they also always, at the same time, configure power relations and are themselves embedded in power relations. For example, social media and smartphones may structure the lives of users in ways that exercise power over them. This might not happen in a "macro" way, in the sense that, for instance, a large corporation or state coerces a person, but, in any case, in a "micro," Foucaultian way, for instance, by limiting the options of use through and in the design of the code and the device, a specific *use-narrative* may be imposed that configures people's lives. Think, for instance, about a running app, which, through its design, encourages a particular way of running and thinking about one's health and physical condition.

Second, we can read Foucault's proposal to do hermeneutics of technologies of the *self* as a hermeneutics that crucially includes not only a linguistic but also a material side, and we can understand material technologies as conditions of possibility for language use and discourse in practices of the self. The use of language that constitutes the self is then seen as crucially connected to the use of technologies. Consider the writing of letters from ancient times until modern times, and the writing and use of visual material on the Internet today. These technologies (material or immaterial, or both at the same time, as in contemporary media and technologies) condition how one exercises power of one's self and over others, how one subjects oneself and others, how one resists (or not), and how one becomes a subject in the process. Subjectivation and becoming a subject are then not only understood as language games and power games but also as what I have called "technology games."

Third, we can use Foucault's earlier *The Order of Things* (1970) (in French: *Les mots et les choses*, The words and the things), which is focused on language and discourse, to further develop the argument about grammar and form of life in relation to power. I propose to read Foucault's concept of *episteme*, the condition of truth and discourse, and the 'field' (xxii) that grounds knowledge, as crucially including material conditions. Episteme can be interpreted as the grammar or form of life that is constituted by the use of material technologies and that shapes our use of words and things. Based on this interpretation, one could study how material technologies shape the field or space of knowledge, and, hence, shape what we regard as true and as knowledge. Such a "grammatical" approach would not focus on criteria, but reveal and critically study the entire material-linguistic space that makes possible a particular knowledge discourse. Indeed, we could emphasize with regard to Foucault, more generally, that he, like the later Wittgenstein and McLuhan, is interested in a transcendental, "grammatical" inquiry. If we use Foucault, we should not only take a historic—or, more precisely: archeological—but also a *transcendental* approach to science. Foucault's episteme concerns the 'conditions of possibility' of knowledge, the ground or space of knowledge (Foucault 1970, xxii). In this sense (and also keeping in mind what I say about McLuhan), Foucault is not only a rhetorician but also a *grammarian*. However, as I explain in my response to Ihde, we need to interpret the transcendental not only in terms of discourse and the social but also as including technologies and materiality.

Finally, using Foucault, one could further explore relations between technology, power, and the *body*, but then pay more attention to the precise ways in which use of words, use of things, and use of the body are connected in the micro-mechanisms of power. This analysis could also connect to feminist uses of Foucault and work on bio-power, but then with a clearer focus on technology. As technologies shape the power in relations and power in practice, these technologies (and not only *discourse*) need to be taken into account in the feminist analysis and in gender studies, in general. To take up McLuhan's example of the anticonception pill: To fully understand contemporary gender issues, a history of the use of birth-control *technologies*, a history of *discourse* on women, and a history of *human* and social relations need to be connected. The same could be done for (other) technologies used in sexual and reproductive practices, for instance, the Viagra pill (to stay in a pharmaceutical-technological context). Feminist analysis of all kinds of reproductive practices has already been done, of course, and, admittedly, has often included technologies. Consider, for instance Terry's (1989) work on 'prenatal surveillance' in the 1980s or, more recently, Takeshita (2011) on how the science related to the intrauterine device constructs women's bodies. However, in Foucaultian analyses of power, the role of the use of *material* technologies, understood as the use of things (not just "technologies" used in a metaphorical way, as Foucault and scholars interested in *discourse* usually use it), usually receives less emphasis than discourse (e.g., scientific discourse, discourse about women, etc.) and the use of language, let alone that there is attention to the precise ways in which the use of words and the use of things interact.

To conclude, I propose to pay more attention to the politics of things, from a Foucaultian perspective or another, but then one that gives us more insight into the relations between uses of words and things, combining thinking about language and discourse with thinking about technology. However, here I do not have the space to further develop these Foucaultian and other political paths.

Embodiment

Finally, the approach developed in this book and the analysis of social media proposed in this chapter is not complete without stressing that use of technologies and performance with technologies is always *embodied*. I already raise the issue concerning embodiment when I note, for instance, that Ihde, inspired by Merleau-Ponty, talks about bodies, embodiment, and technologies. And when responding to Foucault in the previous section, I mention the issue of bio-power. In this section, I wish to further articulate this aspect of embodiment, starting from what I say about *performance and narrativity* in this chapter, and apply it to social media.

Thinking about technology and language, taking seriously the body and embodiment means acknowledging that using technologies and language, and performing with and through technologies and language, is always an embodied affair. With regard to social media, such as Facebook, this could mean that using Facebook should not be seen as an activity that is merely "cognitive" or involves only "mental" processes on the part of the human, but, instead, must be understood as embodied performance. McLuhan was right that we are entirely involved in our contemporary media, and we must add to this insight that, if we consider *use* of technologies and language, and *performance* with technologies and language, we are and *have always been* entirely involved as body-minds (a term I use in order to avoid dualism). Of course, we can be more or less involved in, and engaged with, a particular technology. And perhaps McLuhan was right that contemporary technologies require more involvement. But there is also a general lesson to draw here: *In* use and performance, there is body and mind, mind and body, body-mind. Performance is not only social but also always embodied.

If this is right, then *narrativity*, including technological narrativity, must also be understood as embodied. If the social life and the human life can be compared with theatre, then it is important to stress that, in the theatre understood as social performance, there are not only "characters" and "events" but also concrete, fleshy, and embodied people. The plot organizes characters and events, but also bodies. But, since narrative theory, including Ricoeur and perhaps also Latour, is rooted in the postmodern overemphasis on signs, text, and language, and neglects embodiment, its language does not sufficiently enable us to articulate this dimension of embodiment. The very term "character" already abstracts and alienates from the embodied aspects of narrativity in use and performance. We should make sure that the humans in our framework are real, embodied humans. Otherwise, the framework (and also the virtue-ethics related to it) is irrelevant to human life, human existence, and social and technological performance, which always involve the body. Therefore, thinking about technology in terms of a philosophy of *language* alone is not enough. When it comes to theorizing about the body in use and performance, we have to go beyond Ricoeur, whose narrative theory does not explicitly theorize the relation between narrativity and embodiment. In the previous chapter, I point to some potential intellectual sources we could use to develop thinking about embodiment in relation to language and technology, including Feuerbach and Wittgenstein. One could also consider Virilio, again (Chapter 5), or other postmodern authors who *do* say things about the body. Here, I will limit my discussion to Merleau-Ponty.

While Merleau-Ponty did not write about technologies, the *Phenomenology of Perception* (1945) can and has been read by philosophers of technology as providing suggestions about embodiment and technology. Ihde is known for taking this up, in particular, but not exclusively, in his theory of human-technology relations (Ihde 1990); but, also, later philosophers of the empirical turn have studied Merleau-Ponty, for instance Brey (2000).

According to Merleau-Ponty, the body is not experienced as something external, but, rather, is a means by which or through which we perceive and relate to the world. To put this view in a transcendental way: The body is a condition of possibility for perception and movement. It makes possible interaction with the environment. When we relate to the world, we know the body not usually as an object, but as "body schema:" It is a structure that organizes perception and movement in terms of possibilities and potentiality. Merleau-Ponty compares bodily space to 'the darkness needed in the theatre to show up the performance,' a kind of background against which things stand out (Merleau-Ponty 1945, 115). Usually, we do not think about this. We know that our hand can grasp things, for instance, but this knowing is implicit. One could say that it is a kind of know-how, in particular, a know-how about possibilities for movement. Moreover, a body schema can change when we learn skills, including perceptual skills and skills to handle objects. When we learn to handle objects, objects become incorporated into the body schema. Merleau-Ponty's "technological" examples include that of a feathered hat, a blind man's stick, and the typewriter: They come to be experienced as part of the body; the feather, the stick, and the typewriter become extensions:

'The blind man's stick has ceased to be an object for him, and is no longer perceived for itself; its point has become an area of sensitivity, extending the scope and active radius of touch, and providing a parallel to sight.'

(Merleau-Ponty 1945, 165)

This only happens if we are prepared to learn and acquire the habit, it is a process of skill acquisition:

> 'To get used to a hat, a car or a stick is to be transplanted into them, or conversely, to incorporate them into the bulk of our own body. . . . It is possible to know how to type without being able to say where the letters which make the words are to be found on the banks of keys. . . . It is knowledge in the hands, which is forthcoming only when bodily effort is made, and cannot be formulated in detachment from that effort. The subject knows where the letters are on the typewriter as we know where one of our limbs is'
>
> (Merleau-Ponty 1945, 166)

Thus, it is a process of learning implicit knowledge, it is a process of acquiring know-how. The knowledge is *embodied*.

We can apply this idea to social-media use: In use, my smartphone becomes an extension of my body (in particular, the arm and hand). Moreover, using a device or using a social medium presupposes implicit knowledge, a know-how, which requires an effort. But, once we have the habit, our knowledge of the device is in our hands. Ihde calls this embodiment relations to technology: Like our use of glasses, we do not usually notice that we use these kinds of artefacts; they mediate our experience and are experienced as part of our body, if experienced at all. But we can also use Merleau-Ponty to make the more general point that any kind of use of, and performance with, technologies involves the body, since it always mediates our relations to technologies and to the world, and not only in the way of an "embodiment relation" in Ihde's sense. Even if the medium is not experienced as part of the body, we relate to it and work "in" it *through* the body. For instance, when I use social media and use a smartphone for this purpose, I use my senses to read the screen and I use my hands and fingers to handle the smartphone, often via the screen. Learning to use the technology means acquiring perceptual and motor skills, skills specific to the technology. Using social media requires knowing how to use a smartphone, which involves particular skills, such as motor skills, which add up to the skill "knowing how to use a smartphone" and, in the end, "knowing how to use social media."

The latter, however, also involves language skills. I need to know how to move and how to perform, next to know-how related to language. I need what Merleau-Ponty calls 'praktognosia' (162), a way of accessing the world via the body-in-performance. In performance, space and time are not external; instead, 'my body combines with them and includes them' (162). Thus, in the view of technology use and body use that I am developing now, there is not only "extension" in a static sense, there are also dynamics, there is performance as movement in time, in which technology use is bound up with skilled practice. In this performance, the body knows without a need for use of symbols:

'My body has its world, or understands its world, without having to make use of my "symbolic" or "objectifying function".'

(Merleau-Ponty 1945, 162)

I take this to mean that we do not need the use of words to acquire the kind of knowledge Merleau-Ponty talks about in cases such as the hat. However, while we do not *need* language for acquiring this kind of knowledge, this account of performance involving body and technology *in time* does not mean that language cannot play a role. Starting from Merleau-Ponty, we can also make a connection to language and thinking, linking the knowing-how to move to the knowing-how to use words and think. The idea, then, is not so much to "add" language—by saying that sometimes we also use words when we perform; for instance, in the case of the typewriter, we clearly *also* use words—but, rather, to make a direct link between use of language and use of the body and technology. If the body plays the important epistemological role Merleau-Ponty thinks it has, and, if the body gives me implicit knowledge and if I am usually not aware of it, if these are pre-reflective skills, then, perhaps, without my knowing it, these skills and performances also shape my reflection, understood as use of *language*. Indeed, then, even the *language* I use, even my so-called "linguistic" performance (which is always connected to other aspects of one and the same performance), even my *message* and, indeed, my *thinking* is framed and structured by the body and by bodily skills (and, indeed, by technology). But, to make this move, we need the resources of phenomenology rather than postphenomenology; we need to go beyond Ihde. We need Merleau-Ponty's own view and we can also link to cognitive science. Let me explain links between embodiment, technology, language, and thinking, especially with a view to understand our use of social media.

With regard to use of social media, recognizing the dimension of embodiment means recognizing that, in spite, of the language of the "online" and the "virtual," which today is *already* outmoded, *as users and performers*, and *in* our use and performance with social media, we are always acting and performing as embodied human beings. We engage with the technology as embodied beings; we do not leave the body "behind" when we use the medium, since we need embodied knowledge and skills to see and handle the technology. But there is more. The body "mediates" our experience and practice, not only since our senses are "in between" us and the medium and we need them to see, or since we need motor skills to use a computer or a smartphone, but, more profoundly, since social-media messages, performances, and narratives are themselves made through the body. Our use of language and our message are shaped by the body as medium. Subject and object emerge and are constituted in the process. To give this argument a transcendental twist: The stories we narrate, together with language and technology, are *made possible* and shaped by us having and being a body, which functions as a horizon, a (back)ground, a darkness. When we use language on social media, there is already the grammar of a given language and, connected with this, a *bodily grammar*. Any meaning generated in the context of social media only makes sense on the basis of embodiment, as a basis of sense-making, as what *makes possible* sense-making on social media. Embodiment can thus be seen as a condition of possibility. This transcendental turn goes back to traditional phenomenology, rather than relying on Ihde, but renders "transcendental" a lot less abstract than in Kant or Husserl, or, indeed, in postmodernism: In this conception, the transcendental is not only about logic or language but has also to do with technologies—often material technologies—and with the *body*.

For the purpose of better understanding the relations between language, technology, and the body, we can also use this transcendental approach in order to interpret and work on embodiment and enactivism in cognitive science. As Lakoff and Johnson (1980) and Varela and others (1991) have shown, mind is embodied and our language is full of bodily metaphors; metaphors shape our perceptions and actions, shape how we think and live (Lakoff and Johnson 1980). This can be interpreted as: The body, through our use of metaphor, is a condition of possibility for thinking, which enables and structures thinking. Also in this sense, we do not leave the body behind when we use language or technology. If our use of language and our technology is embodied, then the body is *also* our language, our medium, and our technology. If, indeed, performances with technologies are not only social but also always embodied, then embodiment is not just a particular kind of experience and human-technology relation, as Ihde's interpretation of Merleau-Ponty suggests, or something that plays a role in locating an artefact, as Brey claims on the basis of his reading of Merleau-Ponty; rather, embodiment is also a condition of possibility of such relations and performances. These particular relations, experiences, and performances with technology are made possible by the body in the first place. Hence, a phenomenology and hermeneutics of technology use, performance, and narrativity must be a phenomenology and hermeneutics of *embodied* use, performance, and narrativity. It must recognize the importance of implicit knowledge and a background, grammar, and form of life that makes possible that temporal and embodied use, performance, and narration, and it must recognize that the body, next to or, rather, with language and technology, is part of that background, grammar, and form of life. The narrative role of technology, then, is only possible through the body. Technologies "write us" and configure our time and narratives, but the body, as a condition of possibility of use and performance, makes possible this writing and narrative configuration.

For understanding social media, this means that we need a phenomenology and hermeneutics centered not only on social but also *embodied* use and performance, without reducing embodiment to a particular human-technological relation, and while recognizing not only the mediating but also the transcendental role of the body. Of course, more needs to be said on the relation between narrativity and embodiment, in social media and elsewhere. On the basis of this chapter, we can provisionally conclude that our technological use and performance is always social in a narrative sense, but also that this narration and this structuring of time in which technology and language play a crucial role is always also crucially dependent on embodiment. Our social-media activities, such as checking, liking, and posting, for instance, are "linguistic" and "technological," but always also involve uses of the body (e.g., moving one's finger to a particular spot on the screen, handling the computer mouse, etc.) and are dependent on the body in the way(s) identified here—in particular: They are dependent on the body as a transcendental condition of thinking and experience, linking us up to space and time. We are only able to respond to a Facebook post with a humorous text, for instance, if we understand the metaphors in the text. And this understanding of metaphors, in turn, depends on "the body," that is, on our lived experience as embodied and social beings. It is only because our thinking is already embodied that we can then use language and technologies in particular ways. More generally, our uses of technology depend on experience and the lifeworld: concrete and personal experiences, and the crystallization of such experiences in social and cultural forms, such as game rules and a form of life. With regard to narrativity, this means that the narratives I "write" as a social-media user, for instance, are also dependent on my embodied experience and existence, and the embodied experiences and existence of others. Moreover, if the narratives themselves are lived, then they must be understood in terms of embodied performance. The characters and the narratives are embodied, and the narratives only exist if they are performed, which requires embodiment as a background condition.

This emphasis on embodiment and lived experienced can be further supported by returning to another work by Merleau-Ponty. In his article 'What is phenomenology?' (1956), Merleau-Ponty does not take a lot of distance from Husserl and still seems to endorse Husserl's attention to essences and exact science, but, nevertheless, stresses that there is also another side to phenomenology's method, one that he emphasized in his work: phenomenology as a philosophy 'for which the world is always "already there" as an inalienable presence which precedes reflection' and as a philosophy that provides an account of space, time, and the world "as lived."' (Merleau-Ponty 1956, 59). Both Merleau-Ponty and Ihde show that embodiment is crucial to that lived experience and the lifeworld, and these are helpful insights for philosophy of technology. The challenge with regard to my project, however, is to think together human use of technology, body, *and language*. The focus on narrative time in this chapter, the present suggestions about relations between embodiment and narrativity, and my particular interpretation of the transcendental method introduced in this and the previous chapters are the steps I propose to meet that challenge.

Note that we can also link this view to the interpretation of *Wittgenstein* I develop in the previous chapters: For Wittgenstein, too, thinking is a corporeal, embodied, and performative process of living beings, in which—so I argue—technology and language play mediating roles and the roles of transcendental conditions. Unfortunately, this is often neglected, since Wittgenstein is read as a philosopher of *language*. But I have shown that we can highlight and develop his remarks on embodiment (directed against Descartes) to arrive at a view of performance that is *embodied*, social, and shaped by language and *technology*.

Finally, another interesting way to connect language and narrativity to embodiment, which has only been hinted at so far, but is entirely compatible with the arguments presented here, is the concept of voice. When we speak, this speaking should not only be studied in linguistic terms. There is also the spoken narrative as voiced narrative. More generally, voice can itself be seen as embodied speech; the concept of speech itself stands in need of *embodiment*. This angle brings in material-physical aspects of speech, which are neglected in postmodern and other language-oriented thinking, which is often "bewitched" by the technology of writing and does not usually study use of language in its embodied voice aspects. Ihde's work on voice (2007) does better in this respect, although a lot more needs to be said about the meaning of "voice" and about the relations between voice, language, and technology. However, I have no more space to further develop this here. Now it is time to inventory and integrate the insights gained so far, and present a framework for conceptualizing relations between humans, language, and technology.

Note

1 Wikipedia https://en.wikipedia.org/wiki/Social_media#Most_popular_sites

References

Allen, Amy. 1996. "Foucault on Power: A Theory for Feminists." In *Feminist Interpretations of Michel Foucault*, edited by Susan J. Hekman, 265–82. University Park, PA: Pennsylvania State University Press.

Brey, Philip. 2000. "Technology and Embodiment in Ihde and Merleau-Ponty." In *Metaphysics, Epistemology, and Technology: Research in Philosophy and Technology 19*, edited by Carl Mitcham, 45–58. London: Elsevier/JAI Press.

Coeckelbergh, Mark. 2012. *Growing Moral Relations: Critique of Moral Status Ascription*. Basingstoke/New York: Palgrave Macmillan.

Coeckelbergh, Mark. 2015. "Language and Technology: Maps, Bridges, and Pathways." *AI & Society* (online first). doi:10.1007/s00146-015-0604-9.

Coeckelbergh, Mark and Wessel Reijers. 2016. "Narrative Technologies: A Philosophical Investigation of the Narrative Capacities of Technologies by Using

Ricoeur's Narrative Theory." *Human Studies* 39(3): 325–46. doi:10.1007/s10746-016-9383-7.

Dorrestijn, Steven. 2011. "Technical Mediation and Subjectivation: Tracing and Extending Foucault's Philosophy of Technology." *Philosophy & Technology* 25(2): 221–41. doi:10.1007/s13347-011-0057-0.

Foucault, Michel. 1975. *Discipline and Punish: The Birth of the Prison*. New York: Random House.

Foucault, Michel. 1980. *Power/Knowledge: Selected Interviews and Other Writings, 1972–1977*. Edited by Colin Gordon. Translated by Colin Gordon and Leo Marshall. New York: Pantheon Books.

Foucault, Michel. 1988. *Technologies of the Self: A Seminar with Michel Foucault*. Edited by Luther H. Martin, Huck Gutman, and Patrick H. Hutton. Amherst: University of Massachusetts Press.

Foucault, Michel. 1994 (1966). *The Order of Things: An Archeology of the Human Sciences. A Translation of Les Mots et les choses*. New York: Vintage Books/Random House.

Foucault, Michel. 1998 (1976). *The History of Sexuality, Vol. 1: The Will to Knowledge*. Translated by Robert Hurley. London: Penguin Books.

Fuchs, Christian. 2014. *Social Media: A Critical Introduction*. London: Sage.

Fuchs, Christian, Kees Boersma, Anders Albrechtslund, and Marisol Sandoval, eds. 2012. *Internet and Surveillance: The Challenges of Web 2.0 and Social Media*. New York and Abingdon: Routledge.

Gerrie, Jim. 2003. "Was Foucault a Philosopher of Technology?" *Techné* 7(2): 66–73. doi:10.5840/techne2003722.

Ihde, Don. 1971. *Hermeneutic Phenomenology: The Philosophy of Paul Ricoeur*. Evanston, IL: Northwestern University Press.

Ihde, Don. 1990. *Technology and the Lifeworld: From Garden to Earth*. Bloomington: Indiana University Press.

Ihde, Don. 2007. *Listening and Voice: Phenomenologies of Sound*, 2nd edition. Albany, NY: State University of New York Press.

Kaplan, David M. 2002. *The Story of Technology*. Drexel University. Retrieved February 18, 2016, http://www.pages.drexel.edu/*pa34/The%20Story%20of%20Technology.pdf.

Kaplan, David M. 2006. "Paul Ricoeur and the Philosophy of Technology." *Journal of French and Francophone Philosophy* 16(1–2): 42–56.

Lakoff, George and Mark Johnson. 2003 (1980). *Metaphors We Live By*. Chicago, IL: University of Chicago Press.

McCallum, Ellen L. 1996. "Technologies of Truth and the Function of Gender in Foucault." In *Feminist Interpretations of Michel Foucault*, edited by Susan J. Hekman, 77–98. University Park, PA: Pennsylvania State University Press.

MacIntyre, Alasdair. 2007. *After Virtue: A Study in Moral Theory*, 3rd edition. Notre Dame, IN: University of Notre Dame Press.

Merleau-Ponty, Marcel. 1956. "What Is Phenomenology?" *CrossCurrents* 6(1): 59–70.

Merleau-Ponty, Marcel. 1962 (1945). *Phenomenology of Perception*. Translated by Colin Smith. New York/London: Routledge.

Ricoeur, Paul. 1980. "Narrative Time." *Critical Inquiry* 7(1): 169–90.

Ricoeur, Paul. 1984 (1983). *Time and Narrative—Volume 1*. Translated by Kathleen McLaughlin and David Pellauer. Chicago, IL: University of Chicago.

Ricoeur, Paul. 1986 (1984). *Time and Narrative—Volume 2*. Translated by Kathleen McLaughlin and David Pellauer. Chicago, IL: University of Chicago.

Ricoeur, Paul. 1988 (1985). *Time and Narrative—Volume 3*. Translated by Kathleen McLaughlin and David Pellauer. Chicago, IL: University of Chicago.

Ricoeur, Paul. 2014. "The Later Wittgenstein and the Later Husserl on Language." *Études Ricoeuriennes/ Ricoeur Studies* 5(1): 28–48. doi:10.5195/errs.2014.245.

Takeshita, Chikako. 2011. *The Global Biopolitics of the IUD: How Science Constructs Contraceptive Users and Women's Bodies*. Cambridge, MA/London: MIT Press.

Terry, Jennifer. 1989. "The Body Invaded: Medical Surveillance of Women as Reproducers." *Socialist Review* 19: 13–43.

Van Den Eede, Yoni. 2011. "In Between Us: On the Transparency and Opacity of Technological Mediation." *Foundations of Science* 16(2): 139–59. doi:10.1007/s10699-010-9190-y.

Varela, Franscisco J., Evan Thomson, and Eleanor Rosch. 1991. *The Embodied Mind: Cognitive Science and Human Experience*. Cambridge, MA/London: MIT Press.

Verbeek, Peter-Paul. 2005. *What Things Do: Philosophical Reflections on Technology, Agency, and Design*. University Park, PA: Pennsylvania University Press.

Part IV

Humans, Language, and Technology Speak (Subjects and Objects Entangled)

8 Using and Performing with Words and Things

8.1. Balancing the Triangle: A Framework for Thinking About Language and Technology

Overview of the Argument and Narrative of this Book

Today, most philosophers of technology claim that technologies are not mere instruments for human purposes but crucially and significantly shape what we perceive, think, do, and are. In the course of this book and further developing my preliminary investigation on language and technology (Coeckelbergh 2015), I try to rethink this claim about the noninstrumentality of technology in a way that accounts for the role of *language* in human technological experience and practice, in particular the use of, and performance with, technologies. I do this by constructing, interpreting, criticizing, and revising three "extreme," ideal-type positions, with regard to who or what "speaks." In the course of these operations, I use and respond to various authors and theories, including Wittgenstein, Heidegger, McLuhan, Dewey, Austin, Searle, Bijker, Lyotard, Derrida, Ihde, Latour, Pickering, Ricoeur, Foucault, and Merleau-Ponty. Let me summarize this narrative and further develop its positions and arguments, in order to construct my synthesis of the three "extreme," ideal-type positions and offer an integrative framework to think about the relations between humans, language, and technology. How can we "balance the triangle" and give humans and technology a voice, humanize technology and language, and materialize humans and language?

The first position is that only humans speak. Technologies are seen as instruments to reach human ends. There is an unbridgeable distance between subject and object, and between humans and things. This is not a strawman position: While many philosophers of technology reject it, among other philosophers and outside philosophy it is rather common to think of technology in this way. With regard to the relation between language and technology, the position implies that humans speak *about* technology. There is a gap between discourse and materiality, culture and technology, and, indeed, humans and technology. Technology is seen as something external to the human. Humans are subjects that have language; technology concerns objects that are mute. Perhaps we can add a layer of meaning to artefacts by means of speech—this is what I take to be Searle's view. This modification already shows that technological objects can also get a meaning beyond that of tool, but it does not significantly reduce the gap between subject and object. Speech and what Austin called 'speech acts' are still entirely on the side of the humans.

Then I start criticizing this instrumentalist position by drawing on Heidegger and the later Wittgenstein. First, these authors help us to close the gap between humans and *language*. According to Heidegger, language is much more than an instrument or a tool; language makes possible that things show up for us in the first place, and that things are revealed in a particular way. Human experience and thinking are shaped by language. In this sense, when humans speak, language speaks. It is the vehicle of thought. It also shapes what we say and pre-configures meaning, also the meaning of artefacts. Both Heidegger and Wittgenstein help us to see why there is no such thing as naked physical fact, as opposed to meanings that are declared and social (a view I attribute to Searle); instead, the world is *already* social and meaningful. According to Wittgenstein, language is woven into our activities, our language games, and, ultimately, our forms of life. By drawing our attention to the use of language, as embedded in language games and forms of life, Wittgenstein helps us to connect language to practical use and to social-cultural contexts. In addition, with Wittgenstein, we can see that use of language is much more varied than Searle's declaration theory and social ontology might suggest (a variety the early Searle would certainly acknowledge).

Second, turning around his metaphor of the toolbox and applying his holistic view of language use to technology, I use Wittgenstein to start closing the gap between humans and *technology*, and between language and technology. Like language, technology is something we use, is part of activities, and, as such, is linked to what I call "technology games" and a form of life. Language itself is one of the technologies we use. We perform (with) language, and this is also true for technologies: We perform (with) technologies. We use words and we use tools. Language and other technologies are embedded in, and constitute, language and technology games, practices, and cultures. Artefacts as objects are not given an aura of meaning through words only; they mainly get their meanings through *use* (which may or may not include the use of words). Thus, the concepts of use, games, and forms of life help here to bridge the gaps between humans, language, and technology. In use, understood as performance and related to practice and activities, there is a tight connection between humans, language, and technologies. I also noted that this view can be supported by Dewey's view of language and its relation to technology and the social.

The latter interpretation, among others, helps me to suggest a Wittgensteinian revision of social constructivism by focusing on use and on the role of language in the social construction of artefacts. It also helps to criticize the second ideal-type position I articulate: the view that only language speaks. Inspired by Heidegger, Wittgenstein, structuralism, semiotics, and related language-oriented theory, 20th century poststructuralist and postmodern philosophers of language focus so much on discourse, signs, grammar, etc. that both humans and technologies become invisible and muted. As language is seen in a way that alienates it from concrete activities and uses, the human subject and the technological object are difficult to discern. By using Wittgenstein (in a different way than postmodern philosophers), however, we can bring life to the dead signs of semiotics, and, indeed, to the dead objects of Searle: They are alive in and through use and performance. There are no signs and there is no meaning apart from practical and active use. A focus on use and performance, thus, help to bring about and reveal a "speaking" of the human and of the material artefact. There is not only collective intentionality but also and especially collective use and performance, which involve the use of words, but also the use of artefacts and tools. Searle's focus on language is one sided, insofar as it neglects this use of artefacts; the same can be said of many other philosophies of language in the so-called "analytic" tradition.

However, many so-called "continental" strands in 20th-century philosophy do not fare better. Indeed, Heidegger can be (mis)interpreted as saying that there is only language, that humans and technology are secondary phenomena. Lévi-Strauss's structuralism and, later, Derrida's poststructuralism overemphasize language. For Baudrillard, there are only images and signs, with no relation to reality (the simulacrum). If objects appear at all (in this later work), they are simulated and enigmatic. That being said, there are elements in poststructuralism and postmodernism that can be used to bridge the gap from language to humans and technology. For example, in Derrida, there is a reference to the bricoleur, which, in Derrida, remains a metaphor, but could be used by authors working in this tradition to bring in technological practice via the back door. In Lyotard, there is a Wittgensteinian interest in language games, which mainly serves Lyotard to focus on language and the linguistic nature of the social bond, but which can be reinterpreted in a material way: There are also what I call "technology games." Lyotard also argues that technological transformations have an impact on knowledge, a claim that is still very relevant today. Baudrillard's simulacrum and hyperreality can be interpreted as made possible by very "real" and material technologies: Our Disneyland and Walt Disney World is only possible on the basis of material objects and infrastructures. And in Virilio, often categorized as postmodern, we meet more concrete technologies and bodies. Mersch's posthermeneutics is also interesting, since he tries to think what is not sign, not text. Through the failure to do this, he acknowledges that there is something that resists. We may wonder if this something could include material artefacts. But, if it does, here, the materiality of things and the body remain in the shadow, in the part of reality we cannot speak of.

Language speaks; the rest is silent, although, occasionally and unexpectedly, it may show up. Finally, I argue that we can also learn from a *transcendental* approach to thinking about language and technology, and that postmodern thinking attends us to issues concerning difference and alterity, and to concrete encounters with technology.

A clearer distance from the "only language speaks" position, however, can be achieved by using Arkich and Latour. What one could call their "post-semiotics" can be interpreted as a French version of the material turn, in response to French semiotics. Akrich and Latour interpret semiotics in way that focuses on non-textual and nonlinguistic meaning-making. Here, technologies start to "speak," in the sense that scripts become attached to artefacts, objects become part of a 'program of action,' and, more generally, language and material objects become linked. But I propose that this should not be interpreted in a Searlean way, by saying that meaning gets attached to objects. Instead, it is more adequate to say that, in this view, meaning moves across humans, language, and technologies. We can explain and develop this thought by using the concept of performance. Both language and technological artefacts have a kind of performativity. The gap between humans, technology, and language is being bridged. Enriched with Wittgensteinian thinking and, perhaps, process philosophy (which Latour endorses, or it is at least compatible with his view; see Harman, again), we can think of the technical object as an outcome of a performative process, which is both linguistic and nonlinguistic. As I argue in Chapter 6, if we interpret Latour, not as offering an ontology, but as giving us an account of a social and political *process*, then, perhaps, both humans and nonhumans are not the "ingredients," but the "cake:" They can be seen as the *outcome* of a material-social-linguistic process, which is a performative and a use process. This highlights the role of humans, but also the role of language.

Social constructivism, then, renders technologies "social" and takes a material and empirical turn. But, apart from "discourse analysis," which goes back in the direction of the two extreme positions by being in danger of being only *about* technology, or of divorcing text and the social from human-technological use, it neglects the role of language. Social constructivism is right to see artefacts as socially constructed, as being the outcome of a social process, but the social should be interpreted as crucially involving language. Yet, this language does not function in a Searlean way, by giving meaning and life to dead physical objects. And neither does it function in the form of signs or discourses that are disconnected from material technologies and their human users, as poststructuralism and postmodernism suggest. Instead, artefacts are already social and meaningful through their use and their embeddedness in a form of life, even "before" there is a particular social-institutional process, with its explicit controversies about the meaning and design of the technologies, and with its meaning-giving by particular groups. It is true that there is interpretative flexibility, but this flexibility is not only created via different groups having an impact on the innovation process; it is also constrained by the use of the technologies (within and outside innovation contexts), and, hence, by the language games, "technology games," and cultural context that is already there. When groups and individuals do things with language (use rhetoric, use language in a performative way to reach certain aims) to give meaning to the technology, this is only part of the picture; there is also use of the (other) technologies, material technologies. In innovation, there is use of words and use of things.

In the next chapters of the book, I construct a third ideal-type position, which highlights this materiality and to which I also respond in order to set up my framework: the position that technology "speaks." Here, I use McLuhan, Ihde, and Latour again, but also Ricoeur.

McLuhan's phrase that the medium is the message takes seriously technology as mediating our perception and thinking, as an "in between," but also as a milieu: For McLuhan, we live in a different environment, or rather media-environment. Media ecology's response to McLuhan (Postman) pays more attention to the medium as an environment. Both media theories can be interpreted in an "extreme" way: One can read McLuhan as a techno-logical determinist, and accuse Postman of a postmodern obsession with symbols (and of dualism). At first sight, it seems that, in their views, humans and concrete technologies disappear; the medium takes over, only the medium still speaks. However, I have also emphasized that, for McLuhan, language is also a technology (and language is often linked to the technology of writing) and, for Postman, there is interaction between human beings and media. Influenced by Stephens's comments on Postman, I argue that a post-humanist media-ecological view is possible that has a more integrated view of humans, technology, and language. Moreover, I show that, in the later work of McLuhan, such integration is further supported, since language and technology are equated. He also offers an interesting "grammatical" approach, which is in line with the transcendental approach I propose.

Like Latour's work, Ihde's postphenomenology can also be understood as an empirical and material turn: a reaction to the postmodern overemphasis on the linguistic and to Heidegger's thinking about technology. In postphe-nomenology (Ihde) and mediation theory (Verbeek), phenomenology and hermeneutics are given a new direction by focusing on how technological artefacts mediate and shape our experience, perception, and action. Like in McLuhan, here, technology "speaks," in the sense that it mediates our experience and action in the lifeworld. It speaks "even" in science, which is also a hermeneutic practice. Here, the technological artefact is neither a mute recipient of meaning, as in Searle, nor a mute ghost banned to the shadows of what we can know, outside the human and linguistic order. Instead, technology "speaks," as it actively shapes meaning and experience. However, this use of the speech metaphor is already my interpretation of the position, reading more focus on *language* into Ihde's artefact-centered hermeneutics than there actually is. While postphenomenology may not mute humans (and hence cannot be reduced to the extreme, ideal-type position that technology speaks, but humans and language are mute), it often does seem to mute *language*. Language does not seem to play any mediating role, and it remains unclear what the role of language is in these world-making mediations. This may surprise, since, in his early work, Ihde engaged with language-oriented hermeneutics (Ricoeur). But, because of postphenomenology's allergy to "old-fashioned" ways of phenomenology and philosophy of technology (by which they mean Heidegger, his followers, and his predecessor, Husserl, mainly) and its material turn *away from language-centered philosophy*, the focus is on the material artefact. The result is that the relations between language and technology remain heavily undertheorized. Moreover, any kind of transcendental argument is rejected, since "transcendental" is assumed to be anti-empirical, abstract, and nonmaterial.

However, this need not be our view, and I suggest that, if one wishes to embrace postphenomenology at all, this does not necessarily require one to reject theorizing the role of language in its hermeneutics and mediations, nor does it require one to give up a transcendental perspective. In my proposed revisions of Ihde, I try to merge a recognition of the material and bodily aspects of hermeneutics with a recognition of the role of language in mediating our perception and action. In my view, there is no hermeneutics of things without a hermeneutics of language. The phenomenology and hermeneutics of things does not take place "outside" materiality, but neither does it take place "outside" language. When technological artefacts shape our world, they do so together with language. Both words and things make us see and do differently; both language and technology mediate our relation to the environment.

Indeed, I show that Ihde's relational view can be revised to account for this mediating role of language. I propose different conceptualizations of I-world relations, taking into account not only the mediating role of technology but also the mediating role of language: the mediating role of language in the human-world relation and in the human-technology relation. I argue that there are various types of human-world relations in which language and technology function in different ways. They can be embodied, they can be experienced in a hermeneutic or an alterity way, and they can be recognized as media. In an integrated view and in analogy to what Ihde and Verbeek say about technology, one could say that *language* and technology (if they are different at all) mediate between humans and their world. One of the schemes I propose to expresses this is, then, simplified: human—language/technology—world. However, often this mediating relation is not recognized, since, in everyday use, they are either embodied or they are experienced in a hermeneutic or an alterity way. In other words, it is precisely *because language and technology are such good mediators* that they are often transparent (since embodied) or that we fail to see them and move beyond an interpretation of them as objects, instruments, or even quasi-others.

In my descriptions of these relations, however, I put more emphasis on *use* than is generally done in postphenomenology. This is in line with Ihde's own pragmatist twist to hermeneutics and phenomenology, but, in Ihde's hermeneutics, use is not as much stressed as it could be, since the focus is on hermeneutics as interpretation and mediation. My adaptation and revision of Ihde as presented here does not directly refer to pragmatism (this could be done, but would need further development, for instance departing from my excursion to Dewey's view earlier in the book), but is mainly inspired by Wittgenstein and McLuhan, who do not make a strict distinction between technology and language. In use (Wittgenstein) and experience/perception (McLuhan), language and technology mediate. This approach helps us to account for the mediating role of language and technology and, indeed, their active role in use and performance; but, at the same time, it also keeps the human as a speaker. It may be that language and technology "speak," but, since use and performance is crucial, the human also speaks and acts. Without human use and performance, language becomes dead signs. Without human use and performance, technologies become dead things. Only in use and performance, language and (other) technologies live and have their mediating and hermeneutic functions. This gives us a more integrated view in which humans, technology, and language all "speak."

Going further beyond Ihde, I also offer a more substantial revision of Ihde by giving a transcendental twist to the postphenomenological framework and by criticizing the focus on mediation. Language and technology are not only "in between" humans and their world—as it were, on the same horizontal plane—but also make possible and shape these human-world relations (including mediations) in the first place. I suggest that the hermeneutic operations identified by Ihde and Verbeek should not be understood in a quasi-causal sense, as in Verbeek's phrase "what things do," but, rather, in terms of transcendental conditions or "grammars" that make possible, structure, and shape our relation to the world and to others. Both technology and language, then, are not to be seen as a kind of pool ball on a pool table that are "between" other balls and "do" things to them, as the mediation language often suggests. If it is true that subject and object are constituted by technology (and, I add: language) at once, if there is, indeed, a 'mutual constitution of subject and object,' as Verbeek rightly says (Verbeek 2005, 130), then it is not clear why we should keep using the language of mediation at all to make that claim. Then it seems that this mutual constitution of subject and object can be more helpfully conceptualized by the language of transcendentalism, which is unduly and unfairly rejected by Ihde and Verbeek for abstracting from concrete material technologies. In my use of a transcendental argument, subject and object are both made possible and shaped by language and technology, as transcendental conditions, as conditions of possibility. However, I have not used Kant's or Husserl's conception of the transcendental. My use of transcendental thinking requires a revision of the traditional notion of transcendental conditions, which, in Kant and Husserl, is still very abstract and immaterial: I propose to understand transcendental conditions in a much more concrete and material way. More work is needed to fully develop this argument, for instance in response to Husserl, but I have shown how it can be *used*, which is, according to the framework presented in this book, *the* way of making sense of it. Moreover, whereas, in postphenomeonlogy, there is sometimes an overemphasis on the material artefact's role, this one-sided focus on "things" is avoided if we stress that humans are a crucial part of these hermeneutic and phenomenological processes and constitutions. Keeping in mind the Wittgensteinian argument that the role of the human is not eroded, humans speak and act as users of technology and language, and they speak and act as *performers* with technology and language. I propose a more holistic, transcendental, and process- and use/performance-centered view.

Furthermore, I use Latour again to further develop this view—but, this time, the Latour of *We Have Never Been Modern*. Ihde already recognizes and is sympathetic to what he helpfully interpreted as Latour's hermeneutics of the laboratory, and, as I argue, Akrich and Latour can be interpreted as arguing that language mediates between humans and artefacts (or that artefacts mediate between humans and language). Yet, the "early," post-semiotic Latour is still mainly focused on the technology of writing (script) and text. At this point in my argument, therefore, I instead turn to Latour's later work (or perhaps one should say "middle" work), in which things get a "voice." Here, Latour turns away from semiotics, leaves the 'Empire of Signs,' and tries to bridge the gap between subject and object by his language of networks and hybridity. In his non-modern and non-dualistic view, which crosses over between nature and culture, human and nonhuman, artefacts speak through humans (spokespersons) and have "speech." Humans and things belong to one collective, which is in the process of continuous expansion. Beyond modern purification, there are natures-cultures. Thus, here artefacts are no longer mute; they have a voice.

Now this can be read as giving technology a voice, to the exclusion of humans and language. But a better interpretation, perhaps revision, of Latour, is that humans, language, and technology all "speak." Combining Latour and posthumanist media ecology, the social can be defined as consisting of humans, material technologies, and language, which all speak, in the sense that they perform as entities that are emergent from, and are constituted by, a larger relational and ecological whole and process, which is human, nonhuman, social, natural, and technological at the same time. Indeed, reading Latour in a "process" way, rather than the static, ontic way, which leads him to make a claim about "what is," about a "state" and its "constitution," gives us a dynamic view of the social. It is also a view that is consistent with posthumanist accounts, since the human is itself the outcome of this process, rather than a fixed thing or given essence. Latour's focus on the social also helps us to address another lacuna in postphenomenology, which under-theorizes not only language but also the social.

To stress and elaborate this process approach, but also to connect it back to the Wittgensteinian view I develop, I use Pickering's view, which focuses on the temporal aspect of practice, on emergence, and, indeed, on human and nonhuman *performances*. This link to performance enables us not only to keep the focus on the social but also to reconnect with *human* use and interpretation, and to bring in the time dimension. What nonhumans do must be linked to what humans do. Human performance and practice always involves hermeneutics; in use and performance, words and things are interpreted in new ways. As Ihde rightly points out, their meaning is not entirely stable. But, now, we can include this insight in a framework that puts more emphasis on use and performance as *processes* that generate this ambiguity and variety. Our experience of technology and language (including words and things as actants and *as performing*) is always an experience that has a temporal aspect; it is an experience linked to a performance *in time*, which has a particular structure.

Indeed, first it is not entirely clear how the Latourian view fits with Ihde's hermeneutics or the Wittgensteinian emphasis on *use* of language and technologies. Human use and interpretation seem to be missing. It is also unclear how much room Latour leaves for individual human experience, given his focus on the collective and his third-person, detached description of the social and political (process). Moreover, so far these attempts to arrive at an integrated position miss an important aspect of language: narrativity.

To account for the latter, and, at the same time, to also construct a further building block for the integrative framework, I turn to Ricoeur. Wittgenstein acknowledges the performative dimension of use, but narrativity (and perhaps also time) seem to be overlooked. In Latour, there are the notions of script, actors, and actants, and other elements that suggest a theatre metaphor, but the concept of narrativity seems absent. In postphenomenology, attention to performativity and narrativity is lacking altogether and seems difficult to include—except, perhaps, if one interpreted "what things do" in terms of performance, as I propose, or if one retrieves a notion of use from Ihde's work and then adds an emphasis on narrativity. Ricoeur, by contrast, offers a hermeneutics focused on narrativity, which enables us to bring in a temporal and narrative dimension via the front door, so to speak, and which enables us to directly connect to the *social* through its metaphors of theatre, dance, and performance. While Ricoeur belongs to a tradition that is centered on humans and language, rather than, and even as opposed to, science and technology, I show that this work can be helpfully used to conceptualize what Reijers and I call "narrative technologies." Congruent with postphenomenology, this term has enabled the development of a view that clarifies the more active, hermeneutic role of technologies, but focuses on narration rather than mediation, and takes into account the *time* dimension (diachronicity) of use of, and performance with, technology—not only experience at a given moment of time (synchronicity) and not divorced from the social. The claim is that technologies are not the mere objects of reading (they can be read like a text, as Kaplan argues), but can also actively co-write our narratives. Here, technology acquires a "voice," in the sense that it plays a role in the configuration of narratives by organizing characters and events into plots. In contrast to Searle's view, artefacts are not dead and passive when it comes to meaning-making and shaping worlds; they have an active semiotic and hermeneutic role as coauthors of narratives. Here the social (and, hence, social meaning) is not understood as emerging from a synchronic whole; instead, it emerges from a diachronic narrative; an as emergent, it is made *in time*. This attention to the temporal and *narrative* dimensions of human existence and the human lifeworld is an important addition to Ihde and Latour, who neglect or downplay this dimension in their (later) work. The proposed conceptualization of what technologies do in terms of time and *narrative* process is compatible with the postphenomenological view that technological mediations can also be normative (see, for instance, Verbeek), but gives a narrative twist to it. Technology's writing is never normatively neutral; technology co-scripts the human role in the narrative and co-scripts the action and experiential possibilities in the narrative/game. Moreover, one could also understand this narrative role of technology in a transcendental way: Technology makes possible and shapes particular narratives, and narrative structures given in our culture and form of life constitute a kind of "grammar" and ground that carries and makes possible particular games, activities, uses, and performances with technology.

However, what seems to be missing in the first version of this "narrative technologies" view is an explicit account of the role of language and perhaps even an account of the role of humans; it could be interpreted as too technology-centered, leaving too little room for what humans do and what language does. Yet, his lacuna can easily be remedied. To acknowledge the role of humans, we can stress "co-writing," "co-scripting," and "co-directing" by humans as "speakers:" Humans also play a role, not only as actors and performers but also as co-writers and co-directors; their speaking is not only linked to their acting but also to shaping of the narrative. Moreover, we can construct an integrative view in which humans, technologies, and language all co-write and co-constitute narrative and narrative time—and. hence. the (time dimension of the) social. Language also shapes the narrative through our use of language and through its transcendental role as making possible and structuring narration. Language, then, not only functions as a "grammar," in the sense of syntax that shapes our use of language as sentences within a narrative, but also as what we can call, with Wittgenstein, a 'depth grammar' that "deeply" structures and shapes performance and narration, in the sense that, as language games and a form of life, it makes possible and shapes the content of the performance and narration itself. Moreover, combined with the Wittgensteinian emphasis on performance, narrative itself can be interpreted as performance: It is, then, to be understood, not so much as a 'text,' but as a temporal process. It can only "make" meaning *as* performance, not as dead text. And, indeed, the social itself is performative and takes place in time. The conception of narrative should not be divorced from praxis and use; it should not be understood as separate from performance, which gives life to the narrative, or, rather, which creates the narrative as narration (rather than as text). A narrative, if it is really definitive of the social, is not a collection of signs (text) alienated from concrete and lived performances and experiences; instead, as entangled with use and performance, it is a living, enacted, and ongoing configuration of lived time and experience. Even if the narrative contains higher-order characters and events (e.g., from the past), it means nothing and does nothing unless it shapes and makes possible time-performance here and now. This brings back the humans, their experience, and their language. With and through technology, we narrate and perform our world, and this also involves language: It can also be said that, with and through language, we narrate and perform our world. Both language and technology play a more active narrating role than is usually presumed; but humans retain a key role. With and through humans, *language and technology* narrate and perform our world. But the human and the subject are not fixed or given. In the performative process, human subjectivity is constituted. In terms of speech and embodiment, one could say that humans acquire a *voice* and "sound through" (per-sona); their voice and "persona" is mediated by language and technologies, mediated by techno-logos. This mediation, however, is, in turn, made possible and shaped by technology games/language games and a form of life. Humanization and personalization are made possible and structured by a grammar that is already given in and with the social. If there is such a thing as personality at all, it can be conceptualized as this process of acquiring voice, which is always embedded in, and emerges from, a larger social and cultural whole. In this sense, there no such thing as individuality; there is only personality. Thus, from this perspective, one could say that there is personal experience; in this sense, the neglect of individual experience in the previous accounts is justified. There are always contexts and what we may call "con-technologies" that shape the process of personalization. Similarly, there is the human. But the human is also emergent; it emerges from the performative process. Humanization requires performance over and over again; it needs to be performed and requires work. And it involves use of language and use of (other) technologies, which are not neutral, but shape the human and its narratives as mediators and as conditions of possibility. In this sense, there are context and con-technology. However, at the same time, this human/nonhuman narrative and this process of acquiring voice is always lived, experienced; it cannot and should not be reduced to text or artefacts—especially if the latter are understood as entirely removed from human and personal experience, use, and performance.

I also point to the power dimension of these experiences, uses, and performances with language and technology and, indeed, to the power dimension of narrative. Moreover, I propose some interpretations of Foucault to further develop the power and political aspects of thinking about language and technology. On the basis of my reading of Foucault as a rhetorician and grammarian, we can conclude that uses and performances with words and things—for instance, in practices of the self, but also in other practices—can be understood as speech acts and "technology acts," which have perlocutionary and normative effects and which are as much embedded in "power grammars" as they are embedded in language and technology grammars. These grammars, linked to a form of life, or what Foucault called 'episteme,' are not necessarily written large (say, in the form of language and technologies used by the state to repress people), but play out in everyday relations, at the micro level, that is, at the level of concrete social relations and performances, where they make possible, shape, and constrain subjectivity and the social. This can well take the form of narratives that shape power relations (and are shaped by them), but then always narratives in connection with concrete performances with language *and* technology. But this is already an interpretation and application of Foucault, which links discourse and the social to *technology*—understood not only as a metaphors or symbols but also as material artefacts and material (infra)structures, which also, in their entanglements with power, function both as things we use and as transcendental conditions.

Moreover, taking into account Merleau-Ponty's and Ihde's work on the body and embodiment, we should add that narrative is not only human, personal, and lived; it is also embodied. Moreover, as Ihde rightly argues, we can have an embodiment relation to technology. But, in contrast to Ihde, I argue that the body always mediates performances with language and technology, and I have given my argument about embodiment a *transcendental* form: The body not only functions in specific technological mediations; more generally, it is always a condition of possibility for our engagement with technologies and, indeed, for our narratives. Whereas Ihde needs to "import" Merleau-Ponty from the outside, so to speak, in my view, the concept of *performance* already includes the dimension of embodiment, by definition: Performance, which has a temporal and narrative structure, always involves movement of the body. The body is itself a ground, a "grammar" from which we relate to the world. The body knows; this is implicit knowledge that forms the (back)ground for our engagement with the world and that makes possible concrete performances with words and things. Moreover, as we can also learn from cognitive science, language itself must be understood as embodied. Hence, we can conclude that, when we use language and technology, not only language and technology "speak" but the body also "speaks." Our messages are shaped by embodied performances through the medium of metaphors, for instance. Embodied performances have crystallized as a form of life, including a language that contains bodily metaphors. Hence, metaphors function as a kind of "body grammar" that is rooted in bodily experience and that shapes our speaking and thinking. But "the body speaks" is also literally true, when we use our voice. Voice can be understood as an embodied form of language use or as a condition of possibility for a particular kind of language use (speech). Indeed, we can point to a particular organization of humans, language, and technology: the voice as an instrument used by humans in order to use language. But one could also say that voice makes possible speaking, that it also provides a kind of grammar. It could also be seen as a medium that shapes the message, a "voice grammar" that shapes not only how we say things but also what we say. And there are combinations between use of voice and use of technology. Ihde has studied the human voice, but more philosophical work needs to be done on the exact relations between *language*, voice, and technology. Understanding voice itself as a technology (Young 2015; see also recent work by Peter Rantasa) can, for example, contribute to this project.

In order to show what my approach can do (indeed, to show its use and performance), I interpret contemporary ICT-mediated social media by using some of the concepts discussed so far. I argue that the use of words in social media enables language/culture to speak through human users, that even "virtual" artefacts are embedded within a larger meaningful whole, that social media are not mere worlds of signs but also involve materiality, that social media have various hermeneutic effects and shape our world, that social media have scripts and are full of nonhuman actants that talk to us and compete for attention, and that social media have a narrative and performative dimension. We tell and live stories, but technology co-writes these stories and, thus, also organizes characters and events in plots. As such, they become part of our lives and (hi)stories. Social media structure our personal histories and configure our (narrative) time and our performances—online and offline. Taking into account what I write about personality, one could also say that social media play a key role in personalization processes, in the emergence and construction of personality in and through use, practice, and performance, which is part of and made possible by a larger temporal-narrative and structural whole that includes both linguistic and technological structures and infrastructures. Moreover, taking into account what I say about power and embodiment, we must add that our use of social media is always embodied (even when we believe we are in "virtual" worlds), since our performance and engagement with these media is embodied and shaped by the body, including our use of language, and that our use of social media, understood as use of language and use of technology, always has power aspects and effects, and must be understood as being embedded in power structures and grammars that structure that use. Power and embodiment, thus, must be taken into account when studying the rhetoric and grammar of social media.

I conclude that social media reveal the performative, narrative, and theatrical character of the social life and, indeed, of human existence. To exist as a social being is to perform. That performance, however, is not only a matter of human performance (indeed, we are all actors on the world's stage) but is always technologically mediated and co-performed by technologies as actants—to use Latour's term. It is also mediated and made possible by language and its rules and games, as Wittgenstein suggests. In other words, the "stage" for human acting and playing is linguistic and technological, is not a neutral platform but structures what is said and done, and is itself made and emergent, the result of performance and process. In concrete use, experience, and performance, stage, words, and actors merge. There is use and performance. There are characters and events. There are narratives. These constitute the social, understood in performative and narrative terms, and understood not only in structural but also in temporal terms. What we say and what we do is part of that social process, structure, and whole, and is, at the same time, structured and made possible by it. This does not mean that media, technology, and language determine us—as I say, we are co-writers and co-actors, co-performers and codirectors—but, at the same time, *and because of the participation of language and (other) technologies in this process*, we lack full control and choice when it comes to directing this human-linguistic-technological and performative process, these performances, and these narratives. At most, we co-direct. The same is true for the person: We codirect our performances and narrative, which are enmeshed with, and constituted by, our use of words and things. Therefore, social and personal change can only happen in promiscuity with language and technology, in which we are always entangled as users and performers, and without which there would be no such thing as the social. (I will say more about the concept of entanglement below.)

Note that, starting from this performative-narrative approach, it would be interesting to further discuss power issues. Who has the power to direct or to get a particular script accepted? Who or what plays, writes, and directs what kind of game? What is the performative and authoring power of specific media and technologies as compared to others? Furthermore, the embodiment dimension of technology use and performance with technology remain undertheorized so far. What are the relations between use of body(-mind), language, and technology in performance and narration? What exactly is voice, and how should we understand the "voice" of new technologies? What about the illocutionary and perlocutionary aspects of voice use (Rantasa proposes the term: 'voice acts'): Could new technologies use language and voice in this way, and how are these uses of voice and performances with voice related to power issues? These are questions that I hope to further discuss elsewhere; it is now time to finish this book by presenting a conceptual framework that harvests most of the conceptual work done so far.

Conceptual Framework: Harvesting, Integrating, and Developing the Building Blocks

The investigations in the previous chapters have yielded some building elements (one could say: words and tools) for a framework to think about humans, language, and technology, which does not exclusively focus on one of these elements, but assembles, merges, integrates, and develops them as different aspects of a unified framework. I offer at least the following *approaches*, *arguments*, and *cases* that can be used to conceptualize the relations between humans, language, and technology:

- a Heideggerian criticism of the position that humans speak: Language and technology also frame what we say and what we think;
- a Wittgensteinian criticism of the position that humans speak, but also and especially *an application of Wittgenstein's view of language to technology*—therefore, also criticism of the position that (only) language speaks; this has yielded a focus on *use* and *performance* and, following the social-cultural interpretation of Wittgenstein, a holistic view, according to which this use and performance is embedded in a social-cultural context; I have argued that use of language and technology must be understood as part of language games and technology games, and a form of life;
- a Deweyan recognition of the deeply social and cultural nature of both language use and technology use; both language and technology are instruments for social cooperation and make possible the social life; both technologies (as tools) and language, as the tool of tools, create meaning and the social;
- a Searlean attention to the relation between language and the social, but without borrowing his dualism, with an emphasis on the use of technology, and with an expansion of his social ontology from declaration to all kinds of performances; there is not only collective intentionality but also collective use and performance, which involves both the use of words and the use of things;
- a social-constructivist attention to innovation dynamics, which helpfully brings in the time dimension and thinks of technology in relation to social processes, but then is revised in order to account for the role of language and to add a focus on use (also, on innovation);
- some elements retrieved from poststructuralism and postmodernism, if cured by more attention to human use and material technologies; for example posthermeneutics reminds us of the limits of what words can do; perhaps we also need art, dance, etc. to better understand how we use and relate to technology (see the last section of this chapter);
- a *transcendental* argument, saved from the phenomenological tradition, but then interpreted in a less abstract way as including technological and material next to logical or linguistic aspects; Wittgenstein's view of language can also be interpreted in a transcendental way, hence, my Wittgensteinian argument about technology and language: Their use and performance is embedded in a transcendental grammar, which shapes and makes possible that use and performance; moreover, technology and language are also part of that transcendental ground, that form of life;

- inspiration from McLuhan and posthumanist media ecology, which intimately connect language and technology (for McLuhan, language is a technology) and which add an ecological perspective; mediation should not only be understood as "in between" but also as "milieu;" ultimately, language is a technology; a "grammatical" approach can help us to understand how media and technologies shape us;
- use of postphenomenology and mediation theory, in particular Ihde, but then revised and developed in order to account for the role of *language*: a phenomenology and a hermeneutics that theorizes the mediating role of technology *and* language;
- a more fundamental revision of postphenomenology and mediation theory, which goes beyond mediation by giving the framework a *transcendental* twist: Technologies are not only mediators but also constitute a transcendental condition that makes possible particular uses and mediations; this transcendental approach does not render the approach less "empirical," at least if we interpret transcendental structures and processes as including the role of material artefacts, bodies, etc.;
- interpretations and revisions of Latour's bridging of gaps between humans, language, and technology, which draw on his work with Akrich and his writings about the nonmodern, including his use of the theatre metaphor, but emphasizing *language, process*, and *human* interpretation, use, and *performance*;
- narrativity theory inspired by Ricoeur, to bring in the temporal dimension and narrative structure of human existence and use, but then applied to technology, rather than only focusing on language; this leads to a view in which technology and language do not only mediate synchronically but also diachronically as they cowrite and codirect our narratives and performances; this conceptualization of the active hermeneutic role of technologies and language goes beyond the idea of language and narrative *about* technology; like Latour's work, it also puts more emphasis on the social, than, for instance, Ihde; the narrative role of technologies and narrative structures (as language) can also be interpreted in a transcendental way, as *making possible* uses and performances;
- social media as an interesting case to study this narrative-hermeneutic role of technologies;
- an explicit *process* approach added to the Wittgensteinian account of use and performance with technologies, and to the revised postphenomenological, Latourian, and narrative approaches;
- use of *voice*—with or without (other) technologies and media—as an interesting case to reflect on relations between humans, language, and technology; voice can be understood as embodied speech, but also as embodied or innate technology; per-sona as sounding through, an emergence of voice;

- acknowledgment of the political and *power* dimension of the use and performance with technology: There is a power dimension in the use of technology as social performance, in mediation by technology, in transcendental structuring of technology, and in narrativity; for example, if we ask who or what gets a voice, this is also a political question; the use of voice and speech has power dimensions, and structures in language and technology are always related to power structures;

- use of Foucault to discuss the rhetoric and grammar of power, language, and technology; our use of words and things has power effects and is itself dependent on, constituted by, and made possible by an already-given "power grammar" (as I call it), in which both language and technology play a role in the everyday micro level; power grammars are part of a form of life (Wittgenstein) or what Foucault calls 'episteme;'

- an acknowledgment of *embodiment*: Performance is always embodied; the subject in Wittgenstein, Latour, or, indeed, Ricoeur should be interpreted as a living being, rather than an abstract rule follower, node in a network, or abstract actor-narrator; our bodily existence is a condition of possibility for experience, action, language use, and thinking; the body structures and makes possible use of, and performance with, technology;

- use of Merleau-Ponty to suggest that the body also "speaks," in the sense that our uses and performance are always embodied and in the sense that the body *also* provides a grammar (which I call "grammar of the body" or "body grammar"), a transcendental structure or ground that makes possible and frames our particular uses and performances with language and technology; as cognitive science also shows, language is itself deeply connected to the body via our use of metaphor, and, hence, the body also shapes our speaking and thinking;

- a realization that the social life and existence can indeed be understood by using the metaphors of *theatre*, performance, and so on; but an insistence that, according to the approach(es) presented, acting and performing should not only be understood as the result of human intention but as also shaped by process and as made possible by linguistic and (other) technological grammars; this means that, in concrete use and performance, stage, actors, and words merge;

- an acknowledgment that, in structural, network, narrative, and process approaches, there should still be enough room for *human* and *personal* subjectivity, experience, use, and performance; however, humanity and personality are also emergent from process, use, and performance—they do not preexist and precede process and performance, but live and exist in it and by it; being human and being a person are becomings: Like the worlds and meanings and, indeed, the words and things with which they are enmeshed and entangled, they need to be performed over and over again.

These approaches and arguments give us the following *concepts* as tools to reflect on and conceptualize relations between humans, language, and technology:

- human
- language
- technology
- use
- performance
- the social
- innovation
- collective use
- collective performance
- game
- form of life
- transcendental conditions
- grammar
- episteme
- script
- collective
- individual
- hybridity
- mediation
- embodiment
- narrative
- time
- process
- power
- politics
- voice
- subjectivity
- personality
- entanglement

How can these approaches and concepts be integrated? Although, in my descriptions of the elements, I include many formulations that express a more integrated view, I will soon attempt to further articulate a more integrated view, and, later in this chapter, I will use an additional concept, entanglement, to summarize and consolidate it. However, in order to retain a critical-philosophical angle and at least *pause* my constructive-theoretical narrative, let me first recognize and briefly address some problems related to *this very project of integration* and my *use* of the *concepts* presented.

First, there remain tensions between the theories and concepts discussed in this book and, hence, also between those presented in this integrative framework. Sometimes I mention or discuss these tensions, but perhaps they receive insufficient attention, since the focus is on constructing an approach, arguments, and concepts. Therefore, let me at least highlight a few. For example, Wittgenstein's focus on the *use* of language in the *Investigations* is different from Ricoeur's view of language, which, in its emphasis on narrative, seems to lose track of the language user. This has implications for the project of integrating both approaches, for thinking about technology use in Wittgensteinian *and* narrative terms. The challenge here is to think about narrative technologies in a way that does not lose sight of the insight that technologies and, indeed, *narratives* can only have the roles of configurators, mediators, and grammars, since and as they are used by humans. Consider, also, postphenomenology and Latour, which seem to be in tension with human-centered approaches: At first sight, in their approach, humans seem to lose out to technology when it comes to having their "speaking" and agency recognized. I have already discussed this problem. But, in general, there is always the danger that one of the corners of the "triangle" I introduce in the first chapter—humans, language, and technology—gets too little conceptual weight. While, in the course of the book and in this chapter, I have tried to resolve some of these tensions and have shown that there is also a lot of compatibility, it is important to recognize and address the tensions and problems—here and in further work.

Second, when we use authors and a particular tradition, we also borrow all kinds of problems; this is also applicable to this book. For example, Wittgenstein is often read as antitheoretical and I sympathize with this interpretation and approach. Yet, in this book, I use Wittgenstein to develop a conceptual framework. Is this theory? Is the very project of this book intrinsically anti-Wittgensteinian? And perhaps also: anti-Deweyan? And, if so, is this a problem? Or does its "damage" and fruitfulness depend on how the framework is *used*, i.e., how it *performs* in practical efforts to better understand and evaluate, and to better *use* technologies and language? Other issues arise with the interpretations of, for instance, McLuhan, Ricoeur, Heidegger, and Latour. In the course of my book, I discuss the work of these thinkers, but I could not pause too much to extensively discuss their work and reception history, since this would have taken too much space and would have distracted too much from the main purpose of my book. However, if one wishes to further develop the framework presented here, more work is needed to deal with issues within their work. For example, one could further study and discuss interpretations of Wittgenstein that support the arguments of this book and one could further discuss the "process" interpretation of Latour by engaging with recent work on this, including Harman's work. One could also further read and interpret authors that are merely mentioned or only briefly discussed in the book, but that deserve more attention. For example, one could learn more from Foucault, Merleau-Ponty, Arendt, and Husserl. (And, no doubt, there are many authors that are entirely absent in this book, but could be very helpful with regard to my project.)

The same is true for the concepts that make up the framework. These concepts themselves can always use *more* philosophical attention in order to render the framework more critical and more philosophical. Their meaning is not obvious, clear, or uncontested, and, often, they also bring with them the luggage of the philosophical traditions they come from and, hence, their problems. For instance, one could further discuss the meaning and my use of central concepts in this book, such as technology, language, form of life, mediation, performance, and process. Part of this work may require relating my use of these terms to their meaning in different approaches and philosophical traditions, thus further relating them to various philosophical grammars and academic contexts. While I have of course done some of this work in the course of this book when developing my views and arguments, more critical attention to these concepts and their contexts could help to further use, develop, and critically discuss the framework presented here.

Indeed, I have often taken the liberty to decontextualize concepts: I divorce them from the "home" of the philosophical works, authors, and traditions they stem from in order to serve in the arguments and framework presented. To some philosophers, this amounts to heresy or abuse, to others this is not a problem at all, or is even a virtue; in any case, this operation is far from uncontroversial. My own view on this matter is that, on the one hand, philosophers should acknowledge where concepts come from and be aware of problems that may come with those origins (this includes the politics of concepts, e.g., one may discuss Heidegger's political views and their relation to his writings on technology), but that, on the other hand, philosophers should also be free to use concepts in new ways. Influenced by Dewey and Wittgenstein, I see philosophical language, like any other language, as a technology, consisting of various tools. Tools are part of larger games and of a form of life; at the same time, they can, are, and should be adapted, used, and so on, in order to reach certain purposes. Taking this distance and this liberty is necessary in order to move beyond mere analysis of philosophical texts to investigate and address philosophical problems at hand—here, the question concerning the relations between humans, language, and technology. It may also help us to arrive at a conceptual apparatus we can use to cope with the societal and technological problems we face today. Concepts, theories, and other technologies always have many meanings, contexts, origins, etc.—meanings that, like all meanings, are related to their origin and history. In order to make good and new use of them, these origins and histories should be studied and discussed (indeed, by philosophers of technology, among others). But this should not prevent us from *using* the concepts and theories, performing with them, and experimenting with them, in order to improve thinking about technology and society.

After these remarks, we can now proceed with the main task of this chapter: How can these building blocks (approaches, arguments, cases, concepts) be *used* in a way that integrates the insights gained in this book? In the previous pages, I offer some claims that work towards such an integration, but now I will further tie these together. While the conclusion of this book is not so much a particular view, but, rather, a "framework," which consists of the elements and building blocks outlined that can be used and adapted in various ways, and which must retain an openness in this respect, let me attempt to articulate a more general, integrated view, which emerges from the previous chapters and pages. The following conclusion is my response to the main question of this book regarding the relations of humans, language, and technology: What does it mean to say that humans, language, and technology "speak"? Based on my interpretations and discussions of the approaches and theories presented in this book, I conclude that humans, technologies, and language "speak" in the sense that:

- As human users, performers, and interpreters, we use language and technology, but this role of language and technology is not merely instrumental; language and technology shape our use, performance, and interpretation; from this performative-hermeneutic process, meaning emerges and meaning is made;
- Human use, interpretation, and performance are always lived and living, they are experienced by the human user and involve interaction with a living cultural-environmental whole; this is the case in everyday life, but it is also the case in science and in innovation processes;
- The user can be constructed as a person (rather than an individual), but use always happens within a larger social and cultural whole, which is always given, to some extent, and structures that use and that person; there are all kinds of "grammars," including language games and what I call "technology games," there are narratives and processes, and there is a form of life in which these games and narratives are embedded and which is, at the same time, constituted by them; personality also emerges from this process and this larger whole; there is individual, but also collective use and performance;
- In use, language, technology, and the social are intimately bound up; the social whole and process consists of humans and nonhumans: It is full of texts and tools, languages and technologies, which create a semantic-performative field and environment to which we attune, within which we act and perform with others, and within which we do things with words, do things with things, and do things with others; the social is a theatre in which use of words, use of things, and bodies (see below) combine and unite in performances, understood as processes and narratives;
- As they mediate our relation to the world and make our world, languages and other technologies shape our spatial and temporal hermeneutic and social relations: They form the relation we have to the world and to others, and they co-write our scripts and narratives; they co-create our use-world, experience-world, and acting/performance-world;

- Language also mediates our relation to technology and our uses of technology, for instance when we write and think about technology or when we speak about a new technology;
- Language and technology do not only mediate; language and technology are also themselves transcendental conditions (conditions of possibility) that structure, shape, and constrain our experience, use, and performance; in other words, language and technology are not only tools or mediators but are also part of the whole that makes possible and shapes our use and performance (a whole that is, in turn, shaped and constituted by our uses and performances); language and technology are shaped by grammars, but are also themselves (part of) grammars; these grammars are structures and processes, which can have a narrative character;
- Using the metaphor of speech, we can formulate a more integrative view as an answer to the question: *Who or what speaks?* When we speak through technologies, language and technology also speak; when we use words, language and technology also speak; when technologies and language speak, humans speak;
- The interpretation and making of 'world' happens in processes: Meaning emerges from processes in which humans, languages, and technologies act; there are processes of co-performance by humans, languages, and technologies;
- In these use and performance processes, in these processes of meaning-making, subject and object emerge: They should not be understood as fixed on beforehand, but as emerging from processes of use, performance, interpretation, and experience; this is the case in everyday life, science, and elsewhere;
- Use and performance with technologies and with language are themselves always political, in the sense that they are also a way of exercising power (e.g., of making people do things); humans use words and language to perform and to make others perform; hence, use of language and technologies in everyday life is always political; science and innovation are also political;
- However, there is not only the power exercised in the form of rhetoric of language use and technology use; there is also the power that is in the games and form of life: the "power grammar" in which these uses and performances are embedded and which plays out in micro social relations and situations, and is embodied in institutions; use and performance with technologies and language are always already social, since they are part of a larger social-cultural whole, which structures that use and performance, and are therefore never politically neutral, but also change, interact with, and depend on, power structures, which are also linguistic and technological;
- Use and performance with technologies is also always embodied: Use and performance depends on the body as mediator and transcendental structure; another way of saying this is that, when humans, language,

and technology speak, "the body also speaks," as mediator and as "body grammar;" in this grammar, metaphor plays an important role, since it connects use of the body to use of language (and to use of technology); use of voice is one kind of performance in which humans, language, and technology meet and which shows embodiment of speech; voice can also be considered as a technology; there are individual voices and perhaps there are collective voices;

- Language can itself be defined as a technology, since the use of words, sentences, and other linguistic elements is so similar to the use of tools, and their grammatical, transcendental role is also very similar, since both have syntax (surface grammar), are part of a form of life or episteme (depth grammar), and condition particular uses and performances with words and things;

- This is also true for philosophical concepts: They are linguistic and they are also tools; in addition, writing, word processing, and computers are technologies that are used to support the development of conceptual tools; hence, this "framework" must itself be understood as a tool or a toolbox (but not a "mere" tool or toolbox, given what has been said about technologies and tools in this book), which, in turn, depends on various (other) technologies, such as writing technologies, and which is *used* and *performed* (with) in this book in order to advance thinking about technology and its relation to language;

- Philosophy of technology, thinking about technology, is itself mediated, shaped, conditioned. and made possible by language; as philosophers of technology, we should become aware of these mediations and conditionings by language; as philosophers, we should attend to the elements, forms, structures, and limits of language and language use in contemporary philosophy of technology, and explore other uses and other languages;

- Beyond language, we should also explore other ways of gaining new knowledge, experience, and wisdom concerning our relations to technologies and to each other; in use and performance, there are always encounters and experiences that will resist our general and linguistically mediated descriptions; perhaps collaboration with, and engaging in, scientific-technological and artistic practices can help us to gain different kinds of knowledge and wisdom, or, rather, to gain know-how and wisdom in different ways, through different (kinds of) uses and performances (below, I will say more about art);

- This book is also a use of words and a performance with words, which is embedded in a wider field of language (English, but also specific philosophical languages) and made possible by language and technologies (e.g., writing and writing technology), and which should therefore be read critically by focusing on my language use, my use of technologies, and the problems and limitations that follow from these uses and performances—next to what is opened up and made possible by them.

This framework and integrated view of humans, language, and technology gives us a tool to better understand why technologies are instruments, in the sense that we use them, but also more than mere instruments. A noninstrumental understanding of technology is already offered by Heidegger and is widely accepted among contemporary philosophers of technology. For instance, it is also a central claim and insight in work by philosophers from the postphenomenological school and philosophers inspired by Latour's work. However, in contrast to postphenomenology, this framework enables us to theorize the role of *language* in relation to this technology-as-more-than-technology. Compared to Ihde's and Latour's view, the framework also gives us a more *use*-oriented view (inspired by Wittgenstein), a *performance*-centered view, and a more dynamic, *process*-oriented approach that takes into account the temporal and narrative aspects of the phenomenology and hermeneutics of how we relate to the world and to others through technology and language. Compared to Latour's and Ihde's view, it is more focused on, and better connected to, use, performance, and practice, and to philosophy of language. This does not render their approaches obsolete; rather, we can revise them in this direction. I propose to read *language*—with which these authors engaged in their early work—back into their later work, into their nonmodern (Latour) and postphenomenological-hermeneutical (Ihde) approach. But I also do more; I move beyond Ihde and Latour. I also use Wittgenstein *and* move beyond him. Compared to Wittgenstein's view, the proposed framework offers an explicit account of the use of *technology* (which is only implicit in his writings), stresses performance with language and technology, adds a temporal and *narrative* dimension and an explicitly phenomenological, hermeneutic, and *transcendental* dimension, and emphasizes the *embodied* dimension. And compared to Ricoeur's view, it recognizes that, not only language but also *technology*, plays an important hermeneutic role in shaping our narratives, showing the possibility of a narrative theory and hermeneutics that includes *human and nonhuman* narrators, directors, and performers.

Finally, this books shows how much humans, technology, and language are related. It turns out that, when we try to define one of the terms, we need at least the other two. If the present exercise and reflection contributes to a better understanding of the relations between humans, language, and technology, based on various conceptualizations of language such as Wittgenstein's (use) and Ricoeur's (narrativity), this also has implications of how we conceptualize the three terms themselves. The framework and approach I present leads to a conception of the *human* as user, performer, and narrator with language and technology, whereby the human subject itself is understood as emergent from and constituted by language and technology, rather than fixed on them beforehand. In this sense, there is no human "before" the human-technology and human-language relations and interactions nor "before" the use and performance. Moreover, this view also enables us to re-understand and reconceptualize *technology itself*, in terms of use, performance, "speaking," and narrativity, with "use of technology" and "performance with technology" understood in analogy with use of language. By (re)connecting conceptions of technology to conceptions of language, it thus contributes in a novel way to a central question in philosophy of technology: What is technology? Similarly, my discussion raises the question regarding the nature of *language*: Is language "something" distinct from technology or is it itself a technology? Based on the present investigation, in particular, my interpretations of Wittgenstein and McLuhan, I suggest that it makes sense to regard language as technology. But, whether or not we define language itself as a technology, it is clear that, if we consider their use/performance and the emergence of meaning from within a social-cultural-material context and process, and if we consider their very similar roles as "grammars" linked to forms of life, there are deeper structural similarities between language and technology. Let me end with a further reflection on the relation between language and technology. What can we conclude from the previous chapters about the relation between language and technology, if these terms have a different meaning at all? I will also say more about the limits of philosophical (and, hence, linguistic) inquiry into this problem, and suggest some ways to overcome it.

8.2. Conclusion: Entanglement, Touch, and Art

Conceptualizing the Relation Between Language and Technology: Entanglement, Process, and Touch

Usually, we think of technology as belonging to a different sphere than language: We talk about technology as if there is a scientific-technological sphere, as opposed to a human-cultural sphere, with artefacts separated from humans, objects separated from subjects, and so on. But I argue that this dualistic way of thinking is unfruitful when it comes to understanding technology and its relation to language. The integrative view I present in this book shows that, in processes of use, performance, interpretation, and narration, the human and the nonhuman, but also language and technology, are much more interrelated than assumed in a modern, dualistic view. How can we further conceptualize this interrelatedness?

In addition to the concepts presented previously, I propose to use the metaphor of *entanglement*. In common usage, "entanglement" means that two elements are mixed up and twisted together, or that one is deeply involved in something and cannot escape from it: One is entrapped in something, enmeshed in something. For example, two threads are twisted together, or a person is entrapped in a complicated romantic relationship. In quantum physics, entanglement means that particles interact in such a way that they are correlated, but remain spatially separated: The quantum state of one particle cannot be described independently from the state of other particles; there is, thus, only one quantum state, for the system as a whole.[1]

These qualities and meanings of being mixed up, being twisted together, complication, involvement without escape, interdependence, correlation, and holism render entanglement an interesting metaphor for our present purpose. One could conclude from the investigations in this book that humans, language, and technology also interact in ways that make it impossible to describe them independently. Twisted together in use, performance, and experience, they are always mixed up with the other elements, entrapped in each other. If we describe "the human" independently from the use of words and the use of tools, and independent from performances with technology and language, what do we have left? A ghostly and immobile mind or mute body, unrelated to its environment. If we describe language independently from language users and from language technologies, what do we have left? Dead words that are no longer signs, that cannot be used, and that mean nothing. And if we describe technology independently from human users and from language used to talk about, with, and to technology, what do we have left? Meaningless and, indeed, *useless* artefacts.

However, this does not necessarily mean that the three elements are the *same* or totally overlap, they may still be different and separate; at the same time, one needs to express that they are interrelated and form a whole. The metaphor of entanglement enables us to express this. If we say that humans, language, and technology are entangled, it means that there are co-relations between humans, language, and technology in ways that lead to one speech-performance/technology-performance/human-performance. When they are correlated in a concrete use and performance process, then, at a given time, there is only one speech "state," which is also a technology state and a human state: As they are mixed up and twisted in concrete and lived use and performance, humans, language, and technology speak together. Enmeshed and entrapped in a complicated relationship, due to their roles in use and performance processes, humans, language, and technology are forced to interact and mix. Moreover, together, they are part of a larger whole. When I use technologies and perform with technologies, it is not only me who speaks; the whole also speaks, the grammar speaks, and the form of life speaks. This whole includes linguistic and technological, and human and nonhuman elements. There is one state of speech, or, rather, there is one *process* of entanglement, *entangling*, which is, in turn, related to a larger whole and process that makes possible the concrete entanglement state and performance process. Finally, in this whole and in this process of entangling, nothing is left out and nothing is isolated; the process is radically relational. In the process, human, language, and technology are fatally involved with one another, in the sense that they cannot gain stability on their own (if at all) or retain their own state or border. This is also the case, since, although we can try to grasp a particular "state" at a given time, there is also a sense in which there is no such thing as a "state," since language, humans, and technology cannot escape temporality, and mean nothing outside of temporality; in this sense, there is only process and change.

Indeed, responding to the question, "What is left?," one could say that there is no "is" that is left. Human existence, meaning, and experience, understood as use, performance, and narration, have a temporal and process character; we cannot and should not define it as an "is." If there "is" anything here, it is a becoming, a continuous becoming and continuous change. In this process of becoming, objects and subjects emerge, appear, interact, and entangle. Personality and "individual" and "collective" narratives also emerge. As we use them and as we perform with them, words and tools make us what we "are," or, rather, what we become. There is a speaking; there is a performance. And both speech and the speaker are hybrid human/nonhuman, or, at least, there is always co-speaking. There is no pure subject when it comes to speech, understood as use and performance. Try to find a subject and it will be contaminated by objectivity. In use and performance, the human is already posthuman, contaminated, enmeshed, and entangled with words and things. But neither is there a pure object: words and tools—*as* words and tools—mean nothing outside human use. Try to find an object and it will be contaminated by subjectivity. It is already contaminated once you think of it and use words to describe it, even before you look for or at it. It is entangled with subjectivity once it is used and performed with.

Language, then, is a tool of contamination: It is a tool to come into contact, to communicate, to touch. It mixes and meshes things up. If we must have an ontology at all, it is one we must seek beyond modern mysophobia. We may find this ontology once we embrace the ecology and epidemiology of dirty and contaminated subject-objects, and dirty and contaminated objects-subjects, which are entangled in a dirty embrace. But, better than an ontology, we can think of human beings and of reality in terms of process and communication. There are processes of contacting and touching, processes in which subject and object touch, entangle, and contaminate one another. Words and tools, language and technology, are part of these processes, mediate these processes, and make possible these processes. At the same time, they also play the role of subject-objects we want to get in touch with, that is, insofar as they are also human, insofar as they also speak. Humans speak through using and performing with words and tools. This use and performance contaminates words and tools with a human touch. Posthumanist thinking accepts this contamination and touch. Of course, it is important, as humanists continue to remind us, to get in touch and to stay in touch. It is also important to communicate, to make *communis*, to make the common. But, an important insight we can gain from the traditions of philosophical thinking about language and technology, is that we cannot do this as pure subjects. We need tools and media in order to get in touch and stay in touch. We need words to get in touch and stay in touch. And we need bodies to get in touch and stay in touch. We also need tools, words, and bodies to make the common. At the same time, this making presupposes the social that is already given, and that includes tools, words, and bodies—in particular, their structures and grammars that constitute a form of life, on which basis touching and communication can happen. A lot more needs to be said about this. But, here, I conclude that we must conceptualize getting in touch and staying in touch as a performance involving the use of language and the use of technology, during which meanings emerge and during which not only relations and the common(s) but also subjects and objects, humans and nonhumans, are mutually constituted.

This view of language and technology takes us far beyond instrumentality, beyond the instrumentality of language and technology. However, as I note, there is something paradoxical about this. When articulating this noninstrumentality in the course of this book, I have been influenced by a Wittgensteinian focus on use and performance (which was further supported by Dewey), which, in a sense, is an "instrumental" view, if "instrumental" is understood in terms of use and performance. I thus started with recognizing the instrumentality of technology, in analogy with the instrumentality of language: Just as language is about the use of words as instruments, technology is about the use of tools. But following this path in various directions has led us to recognize that language and technology also are and do much more. Considering their mediating, narrative, and transcendental, "grammatical" roles, it turns out that they also shape and structure our uses, performances, experiences, perceptions, interpretations, and thinking. They actively intervene in processes of meaning-making and processes of meaningful and meaning-shaping performance. But this is possible *because we use them* and perform with them, using them as tools. Language and technology are, thus, both instrumental and noninstrumental, or, rather, they are noninstrumental because of, or *in* their instrumentality, that is, in their use and in their roles in performances. It is because we use words and tools as performing and embodied beings, it is *in* our use of words and tools, that meaning is made. It is because, and insofar as, they are such use-full instruments that they have the mediating, performative, narrative-hermeneutic, and power-full roles ascribed to them. It is because they are in-betweens *in* use, because they constitute milieus *for* use, and because they *make possible* that use, that technologies are more than what instrumentalists suppose they are. Hence, usefulness is not necessarily opposed to meaningfulness at all; the meaning is not only in the aim but also in the means. The means also means. Meaning is a holistic matter, emerging from "states" and processes of entanglement, contamination, and touch between humans, language, and technology. In the view I develop, it is in their usefulness that language and technology are so actively involved in the human(/nonhuman) making of meaning. It is their use in performances, activities, games, and practices, within a larger social and cultural form of life that brings us beyond the useful and to the meaningful, that enables us to build the common(s) and to get in touch with and via language and technology.

One kind of use and performance from which meaning emerges is, of course, writing, including writing as philosophers of technology. It is true that, as Mitcham (1994) argued, we need bridges between, on the one hand, the world of technology and engineering, and, on the other hand, the world of letters, the world of the humanities. We need bridges between technology and language, and between technology and the human. Now it turns out that there was already a bridge, but we (philosophers of technology) did not see it because it was transparent in use: philosophical writing, understood as the use of words and performance with words, and as crucially involving language and technologies—indeed, as constituting concrete entanglements of language and technology. Philosophical writing is techno-logos as process and performance. We used it, we performed with it, and, therefore, it remained transparent and unexposed to critical philosophical attention.

But, if this is the case, then we should also be aware of the limitations of our tool(s). Perhaps there are other things to say, to speak out, about technology and other matters—things that cannot be said with language and with writing technology, that need a different technology. And perhaps there are things that cannot be said at all, as Wittgenstein suggests in the *Tractatus*. Then we could touch, without words or things—if this makes sense at all. We could also use words and things to explore or "touch" new possibilities. In any case, touch also offers an interesting metaphor that can be connected with thinking about change and structures, in line with the "grammatical" inquiries proposed by Wittgenstein and McLuhan. Consider what the McLuhans say about touch:

> 'Touch, as the resonant interval or frontier of change and process, is indispensable to the study of structures. It involves also the idea of 'play' . . . as the basis of human communication.'
>
> (McLuhan and McLuhan 1988, 102)

Perhaps thinking about new and changing forms of life, structures, and grammars, then, is a matter of touch and exploration. It is itself a kind of play; it may well require playing around with words and things. And perhaps touch can even be used as a metaphor for thinking itself, for philosophical thinking, especially: When we think, do we not touch around? Do we not try to grasp and touch? 'We go towards the thing we mean,' Wittgenstein wrote in the *Investigations* (§455, 140e). And when we go toward the thing we mean, we touch and explore meanings. We play with meanings. And we communicate.

But what if the use of words is insufficient? What if we need different experiences in order to *move on*? Realizing the limitations of what can be said by means of written discourse makes me move towards (other) technologies and (other) *arts*.

Other Tools: Technological Practices and the Role of Art

One way of taking a critical stance towards philosophy of technology as use of language and as writing performance is to study and intervene in *discourses* about technology—inside and outside the philosophy of technology, and inside and outside academia. This is, of course, an important task for philosophers, as language users and as writers. It is a classic humanistic duty to write and interact with a community of writers, indeed, to build a *communis* by means of the use of words. And it is a central task of philosophers to do conceptual work, that is, to use language. To do this in a critical way means to use language in a way that renders oneself and others aware of language use itself and its implications, and to use language and write in a way that critically evaluates and adapts use of language (words) in order to address problems—including the use of language in order to question what the problem is, and what should be framed as a problem at all.

Another way of taking a critical stance, however, is to recognize that language and writing are not the only tools we can use in order to better relate to technology, to better understand technology, to gain more wisdom about it, and to evaluate and intervene in what happens with technologies. Wittgenstein (this time the *Tractatus*: there are things one cannot speak about) and postmodern reflections on language and the real (e.g., Mersch) remind us that there is an "outside" of language. This issue concerning the limits of language touches, of course, upon a long-standing discussion in philosophy. For my present purposes and in the light of thinking about technology, we can narrow down the scope of this discussion and reconceptualize the claim concerning the limits of language as: *Language is not the only tool*. There are other tools, and it seems that we may well need them to achieve more understanding and wisdom with regard to technology. Given the limits of what words can do, it is important to use other tools than words to try to understand, reflect on, and experiment with, technology. So *which other tools could be used for understanding technology?*

First, when we leave the study room, we find other technological practices from which we can learn, which may involve use of words, but also more than words. There are not only writing practices but also what are commonly regarded as "technological" practices: practices that involve use and performance with technologies more narrowly defined in terms of material artefacts (defined as not including linguistic elements). For instance, there is the use of technologies in scientific practices. We can go to the lab. There is also the use of technologies in DIY practices. We can go to the garage or the workshop. We can also look at how technology is used, produced, and designed in industry. We can go to the factory. And in all these practices we can watch, but we can also participate, use, and develop. Use and development of technologies can then itself be seen as a path towards understanding technologies and gaining a critical relation to them. But what constitutes *critical* use and development of technologies? It seems that this can only happen under certain conditions. Usually (literally), we cannot do this, since a particular technology is transparent when we use it. We usually do not think about our use. But thinking is made possible by language and involves use of language. Should language and the use of language therefore be seen as one of the (necessary) conditions for critical use of technologies and critical thinking *through* technologies? While as yet I do not have a clear and comprehensive idea about what these conditions could be, we can observe that some people in what is considered as "the world of technology" reflect on technology and their use of technology, and gain a critical relation to it, by using technology and by using language, but based on, and *in* and through their experience with developing and engaging with the technology—in and through their *performances* with technology. For instance, some hackers/programmers use technology (code) in a way that is intended as a critical gesture, e.g., a particular use of code to hack a particular system (say, of a large corporation or a government agency) may reveal something about our surveillance society and, in this sense, "reflects" on security, "asks questions." But the latter, language-based metaphor is perhaps not the best or, at least, not the only metaphor we can use to describe this, since the critical performance is not *mainly* done by using language— at least not ordinary language. It is mainly a technological performance, which uses programming language and other tools and which intervenes, "comments" (again, a metaphor from language use) on what goes on in the world of technology. But, next to the link with programming language as *language* (code is "technological," but also linguistic at the same time), this metaphorical connection—which is as much linguistic as it is practical and technological—is interesting in itself. Given language-technology entanglements in concrete performance and experience with technology, it seems difficult, if not impossible, to disentangle language and technology when attempting to articulate what goes on here. Again, it becomes clear that there is entanglement and that the "narrow,", instrumental conception of technology does not work. But, in any case, my example suggests that doing something with technology can have an important critical function, with *technology* playing an important role next to language—even if use of language, by the person(s) doing this and by (other) interpreters of the person's performance with technology, seems necessary to *reveal* the critical function of the "technological" act, gesture, and performance. Moreover, one may need to know(-how) particular games, including the game of so-called "white hat" hacking, for instance, which is aimed at exposing vulnerabilities for the sake of improving a system. There may also be games of resistance at play. But these games are not only linguistic or "cultural;" they are also technological at the same time. The act/performance and its interpretation seem to presuppose both linguistic and technological grammars, and these grammars seem to have a lot of overlap.

Second, we can also go to the studio. Art is also a technological practice, using language, but also different technologies. As Heidegger and McLuhan also recognized, art and artists can offer modes of thinking and knowledge, perspectives and sensitivities that can help us to understand our relation to technology and media. McLuhan thinks that artists perceive differently and can make us perceive differently. Influenced by Ezra Pound's phrase that 'Artists are the antennae of the race,' he argued that artists are a kind of "perception experts" that are sensitive to transformations of experience made possible by new media and technology:

> 'The effects of technology do not occur at the level of opinions or concepts, but alter sense ratios or patterns of perception steadily and without any resistance. The serious artist is the only person able to encounter technology with impunity, just because he is an expert aware of the changes in sense perception.'
>
> (McLuhan 1964, 19)

Art can then function as a kind of early-warning system and as a way of preparing for, and coping with, the changes at hand:

> 'The artist picks up the message of cultural and technological challenge decades before its transforming impact occurs. He, then, builds models or Noah's arks for facing the change that is at hand. . . . the artist is indispensable in the shaping and analysis and understanding of the life of forms, and structures created by electric technology.'
>
> (McLuhan 1964, 71)

Thus, according to McLuhan, artists can probe and detect changes in our society and culture, changes most of us do not really see, let alone are prepared to confront. But we need to understand what is happening and what is changing in order to prepare for the future. Paradoxically, in order to best prepare for the future, we have to try to better cope with the present. McLuhan thinks artists are particularly good at facing the problems of the present: 'Only the dedicated artist seems to have the power for encountering the present actuality' (77), and for accepting and coping with its problems, for 'taking it on the chin,' as McLuhan puts it (73). The artist must then 'play and experiment with new means of arranging experience' (276). McLuhan thought that art could lead to a new, more integral form of consciousness and awareness. In addition, in the *Laws of Media*, the McLuhans write that it can create a new, liberating environment, a 'counter-environment:'

> 'All serious art, to use Pound's phrase, functions satirically as a mirror or counter-environment to exempt the user from tyranny by his self-imposed environment, just as Perseus's shield enabled him to escape stupefaction by the Gorgon.'
>
> (McLuhan and McLuhan 1988, 226)

We may also consider Dewey's view that art embodies 'possibilities' that are 'not elsewhere actualized' (268) and that art 'is the living and concrete proof that man is capable of restoring consciously, and thus on the plane of meaning, the union of sense, need, impulse and action characteristic of the live creature' (Dewey 1980, 25). Dewey thus connects art to the 'activities of a live creature in its environment' and sees it as realizing a 'union of material and ideal' (27), mind and body, nature and spirit. Art, then, becomes a tool for achieving a non-dualist awareness and (inter)action, perhaps for achieving an awareness also of the entanglements and touching processes I have described in this chapter. But whether or not we agree with McLuhan's specific claims about consciousness and environment, and with Dewey's definitions of art and hopes for what art may achieve, it is clear that artists—next to others—can have an important role in helping us to address what Heidegger calls 'the question concerning technology.'

"Art" should, of course, not be reduced to painting, or even to visual art. "The arts," in the broad sense, must be considered here, including fine arts/visual arts, crafts, and architecture, but also—keeping in mind the conceptual focus on performance that emerged in this book—*performing* arts, such as music, theatre, dance, and performance (here, narrowly understood as a specific form of art), which may reveal our use of words and tools *as performance* and process, and help us to rethink, reembody, move, remove, and redirect what we are doing. Art may also offer other tools we can use to *touch* and get in touch with one another, and to create the common(s). Writing may help. For instance, some of the words and sentences in this book and in other writings may be used in making this connection with art. But, here, I wish to emphasize the value of nonlinguistic tools and media, of attending more directly to the performance and the touch. Art in its nonlinguistic aspects can make us aware of the performative and touch dimension of what we are and do, including what we are and do with technology. And as Mersch argues (see Chapter 5) and, keeping in mind the postphenomenological and "empirical turn" interest in materiality: In art, materiality also shows itself, often in a way that resists integration in language. Consider also, again, Wittgenstein's view in the *Tractatus* that there are limits to what language can do. This is also true for understanding technology and our relation to technology. The performative, embodied, touch, and material dimensions of technology use and performance are not always (made) clear in philosophy or, in science, are understood as theory-based, language-based, and logic-based practices. Practitioners in these fields are either unaware of these limits (at least *as* scientists) or actively suppress the concrete, the particular, the embodied, the material, etc. (this often happens in philosophy). In these respects, artistic approaches may well be a necessary complement to scientific and philosophical approaches to the question concerning technology.

McLuhan, however, does not sharply distinguish between art and science, or between art and research, and neither should we:

> 'The artist is the man in any field, scientific or humanistic, who grasps the implications of his actions and of new knowledge in his own time. He is the man of integral awareness.'
>
> (McLuhan 1964, 72)

This broad definition of the artist suggests that science, art, and philosophy can and should be combined to cope with our problems, and that a discourse that keeps these domains and approaches entirely separate is unhelpful. Another argument for combining art, science, and philosophy could be that, although each is different, each has unique perspectives to contribute, that a combination could lead to a more integral perspective, and that focusing on one of them to the exclusion of the other(s) is therefore problematic. If this is true, it implies that philosophy of technology should not be done in isolation from science and art, but should engage in collaborations and crossovers between these fields and domains. Moreover, there are many artists today that are artists and scientists at the same time, and philosophers can learn from their work and their approach. Some philosophers also engage with art and/or science, and there are many scientists that are interested in art and philosophy. Such stimulating crossovers could be very helpful for understanding technology, among other things.

Yet, McLuhan also sees a problem that is increasingly relevant today: The speed of technological development makes it increasingly difficult for artists (or for philosophers, for that matter) to keep up with what is happening. 'For in the electric age there is no longer any sense in talking about the artist's being ahead of his time. Our technology is, also, ahead of its time' (McLuhan 1964, 71). Indeed, developments in new media and technology, and the related patterns of human experience, change very fast. Can we keep up, in our philosophical and artistic performances, with the rapidly developing technologies and media? Can we still touch before we are being touched? Can we reflect on who or what is speaking when we *are already spoken* before we can hear our own voice? Are we spoken *to* – with the "we" being fixed in advance of the speaking – or does the speaking constitute us, bring the "we" into being? Do we still have a chance to co-write our script, codirect our performance in the global theatre, or does technology—in the form of machines—become the only author and director?

Perhaps this "keeping up" narrative suggests too much a battle between humans and machines. Perhaps the main problem is not to "keep up" with the machine, but to "keep going" with the dialogue *in* the machine while performing and moving *with* the machine—to make sure we perform, touch, and speak with others, even if we are and remain part of the global world of speed and technology.

In 'Roles, Masks and Performances,' McLuhan (1971) uses the metaphor of a taxi ride. Influenced by Plato, he points to the role of the stranger to provoke dialogue, with the philosopher taking the role of the cab driver; he thinks this talking with the stranger leads to global thinking and helps us to cope with the new world. Reflecting on this metaphor in light of this book, we can highlight the relations between technology, humans, and language, and, indeed, the transcendental role of technology in this metaphor: It is the technology, the car, which makes possible dialogue. But we also need humans, including philosophers and (other) strangers. And we need the talking, the use of words, and especially the dialogue, which is made possible by this amazing ancient technology: language. In this picture, there is, of course, power and there is politics, and there is the technology that "speaks." But, perhaps, when faced with crisis, it is less important who or what speaks; the important thing is to keep talking to one another while using language and while working and performing with the technology—to keep the dialogue and the car going, and to keep listening to the stranger, even and especially in the face of fast and sweeping, all-embracing technological, political, societal, and environmental changes in the global theatre McLuhan talks about. "What is left," then, are words and things, or, rather: Our social and embodied uses, performances, and games with words and things, perhaps involving touch and play. What else can we humans usefully, mind-body-fully, and meaningfully do?

Note

1 I assembled these meanings from various online English dictionaries, including Google's dictionary and, for instance, www.thefreedictionary.com/entanglement, under the assumption that these online dictionaries give us some idea about the common usage of the term. This was sufficient for my purpose here, which is not to provide a definition of entanglement, but to use it as a metaphor to say something about the relations between humans, language, and technology.

References

Coeckelbergh, Mark. 2015. "Language and Technology: Maps, Bridges, and Pathways." *AI & Society* (online first). doi:10.1007/s00146-015-0604-9.

Dewey, John. 1980 (1934). *Art as Experience*. New York: Perigee.

McLuhan, Marshall. 1971. "Roles, Masks, and Performances." *New Literary History* 2(3): 517–31.

McLuhan, Marshall. 2001 (1964). *Understanding Media: The Extensions of Man*. London/New York: Routledge.

McLuhan, Marshall and Eric McLuhan. 1992. *Laws of Media: The New Science*. Toronto: University of Toronto Press.

Mitcham, Carl. 1994. *Thinking Through Technology: The Path Between Engineering and Philosophy*. Chicago, IL: University of Chicago Press.

Index

actor-network theory (ANT) 195
actor theory 102
agency 35, 107–8 *See also* human
  agency
Akrich, Madeleine 100
Allen, Amy 238
alterity 129–31, 157–8, 173, 185,
  256, 258
Arendt, Hannah 4, 196
Aristotle 2, 3, 216
art/artist, as technological practice
  282–6
artefacts: baptism of 77; design
  flexibility 86–7; design process
  as social process 86; as emergent
  outcome of performative process
  103–4; as extensions of man 149;
  hermeneutical role of 142; hyperreal
  118; language use contributing
  to meaning of 90; meaningless
  278; meaning of, not related to its
  use 89; meanings attached to 86;
  mediating between humans and
  192; as mediators 142; perception
  and 159; performative 73; role of
  64–5; semiotic functions of 153–4;
  shaping our experience 155; social
  construction of 83, 85–90; social
  facts and 56; social-linguistic
  construction of 94; social ontology of
  53–6, 83; technology acts and 114;
  translating 149; virtual and digital
  55, 57, 229
Augustine 2
aura, concept of 121
Austin, John Langshaw 51, 69–71

background relations to technology 158
Barinaga, Ester 102

Baudrillard, Jean 116–19
Being and Time (Heidegger) 28, 61–2
Bijker, Wiebe, E. 85, 87
Blue and Brown Books, The
  (Wittgenstein) 75
body grammar 246, 269
body-in-performance 244–5
Bölker, Michael 88
boundary games 64–5
bricoleur 111, 255
Burbules, Nicholas C. 113

care robot 95–6
collective, defined 196
collective consciousness 143–4
collective intentionality 51–5, 58,
  64–5, 229
collective performance 66–7
collective use 66–7
Collins, Harold Maurice 98
commissives 70
"Computer as Component, The"
  (Heim) 45
computers: humanization of 37–8; as
  opponent 45–6
construction, linguistic-social 83–4
Construction of Social Reality, The
  (Searle) 51, 64
context principle 2–3
counter-environment 284
Cratylus (Plato) 1
cultural hermeneutics 159–60
culture, surrendering to technology
  145

Davidson, Donald 2–3
declaration 52, 69
Deely, John 108
De Interpretatione (On Interpretation) 2

deontic powers 52, 56
Derrida, Jacques 110–12, 255
de Saussure, Ferdinand 150
design-and-innovation technologies 92–3
design process, as social process 86
determinism 132, 224
Dewey, John 33–7, 189, 284–5
difference, philosophical tradition of 128–9
disciplining 238
discourse analysis method 88–9, 256
Dreyfus, Hubert L. 177
Dreyfus, Stuart E. 177

ecological thinking 144–6
electric technologies 149
embodiment: defined 8; hermeneutics and 246–7; human-technology relations 171–2; human-world relations 170–1; Ihde's focus on 175–6; language-technology relations 172–4; linguistic performance and 165, 245; lived experience and 247; of the mind 246; narrativity and 242–3; performance and 269; relation to technology 156–7, 159; social media and 242, 244; transcendental approach to 264
emplotment 216
engineering, humanities and 205
entanglement 277–8
episteme 41–2, 241, 264, 269
ethical discourse 185
everyday speaking 41–2
Expanded Hermeneutics (Ihde) 157
Experience and Nature (Dewey) 33

forms of life: artefacts and 90; boundary games as 65; cultural embeddedness and 160–1; defined 96; grammar as 73, 84; language as 106–7; language games as 26, 42–3, 109, 127, 215; music technology and 177–8; the social as 87–8; social-communication games and 67; as social practices 97–8; speaking of a language as 37; technology games as 11, 44; transcendental approach to 182; use, games and 186. *See also* technology games
Foucault, Michel 127, 237
Frankenstein narrative 41
Frege, Gottleb 3
French semiotics 256

gaming technology 218–20, 222
Gerrie, Jim 239
gestures, suppression of 120
Gier, Nicholas F. 182
givenness: influencing existing games and forms of life 95; of linguistic-technological form of life 43; music-technology and 179; of the social 51
grammar: of the body 269; as interpretation of texts and patterns 151–2; *Laws of Media* and 147–53; power 264; power and 240–1; of speech acts 73; of technology acts 73–4; transcendental conditions and 182
Growing Moral Relations (Coeckelbergh) 125

Harman, Graham 199
Heidegger, Martin: *Being and Time* 28, 61–2; on instrumental conception of technology 24; on language 106–7; on language machines 45–7; on present-at-hand 171; 'Question Concerning Technology' 180–1; on technology 107
Heim, Michael 45
hermeneutic phenomenology 164
hermeneutics: bridging gap between technology and 166; embodiment and 246–7; human-technology relations 172; human-world relations 171; as an interpretative activity 162; in Jewish-Christian biblical tradition 164; in the laboratory 191, 260; language and text as central to 164; language-technology relations 173–4; narrative technologies and 222–3, 226–7; relation to technology 157; semiotics and 192–3; social media's role 233; technoscientific praxis 162; of things 167–8; visually oriented 162–3; vocabulary 188–9
Hørstaker, Roar 102
hubris, concept of 41
human agency: Searle's view of 57–8; semiotics and 200–1; technology as 23
Human Condition, The (Arendt) 4
'Humanism of Media Ecology, The' (Postman) 144
humanities 7–8, 205
humanization of technology 37–8, 263
human/nonhuman performances 100–1

humans, language, and technology: concepts as tools between 270; conceptualizing relations between 267–9; integrated view of 273–6
humans shaping language 43
human-technology relations 171–2
human-technology-world relations 159, 230–1
human-world relations 170–1
hyperreality 117

ideal types 9
Idea of a Social Science, The (Winch) 97
Ihde, Don: *Expanded Hermeneutics* 157; hermeneutics of science 161–70; human-technology-world relations 158; I-world relations 158; material and relational phenomenology 155–61; on multistability 27–8, 160; on postmodern hermeneutics of things 191; *Technology and the Lifeworld* 161–2; on technomyth 168
illocutionary acts 70–1
imagination, role of, in technological practices 88
induction theory 27
innovation 85, 90, 96–7
instrumentalism 23–5, 280
interpretative flexibility 85–7
Investigations (Wittgenstein) 125
I-world relations 158, 170, 258

Johnson, Mark 246

Kellner, Douglass 108
knowledge: as commodity 116; consisting of tacit rules 98; narrative and 115; social practice and 34; technological transformations' impact on 115–16, 255; technology changing nature of 116

Lakoff, George 246
language: anthropological view of 24; anti-nominalist view of 34; artefact's meaning contributed by 90; as condition of possibility 184–5; diversity of tools of 59–60; embodied natures of use of 165; fall and salvation narrative 143–4; as form of discourse-text 111; framed by computers 38; as house of being 106; humans shaping 43; innovation 90; instrumentality of 29, 280; instrumental view of 24; Latour's view on 100; learning 26, 31; mastering a 27; measurement instruments 30; mediating role 170; as a medium 109; as medium between humans and technology 171–2; as medium between humans and world 170–1; naming ritual 63; neglect of issues concerning 8; performative 51–3; perlocutionary force of 35; politics and 199; rhetorical closure 87; Searle's view of 51–3; shaping use of technology 74–7; in social constructivism 91–2; social-cultural interpretation of 32–3; as social instrument 33–5; in social robotics innovation 92; structuralist theories of 110–12; subjectivity in 4–5; as supplement to technology 36; as a technology 29; as text 101–2; thinking and 60, 75; as a tool 25–6, 31; as tool of contamination 279; as totality of words 61; transcendental role of 184; use of 29; as vehicle of thought 40; Wittgenstein's view of 31–3. *See also* language games
language games: defined 113; feelings and thoughts to interpret 32; induction theory and 27; metanarratives and 112–13; Searle on 57; social bond and 113–14; social media and 236; speech act of promising as 70–1; transcendental approach to 127, 181; Wittgenstein on 26
language learning 4
language machines 45–7
language speaks 109–10, 254–5
language-technology relations 172–4
language use, holistic view of 58–9
Latour, Bruno 5, 88, 100–2, 141, 192–200
Laws of Media (McLuhan) 147–53, 239–40, 284
Leviathan and the Air-Pump (Shapin) 99–100
Lévi-Strauss, Claude 111–12
linguistic determinism 132
linguistic-social construction 83–4
linguistic-technological performances 72
Linvingston, Paul M. 61
Logic (Dewey) 36
logistics of war 119–20
logos 149–50
Lyotard, Jean-François 112–13, 116, 255

machine alterity 129–30
machines shaping language 47 *See also*
    technology
Mangle of Practice, The (Pickering) 200
material, social constructivist's
    emphasis on 92
material agency 201
material phenomenology 155–61
mathematization in science 167, 176
McLuhan, Eric 148, 150–2, 281
McLuhan, Marshall 24, 108, 143–4,
    146–8, 150–2, 281, 283–4
meaning: attached to artefacts 86;
    context principle 2–3; control of
    95–6; holistic view of 2, 75–6;
    representational view of 102; shared
    experience with using technology
    gives 30–1; of virtual and digital
    technological objects 57; way words
    are used 26
measurement instruments 30
media, as grammatical inquiry 147–53
media ecology 144–6, 257
mediating, defined 170
mediation 171–4, 223–4, 232–3, 268
mediation theory 179, 183–4, 226
mediators, artefacts as 142
Merleau-Ponty, Maurice 159, 243–5
Mersch, Dieter 121–3, 255–6
metallic body 119–20
metanarratives 112–13
metaphors 149–52, 183, 206, 255,
    264–5, 286–7
mimesis 216, 219
Mitcham, Carl 7–8, 280–1
multistability of technology 27–8,
    44, 160
music-language 177–8
music-technology 177–8
muting language 144

name, as an instrument 1–2
naming ritual 63
narrative: fall and salvation 143–4;
    knowledge and 115; reconfigurative
    role of 216; temporality and 215;
    theories on 102; as transcendental
    condition 207
narrative technologies: capacity to
    change human lifeworld 220;
    computer games as 218–20;
    configuring understanding of the
    world 225–6; defined 213, 261;
    fantasy game and 222; hermeneutics

and 222–3, 226–7; mediation and
    223–4; normative dimensions 221;
    performance and 224–8, 233–5;
    in postphenomenology 214, 261;
    reconfigurative role of 216–17; role
    of humans and 262; the social 225;
    social media and 233–5; wooden
    bridge as example of 217–18
narrative theories 18, 102, 214, 217,
    220, 242–3, 276
narrative time, scientific time vs. 215
New York School 144
nominalism 34
noninstrumental view of technology
    72–3
nonlinguistic reality 103
non-mediated perceptions of technology
    158
nonmodern approach to technology
    141, 142, 193–200
nonmodernism 193–5, 198
nouns, meaning of 2

object. *See* artefacts
On Certainty (Wittgenstein) 26–8,
    73–4
Ong, Walter 5, 143
ontic 126, 132–3, 181, 185–6, 198
ontological level 185–6
Order of Things, The (Foucault) 240–1
Oudshoorn, Nelly 89

perceptual-bodily activity 162
performative-hermeneutic process
    175–6
performative language 51–3
performativity: embodied 264–5;
    holistic analysis 188; human/
    nonhuman 100–1, 261; of language
    as text 101–2; process of 103–4,
    261; representational view of
    meaning vs. 102; social media and
    72; use of language as 187; use of
    technology as 187; variety of speech
    acts and 69–71; words and deeds
    149
perlocutionary acts 70
perlocutionary force of language 35
personalization of technology 263
phenomenology, transcendental
    arguments and 123–8 *See also*
    postphenomenology
Phenomenology of Perception
    (Merleau-Ponty) 243

Philosophical Investigations
(Wittgenstein) 25–6
philosophy of language,
overemphasizing language 8
philosophy-of-technology toolbox,
transcendental approach to 132–3
phonocentrism 110
Pickering, Andrew 200–2
Pinch, Trevor J. 85, 87, 89
Plato 1, 3
player narrative, gaming 219–20
pleasing game 231
Pokémon Go 218–19
politics, language and 199
Politics of Nature (Latour) 195–7
posthermeneutics 121–3, 255–6, 267
posthumanism 145, 190–2, 201, 279–80
Postman, Neil 144–5
Postmodern Condition, The
(Lyotard) 113
postmodernism 108–9, 145, 194
postphenomenology: defined 123–4;
expanding framework of 170;
hermeneutics and 47–8, 257–8;
of Idhe 154–5, 257–8; narrative
technologies 214; posthumanism and
190–2; theoretical attitude to 181;
transcendental approach to 124–8
post-semiotics 256
poststructuralism 123–31
Pound, Ezra 283–4
power: disciplining and 238;
Foucault's approach to 237–8;
micro-mechanisms of 237–8;
omnipresence of 237; relations
between technology, the body and
241; sexual harrassment and 238;
social media and 236
power games 236
power grammars 264, 269
power relations 240
praktognosia 244–5
prenatal surveillance 241
presuppositions, examining 129
private technology 44
process metaphysics 213
Prometheus 41

'Question Concerning Technology, The'
(Heidegger) 180–1

relationality 156
relational phenomenology 155–61
relationism 195

rhetorical arguments 87
Ricoeur, Paul 212, 261
robot ethics 126
Roderick, Ian 88
'Roles, Masks and Performances'
(McLuhan) 286–7
romantic epistemology 121, 159
Rule of Metaphor, The (Ricoeur) 151

Schaffer, Simon 99–100
science: as dance of agency 201–2;
mathematization in 167, 176;
performative understanding of 200;
philosophy of 97–8; sociology of
97–9
Searle, John R.: *Construction of Social
Reality* 64; on technological artefacts
53–6; view of language and social
reality 51–3
selective romanticization 159
semantic agency 96
semantic holism 2, 59
semiotics 153–4, 192–3, 200–1, 256
sensible embodiment 176
sexual harrassment 238
Shapin, Steven 99–100
signs: defined 62; external reality
of 132; language as 111–12;
postmodernity and 108–9;
simulacrum and 117; social media
and 230; technological artefacts
reduced to 84; textual 100–1;
thinking and 75–6; use of 62
Simulacra and Simulation (Baudrillard)
117
simulacrum 117–18
skill acquisition model 177
Smith, Dominic 123–5, 129
social, the: defined 260; as form of life
87–8; language and 207; narrative
technologies and 225; posthumanist
ontology of 231–2; production of
198; transcendental approach to
189
social bond as a language game 113–14
social constructivism: language and
90–7, 254–5; technology and 85–90,
256–7
social context 91–2
social media 72, 178, 228–36, 242,
244–5, 265–6
social reality 51–3
social robotics innovation 92
social robots 93–6

Social Studies of Science and
    Technology (STS) 83, 88, 134
sociology of science 97–9
Socrates 1–2
speech, as performance of thought 164
speech acts 70–1, 73, 91, 150
speech prostheses 196–7
speech subjectivity 198
speech-technological performance 75
speed, violence of 119–20
Speed and Politics (Virilio) 119–20
sphere of freedom and politics 4
sphere of necessity 4
sphere of the household 4
spoken words, as symbols of mental
    experience 3
status declarations 52
status function, declaration as 52
Stephens, Niall P. 145–6
Sterrett, Susan 30
structuralist theories of language
    110–12, 255
subjectivity 4–5
subject-object configurations 226
subject-technology-world relations 226
'Summary of a Convenient Vocabulary
    for the Semiotics of Human and
    Nonhuman Assemblies' (Akrich and
    Latour) 100
symbolic, Postman's stress on 145

taxi ride (metaphor) 286–7
technological determinism 132
technological innovation 85
technological practices 282–6
technological transformations 115–16
technologies of the self 239
technology: alterity and 129–30;
    anthropological definition of 23;
    autonomous view of 124–5; bodies
    and 119–20; changing patterns of
    human activity 67; collaborating
    with 45–6; computers 45–6; as
    condition of possibility 184–5;
    culture surrendering to 145;
    discourse analysis method 88–9;
    discourse and narratives about 41–5;
    empirical orientation of 99; everyday
    speaking about 41–2; as external
    to humans 253–4; grammar within
    59; Heidegger on 23; hermeneutical
    role of 40, 47–8, 155; as human
    agency 23; instrumentality of 23–4,

39, 280; know-how 28–9; material
    dimension of 133–4; as a medium
    117, 143–7; multistability of 27–8,
    44, 160; neutralist view of 57; as new
    worlds 37–8; noninstrumental view
    of 72–3; nonlinguistic realm of 4; as
    performative agents 201–2; shaping
    human consciousness 24; shaping our
    thinking 46; and the social 50–1; as
    a social instrument 35; theoretical
    meaning of 30–1; typewriters 46–7;
    use of 29; world-making 37–9.
    See also artefacts; narrative
    technologies; technology games
technology-act 71–3
technology acts, artefacts and
    114
Technology and the Lifeworld (Ihde)
    161–2
technology games: alterity and 130–1;
    defined 160–1; language shaping
    74–7; transcendental approach
    to 127, 181; variations in 68;
    Wittgenstein on 25–31. *See also*
    forms of life
technomyth 168
Technopoly (Postman) 145
textual signs 100–1
things, hermeneutics of 167–8 *See also*
    technology
thinking: agency in relation to 107–8;
    ecological 144–6; language and 60;
    language use and 75; and operating
    with signs 75–6; posthumanist
    279–80; technology use and 76
tools: cultural approach to 36; human
    capacity for using 65–6; serviceability
    of 62; words and 36. *See also*
    technology
touch 281
transcendental: conditions 133, 180–1,
    259–60; defined 212; empiricism
    123–8; materializing 213

understanding, role of 97
use of language. *See* language games
use of signs, use of tools vs. 62
use of technology. *See* technology
    games

Van Den Eede, Yoni 147
Varela, Franscisco J. 246
verdictives 70

Virilio, Paul 119–21, 255
virtual and digital technological objects 55, 57
virtual artefacts 229

We Have Never Been Modern (Latour) 99, 193–5
'What is phenomenology?' (Merleau-Ponty) 247
whole body perception 162
Winch, Peter 32, 97
Winner, Langdon 37–8, 67, 89
Wittgenstein, Ludwig: *Being and Time* 28; *Blue and Brown Books* 75;

On Certainty 26–7, 73–4; cultural interpretation of 36–8; "hinges" metaphor 27; holistic view of meaning 75–6; *Investigations* 125; on language games 26; Latour and 100–2; *Philosophical Investigations* 25–6; on technology games 25–31; on words 2
Wittgenstein flies a Kite (Sterrett) 30
words, tools and 36
word-things 61–3
world-making 37–9
writing technology 5, 46–7, 110–11, 143, 280–1